For Elliott,
Best wishes,
Ellery
April 1992

PROBABILITY
and
RATIONALITY

Studies on L. Jonathan Cohen's
philosophy of science

POZNAŃ STUDIES
IN THE PHILOSOPHY OF THE SCIENCES AND THE HUMANITIES

VOLUME 21

EDITORS

Jerzy Brzeziński

Andrzej Klawiter

Tomasz Maruszewski

Leszek Nowak (editor-in-chief)

Izabella Nowakowa

Ryszard Stachowski

ADVISORY COMMITTEE

Josef Agassi (Tel-Aviv)

Etienne Balibar (Paris)

Piotr Buczkowski (Poznań)

Mario Bunge (Montreal)

Robert S. Cohen (Boston)

Francesco Coniglione (Catania)

Andrzej Falkiewicz (Wrocław)

Ernest Gellner (Cambridge)

Jaakko Hintikka (Tallahassee)

Jerzy Kmita (Poznań)

Władysław Krajewski (Warszawa)

Krzysztof Łastowski (Poznań)

Theo A. F. Kuipers (Groningen)

Witold Marciszewski (Warszawa)

Ilkka Niiniluoto (Helsinki)

Günter Patzig (Göttingen)

Marian Przełęcki (Warszawa)

Jan Such (Poznań)

Klemens Szaniawski

Jerzy Topolski (Poznań)

Johannes Witt-Hansen

Ryszard Wójcicki (Łódź)

Georg H. von Wright (Helsinki)

Zygmunt Ziembiński (Poznań)

The principal task of the book series "Poznań Studies in the Philosophy of the Sciences and the Humanities" is to promote development of philosophy which would respect both the tradition of great philosophical ideas and the method of philosophical thinking introduced by analytical philosophy. Our aim is to contribute to practicing philosophy as deep as phenomenology or Marxism and as rationally justified as positivism or hypotheticism.

The address: prof. L. Nowak, Cybulskiego 13, 60-247 Poznań, Poland.

PROBABILITY
and
RATIONALITY

Studies on L. Jonathan Cohen's philosophy of science

Edited by

Ellery Eells
&
Tomasz Maruszewski

AMSTERDAM—ATLANTA, GA 1991

ISBN: 90-5183-316-4 (CIP)
©Editions Rodopi B.V., Amsterdam - Atlanta, GA 1991
Printed in The Netherlands

CONTENTS

FOREWORD

Scientific activity on the borderlines of different disciplines may often be fruitful, but it may also lay the ground for various kinds of interdisciplinary conflict. On the one hand, it is possible that such activity will recognize new problems and see new possibilities for the solution of old problems; but on the other hand, representatives from different disciplines may be intolerant of, skeptical about, unaware of, or at best critical about, the methodologies of the other disciplines.

The distinguished British philosopher L. Jonathan Cohen is surely one of the best contemporary examples of an investigator with both interdisciplinary interests and interdisciplinary influence. In this volume, the works of Cohen are discussed, both critically and in the way of exposition of his ideas. He is the author of such much appreciated books as *The Diversity of Meaning* (London, 1962, 1966), *The Implications of Induction* (London, 1970), *The Probable and the Provable* (Oxford, 1977), *The Dialogue of Reason* (Oxford, 1986), and *Introduction to the Philosophy of Induction and Probability* (Oxford, 1989). In his work he has endeavored to include psychological theory and data into broader philosophical perspectives. It is obvious that his work on the nature of *induction* has deep and extensive implications for psychology, both in terms of theory evaluation and in terms of methodology.

Induction, of course, is the mode of reasoning we engage in when premises need not confer an *absolute* guarantee of truth on a conclusion. (*Deduction*, on the other hand, is a "yes or no" kind of reasoning, in which the reasoning is "correct" if the premises absolutely guarantee the truth of the conclusion, and "incorrect" if there is no such absolute guarantee). In induction, the goodness of reasoning comes in degrees, sometimes called *inductive probability*, which may be a kind of (rational) *subjective probability*.

Induction is both a common, everyday kind of reasoning and also an important kind of scientific reasoning. Some reconstructions of this activity, such as the one proposed by *attribution theory* (Kelley, 1973), have stressed the fact that people, when they reason about causes of behaviour, almost always use induction. They use a combination of Mill's Methods of Agreement and Difference (Maruszewski, 1983). Many other uses of this kind of reasoning could be cited. Inductive reasoning is also a part of planning and interpretation in psychological investigations, so that induction is important for psychologists themselves as well as for their subjects. Therefore, knowledge of the rules of induction, of its traps, its advantages and disadvantages, and so on, should be an important factor in further developments of empirical and theoretical studies.

We may distinguish two basic kinds of theories of induction: the *prescriptive* (or *normative*) ones (which are about how we *should* reason), and the *descriptive* ones (which are about how we in fact *do* reason). In the prescriptive theories, an assumption of rationality plays an important role. In the descriptive theories, on the other hand, such an assumption is not so central. Before we begin evaluating a particular theory of induction, therefore, it is important to know just what kind of a theory we are dealing with. As Cohen puts it,

> "... well argued analytical theories about the nature of probability are not always targeted at exactly the same objective. One polar aim for such a theory is to describe, analyse and explain how probability is actually conceived in human judgments. The alternative aim is to prescribe how probability should be conceived and, in particular, how the formalism of the mathematical calculus should be interpreted. But those two objectives, though quite different from one another in principle, are often difficult to hold apart in practice. On the one hand, it is natural for philosophers of science to prefer to describe what they take ideally to be implicit in the best usage they can find among the reputable scientists and mathematicians, rather than whatever is actually implicit in the imprecise and inexpert thoughts of the man-in-the-street. On the other hand, anyone who prescribes how probability should be conceived presumably thinks of himself as understanding and answering a question that relates to the familiar existing practice of judging probabilities: the question is about how this practice should be executed, and so any answer to it must presuppose at least some description of the presumed aims of the practice and of the factors that constrain it" (Cohen, 1989, p. 41).

Without first addressing these issues, having to do with the appropriateness and nature of rationality assumptions, our efforts directed at understanding the nature of induction may get us nowhere.

These preliminaries aside, the reader may, upon opening this book, ask himself why we once more enter into analyses in an area that already has so rich a psychological and philosophical tradition. Of course, everybody would agree that it is possible to collect a weighty library of books and papers already written on the topic of rationality. Why, then, do we need to add this next one?

The main reason is that there are various current "nonstandard" approaches in the field of rationality, induction, and subjective probability (both among professional philosophers and among professional psychologists) that may be fruitful for others (in either official discipline) who deal with these issues. Of course, many investigators may disapprove of these solutions at first sight. We cannot expect instantaneous transformations in the people who are adherents of the various perspectives. But we do think that studying a variety of different perspectives may lead to a deeper and more fruitful understanding of cognitive activity.

We may refer to the Piagetian conception of egocentrism as a kind of analogy that may be useful in understanding our thoughts on the development of the debate about the nature of rationality, subjective probability, and so on. Piaget says that every child in his or her cognitive development has to go through the *primary* stage (1964, pp. 122-123). In this stage, a child sees everything from his or her own perspective only, and cannot appreciate the legitimacy of other perspectives. He or she assumes that everybody should perceive, believe, and know as he or she does. Of course, this egocentricity involves, first of all, the world of perceptions. One of us (T. M.) would like to recall the remark of his three year old son, who said, "Dad, look at this lady. She has a big hole on her knee". He was totally surprised to learn that his dad could not see anything.

We are not going to tell a long story about the development of a person's ability to accept the legitimacy of other people's perspectives, but we would like to say a few words about the last stage of its development. Piaget called it *decentralization* (1964, pp. 122-123). In this stage, an individual is able to realize that other people may see the same thing or event or process from a point of view that is different from his or her own point of view. Of course, this may pertain to perception of concrete items as well as to grasping abstract ideas and relationships.

We feel that what is urgently needed in many debates in the philosophy of science and in psychology is a kind of decentralization — what may be called *scientific decentralization*. We would like to encourage our colleagues to take into consideration the fact that different investigators, in the same or in different disciplines, may have different starting points.

These starting points involve philosophical assumptions about the object of study. The point may be summarized as follows. We do not elaborate ideas that have to do with actual objects, or even actual people (like psychology students), but rather, if our method is the universal method of the sciences and humanities, our efforts pertain to *ideal* objects, and not to people at all (Nowak, 1980). And, of course, the objects of study, as abstract entities, are subject to multiple interpretations.

We note that physics, which is often said to be a model of well developed science, does not apply most directly to real objects such as gases, liquids, and so on: the laws of physics describe the behavior of, for example, *ideal gases*, which have never existed. Likewise in psychology: a theory of decision making may describe the behavior of the *ideally rational* agent, and signal detection theory pertains most directly to the *ideal* observer. Of course, this does not make these theories speculative or metaphysical. Nobody denies that decision making or signal detection can be studied empirically: empirical studies of decision making and signal detection belong to well developed, and developing, areas in psychology.

Let us look for a moment at studies of reasoning and subjective probability. There was a lot of misunderstanding after the publication of Cohen's influential paper, "Can Human Irrationality be Experimentally Demonstrated" (1981). We are not going to provoke further discussion on this (from 1981 to 1984, *The Behavioral and Brain Sciences* published 43 open peer commentaries on this paper, as well as three distinct author's responses). The debate over this paper generated not only light, but also heat. This may be considered as a clear indication of the importance of the topic and of its resistance to standard solutions. And it seemed to manifest a certain amount of intolerance, or lack of decentralization, among the parties concerned, in dealing with the abstract ideas involved.

Most of the contributions in this volume concern rationality itself or its manifestations in human behavior. We would like to consider rationality itself as some kind of "essence" that can have various manifestations, where the manifestations cannot be identified with the "essence". If this distinction can be maintained, then we hope we should be able to avoid the kind of pessimism displayed by Baruch Fischoff (1981), who was seriously afraid of the lack of cooperation between psychologists and philosophers. This book may be seen as testimony to the possibility of at least coexistence of psychologists and philosophers in the study of rationality. May it be that, in the future, this coexistence should turn into cooperation.

A prerequisite to such cooperation is the acceptance of the distinction between empirical facts and theoretical models, where the latter may be either descriptive or prescriptive. Unfortunately, the behaviouristic tradition that has recently dominated psychological thinking tends to obscure the difference between descriptive theoretical models and empirical generalizations. Failure to appreciate this difference was not uncommon in the particular debate alluded to above. Given the empirical data, it is hard to deny that people's behaviour often reveals cognitive biases and illusions. But the behaviour itself is one thing, and the mental processes that issue in the behavior is another. If psychological theories are to address the latter phenomena, then we must put the behavior itself in the proper perspective in relation to the underlying mental processes. Of course, this is not to deny that mental processes are not always working optimally, that subjects may be distracted, or lacking in the relevant basic knowledge, and so on. Despite these possibilities, mental processes have the ability for autocorrection, for learning, for looking for alternative solutions, and so on. We believe it would be a mistake to take into account only isolated occurrences of mental processes: we should take into account, as well, long term patterns, and, to some degree at least, overlook temporary changes, deficiencies, and so on.

At this point, we encounter the distinction between competence and performance. This distinction was used first by Chomsky (1965) in the field of linguistics, and Cohen (1981) applied it to the analysis of rationality and subjective probability. Because this distinction suggests that we should sometimes overlook temporary changes and deficiencies in behavior, it has become the target of many attacks (as in the debate surrounding Cohen's (1981), alluded to above). The critics have maintained that using the distinction in this way may lead to the overlooking of systematic cognitive deficits and biases. We suppose it *might*, but perhaps the question is not correctly posed in the first place, as we now explain.

Let us begin with this fundamental question: What and who is psychology about? For example, is it about what people do or is it about what people can do? Most psychologists choose the first alternative, and as a result, we have many studies on biases, we have clinical psychology, and we have psychoanalysis. Of course, we do not deny the importance, and necessity, of this work, but we should also stress the second alternative mentioned above. Instead of looking only for weak points in human functioning, we may also concentrate on the strong points. This is to see the importance of the question, "What *can* people do?" There are studies of human behaviour and mental activity in an area that is described as a study of human *potential* (Otto, 1981). Although we do not endorse all the details of work in this field, we do think that the orientation of the work is a step in a right direction. Concentrating on human abilities is not a repetition of traditional analyses of individual differences; it involves, instead, a quite new approach for psychological investigation.

This cognitive shift can be illustrated with an example from the traditional area of psychometry. Traditional test theory, as developed by Gulliksen (1950) and by Lord and Novick (1968), assumed that the result (score), x_i, of every test includes two components, x_∞ and e_i, as follows:

$$x_i = x_\infty + e_i,$$

where,

x_i = fallible score earned on one administration of the test,

x_∞ = true score,

and,

e_i = error on that particular administration of the test.

Psychometry assumes that error scores may be either positive or negative, because they are random variables. This assumption seems disputable: it allows for higher results than would correctly correspond to the actual abilities of a particular subject. If we assume that the value of the error component is always negative, then the fallible score will always be less than the true score. This assumption is reasonable only in relation to tests of abilities, skills, or intelligence, and not, for example, in personality tests. In what follows, we will always have in mind the former kind of test. Of course, the smaller (closer to zero) the value of error, the smaller will be the difference between fallible and true scores.

The procedures used in psychometry try to estimate the true score on the basis of many independent measures of particular functions or ability, on a given person on different occasions. The arithmetic mean of these measurements is considered to be an approximation of the true score. If we want to estimate the true score, starting with the assumptions mentioned above, then we should find situations in which the value of the error should be expected to be the smallest. It is in these situations that the person in question achieved the best result in a particular task.

Many years ago, a Polish psychologist — Mieczysław Kreutz (1927, 1933) — proposed to take, as a measure of ability, the highest results achieved by the subject in question at a particular task. We should, in our opinion, extend this suggestion to all kinds of psychological measurements of ability or skill. Of course, this would entail many changes in psychological methodology and the statistical analysis of data. Such a procedure would no longer include "accusational" elements, connected with looking for deficits and biases in people's mental functioning. Instead, we should focus on the best results achieved by subjects, and the conditions under which these best results arise. Another alternative would be to try to estimate the influence of factors that may interfere with optimal performance, and then use this result as a correction for an actual score whenever we suspect the operation of an interfering factor. Have investigators concerned with identifying systematic deficits and biases tried something like this? We are unsure about this, but it seems that they have focused more on the behavior itself, rather than on the actual mental activity underlying the behavior.

From the point of view being urged here, psychology needs more idealization. Of course, we do not have in mind idealization in the sense of ascribing more positive value to something or someone. We have in mind idealization in the technical or methodological sense, as explained by Nowak (1980, and his contribution to this volume). It involves accepting idealizing assumptions to the effect that the strength of certain kinds of factors interfering with optimal performance is zero. Then we may investigate relationships between variables that are the main determinants of behaviour, and we may then find laws governing the influence of these variables. We may then be in a better position to investigate, one by one, the nature of the influence of factors that interfere with optimal performance. And finally, in this way, we may better understand the impact of all the most important variables and factors that operate in natural conditions — and then understand, in this new way, the results of the empirical studies.

Contributors to this volume have been inspired by Cohen's work in various ways, and in connection with a number of areas and issues that Cohen has dealt with in his writing. In Part I, Cohen delineates the development of his thought in various areas, including political philosophy, philosophy of language, philosophy of law and the application of psychological theory here, and our deductive and inductive rationality. In the next three sections, other investigators discuss the relevant issues, and Cohen's contributions, as we describe below.

Part II, containing three contributions, is about Cohen's "method of relevant variables", and other modes of scientific reasoning. Leszek Nowak here investigates the interrelationships between Cohen's method of relevant variables and the idea of idealization, which Nowak considers to be central to the methodology of both the natural sciences and the humanities. Menachem Fisch then gives a critical examination of Cohen's method of relevant variables. Fisch argues that the method has its place in the evaluation a kind of "higher-order" hypothesis in "dynamical" theoretical settings, and not with "first-order" hypotheses in more "static" theoretical settings. Concluding Part II, Maurice Finocchiaro discusses some of the main themes in one of Cohen's most recent books, *The Dialogue of Reason: An Analysis of Analytical Philosophy*. He focuses

14

on Cohen's idea of the "inductive-intuitive method", on Cohen's emphasis on the importance of the idea of norms of reasoning in understanding analytic philosophy, on Cohen's ideas about deductive and inductive methods, and on Cohen's analysis of the role of intuition in analytical philosophical reasoning.

Part III contains four articles that focus on the topic of subjective probability and Cohen's distinction between (what he calls) Pascalian and Baconian conceptions in this connection. Here, David Schum first describes the different kinds of conception of probability, and then elaborates models of applications of the different conceptions to the problem of assessing the credibility of human testimony, when the relevant inference is not direct but rather involves a chain of reasoning that may have multiple intermediate stages. James Logue's paper examines the idea of the "weight" of subjective probabilities — roughly, the degree of confidence we have in our subjective probability assessments, which is, or presumably should be, a function of the amount of evidence that our assessments are based on. Logue analyses weight as a function of second-order probabilities, and applies his analysis to Cohen's arguments for pluralism — especially to the issue of whether a Baconian, rather than a Pascalian, conception of probability is the one that is appropriate in judicial contexts.

Continuing with Part III, Czesław Nosal then focuses on the possibility of neurological correlates to subjective probability assessments, which may be causally responsible for subjects' evaluations of frequencies of events. It is difficult to find the principles behind such evaluations from data obtained from various animal studies, but Nosal suggests that a broad construal of Cohen's idea of the "norm extraction method" is at least a principled way of tracking the data, and more precise than other methods. Finally, Ryszard Stachowski's investigation concerns the measurement of various psychological magnitudes: it turns out that there are interesting parallels between certain standard and nonstandard psychological approaches on the one hand, and Cohen's proposals concerning the Baconian interpretation of probability on the other hand.

Part IV, which contains five papers, is about controversies concerning human rationality, and about methodological pluralism in the evaluation of these debates. Lola Lopes and Gregg Oden begin by arguing that the conventional notion of rationality — involving the idea of expected utility — implies a conception of intelligence that involves competence in deductive logic, in probability and expected utility calculations, and in following "linear", rule oriented, decision procedures. Empirical evidence seems to show that we are not rational in a way that is consistent with this kind of intelligence — rather, we seem to use fallible *heuristics*, such as "representativeness", "anchoring and adjustment", and "availability", which have been described by Kahneman and Tversky (and are summarized in Lopes and Oden's paper). Lopes and Oden question certain assumptions behind the empirical research; they make comparisons with artificial intelligence (AI) research (which sometimes *strives* to incorporate heuristics with properties that the authors argue are shared with the three main heuristics Tversky and Kahneman have identified); and they argue, using recent ideas in AI and in psychology, that using the relevant heuristics embodies a more valuable kind of intelligence than the kind that works in the special conditions that obtain in the empricial research in question.

Next in Part IV, Gerd Gigerenzer addresses the idea that there is an analogy between *visual illusions* and purported *cognitive illusions*. He points to three features of visual illusions: 1) there is only one way things *really* are, 2) illusions persist even when the operative perceptual principles are known to the subject, and 3) the fact that what is perceived depends on certain contextual factors and on prior knowledge. Gigerenzer argues that the analogy fails for features 1) and 2). But he agreees with the analogy for feature 3), and he develops the analogy into an interesting way of seeing how *optimal cognitive function* can be consistent with *systematic cognitive illusions*. This conclusion is similar to those drawn by Lopes and Oden, though the analyses are different. Next, Jonathan Adler comes to conclusions similar to those of Lopes and Oden and of Gigerenzer, but for yet different reasons. Adler argues that considerations of "conversational implicature" (urged by the late British philosopher H. Paul Grice) suggest that subjects can be misled by informal features of instructions given in tests. In some tests, subjects are given information the very supplying of which would seem to "conversationally imply" its *relevance* to the solution of the problem at hand, whereas this information is actually irrelevant to the problem understood in a formally standard way. Adler also explores the limitations of his conversational approach, and he argues that the connection between subjects' responses and the assessment of their beliefs and rationality is not very straightforward. In line with some of the ideas urged above, Adler thinks of rationality as testable only in the long run; and in line with Lopes and Oden, he thinks of cognition as having multiple "objectives", truth being just one of them.

The next two contributions focus on Cohen's norm extraction method. Tomasz Maruszewski's essay stresses the fact that, according to this method, we may distinguish different kinds of human rationality. The author's aim is to transfer the distinction between competence and performance into an analysis of rationality, and this is another application of Cohens's idea that had been used before in his analysis of subjective probability. Finally, Ireneusz Scigała applies the norm extraction method to the question of the nature of mental health. Most previous attempts to understand the idea of mental health were postulative in nature, and often combined prescriptive and descriptive elements. It is argued here that the use of the norm extraction method avoids some pitfalls of previous attempts to understand mental health.

This volume concludes with Part V, in which Cohen offers commentary on the main themes discussed in previous contributions: methodology, induction, and rationality.

We are deeply grateful for financial support from different sources which made this book possible. The first of us (E. E.) would like to thank the Research Committee of the Graduate School of the University of Wisconsin-Madison for financial support. The second of us (T. M.) would like to thank The British Academy for the grant that enabled him a visit to Oxford. The Queen's College provided a whole intellectual community. Preparation of this volume was also supported by Research Project RPBP III. 29.

In conclusion, we would like to thank all contributors to this volume for their efforts. Our hope is that these efforts will result in further interdisciplinary discussion and cooperation in investigating the relevant philosophical and psychological problems. A special debt of gratitude goes to Prof. L. Jonathan Cohen who was a very patient and a very careful reader of all contributions. His cooperation has improved the contents of this volume a lot.

Ellery Eells
Department of Philosophy
University of Wisconsin-Madison
Madison, Wisconsin 53706
USA

Tomasz Maruszewski
Institute of Psychology UAM
Szamarzewskiego 89
60-568 Poznań
Poland

REFERENCES

Chomsky, N. (1965). *Aspects of the theory of syntax*. Cambridge, Mass.: The M. I. T. Press.

Cohen, L. J., (1981). Can human irrationality be experimentally demonstrated. *The Behavioral and Brain Sciences*, **4**, 317-370.

Cohen, L. J. (1989). *An Introduction to the Philosophy of Induction and Probability*. Oxford: Clarendon Press.

Fischoff, B. (1981). Can any statements about human behaviour be empirically validated? *The Behavioral and Brain Sciences*, **4**. 337.

Gulliksen, H. (1950). *Theory of mental tests*. New York: Wiley.

Kelly, H. H. (1973). The processes of causal attribution. *American Psychologist*, **28**, 107-128.

Kreutz, M., *Zmienność rezultatów testów. Część I. Znaczenie zmienności rezultatów dla wartości testów*. Polskie Towarzystwo Filozoficzne, Lwów, 1927, ss. 72. *Część II. Przyczyny zmienności rezultatów testów i konieczna modyfikacja metody testów*. Lwów, 1933 ss. 4 nlb 207 (Variability of test results. Part I. The significance of variability of results for a value of tests. Polish Philosophical Society. Lvov, 1927, pp. 72. Part. II. Causes of variability of tests results and necessary modification of method of tests, Lvov, 1933, pp. 4, extra pp. 207).

Lord, F. M. and Novick, R. M. (1968). *Statistical theories of mental test scores*. Reading, Mass.: Addison-Wesley.

Maruszewski, T., (1983). *Analiza procesów poznawczych jednostki w świetle idealizacyjnej teorii nauki* (An Analysis of the Cognitive Processes of an Individual in the Light of the Idealizational Theory of Science). Poznań: Adam Mickiewicz University Press.

Otto, H. A. (Ed.). (1968). *Human Potentialities*. St. Louis, Missouri.

Nowak L. (1980). *The Structure of Idealization*. Dordrecht: Reidel.

Piaget J. (1964). *The Psychology of Intelligence*. London: Routledge E. Kegan.

PART I

SETTING THE STAGE

came to a sudden end in mid-August, 1945. But when the landing-ship that I was on had got about half-way to Penang the charms of philosophy prevailed over alternative attractions. As the outline of Car Nicobar loomed up through the monsoon rain I took one of those crucial decisions that determine a lifetime. I returned to the U. K. as quickly as possible from Penang, and was demobilised and back at Balliol College, Oxford, by mid-November 1945. And in October 1947, after two more years' study at Oxford University, I became an assistant in the Department of Logic and Metaphysics at Edinburgh University, ready to work for a few years on political philosophy.

But no-one could work seriously in any branch of philosophy in the U. K. at that time without coming to terms with the various linguistic approaches to the subject that emanated from Frege, Russell, Wittgenstein, Carnap, Ryle, Austin and others. Certainly those were the issues that assistants and younger lecturers talked about when they got together in Edinburgh, — and in particular Errol Bedford, Peter Heath, Ernest Gellner (then a passionate champion of Wittgensteinianism and, indeed, the only champion of it in Edinburgh), and myself. So my first published papers (1949, 1950a, 1950b, 1951a, 1951b, 1951c) were all concerned with such fashionable topics as the relativity of philosophical analysis to natural language, the redundancy theory of truth, the meanings of token-reflexive expressions, the structure of purposive explanation, the logic of moral reasoning, and so on. At the same time, through an old school-friend, Michael Samuels, who taught in the English Language department, I came in contact with a group of people — including Angus McIntosh and David Abercrombie — who were interested in the study of language from an empirical point of view. Nor did such an interest preclude the possibility of theory-construction, as I learned by attending a seminar given by the Danish grammarian Hjelmslev who was visiting Edinburgh at that time. And as I browsed fairly widely in the literature of linguistics, and later (having become a lecturer at the Dundee campus of St. Andrews University in 1950) attended meetings of the St. Andrews Linguistic Circle, the impression began to grow that there were flaws, or at best serious oversimplifications, in many of the bold and dogmatic statements that analytical philosophers had made, and were still making, about human language. It was as if the very same thinkers who strongly condemned *a priori* theorising about empirically checkable features of Nature were quite ready to theorise *a priori* about empirically checkable features of language. At the same time my wartime studies in Japanese exercised some influence in the same direction. Or at least they tended to undermine some of the imprudent grammatical generalisations that a classical education had encouraged.

Moreover, just as many analytical philosophers spoke or wrote in apparent ignorance of contemporary general linguistics, so too they were apparently unaware of past rapprochements between linguistic theory and philosophical analysis. The seventeenth century was particularly innovative in this area, and I tried to put the record straight by writing an article about some of the artificial languages that were constructed or projected for various purposes at that time. I had learned about the existence of such languages from the biologist, statistician and polymath, Lancelot Hogben (with whom I had several day-long conversations after we had met through a common involvement in the politics of the world-federalist movement that flourished briefly in the U. K. in the late 40's). And a few months' work in the British Museum library turned up sufficient material to provide a basis for some philosophically relevant comparisons between seventeenth- and twentieth-century work in artificial language (1954a).

Consequently, as the book on political philosophy neared completion (1954b), the idea of a book about meaning began to form itself. A pervasive theme of the book would be the diversity, amid systematic interconnectedness, of the different conceptions of linguistic meaning appropriate to historians of ideas, lexicographers, translators, logicians, etc. In particular it seemed that both the historical or sociological dimension of the subject and also the formal-logical one were considerably undervalued by current orthodoxies at Oxford. Nor did these orthodoxies show a proper respect for Carnap's monumental explorations in the analysis of science. But as yet I knew only the elements of modern logic and it was clear that the best way to remedy this defect at that date was to spend a lot of time at one or two appropriate centres in the U. S. A. So I managed to get a Commonwealth (now called 'Harkness') Fund Fellowship to spend one semester at Princeton and another at Harvard, within the academic year 1952-3. At Princeton I attended — with equal profit — a basic course in Carnapian analysis that was given by George Berry, an advanced seminar in Frege's logic of sense and denotation that was given by Alonzo Church, and a small informal discussion-group at which John Kemeny and Norman Malcolm were among the more active participants. At Harvard I went to three sessions of a seminar given by Charles Morris and reacted adversely to them. They did not seem to fit easily into a movement for exact philosophy. C. I. Lewis's course in epistemology and metaphysics was more interesting and so was Israel Scheffler's course on philosophical analysis. But the main excitement that semester was Quine's course on logic and language which later turned into his book *Word and Object*. Here the modal-logical analyses that Church and Lewis had accepted without question were subjected by Quine to systematic

24

criticism, and at each seminar, after Quine's lecture was over, argument with Quine continued over lunch and long into the afternoon. Noam Chomsky and Ronald Dworkin were flexing their intellectual muscles in these discussions. But at the time the most active and effective contributor among the junior participants was Burton Dreben.

2. *The generalisation of non-extensionality*

Clearly one issue that was well to the fore in East Coast American philosophy of language in the early 1950's was the treatment of non-extensionality, i. e. of linguistic contexts that do not always preserve the same truth-value when terms denoting the same entities are substituted for one another. Indirect discourse is one example of such a context and another is created by the use of modal expressions as when the true proposition:

Necessarily 4 is greater than 2
is converted into the false one.

Necessarily the number of books in Locke's *Essay is* greater than 2. But, so far from being persuaded by Quine's programme for the elimination of non-extensional contexts, I came to think that their structure and variety were still insufficiently explored. The outcome, mostly after my return to Oxford in 1957 as a fellow of Queen's, was chapters VII and VIII of my book about meaning (1962d) and a number of papers on related questions. Some of these papers were polemically oriented and criticised analyses that had been proposed by Arthur Prior (1958b, 1959a, 1963a, 1964b, 1974b) or Peter Geach (1962a, 1968b). Some were concerned to develop new approaches to other familiar problems (1955a, 1955c, 1957b, 1958a, 1960, 1961, 1968a, 1970a). For example, I argued (1957b) that some types of logical truth involving indirect discourse required to be formalised in a way that did not exploit the supposed existence of a Carnapian hierarchy of languages, or of a Churchian hierarchy of intensional entities. One such logical truth was "If the policeman testifies that anything which the prisoner deposes is false, and the prisoner deposes that something which the policeman testifies is true, then something which the policeman testifies is false and something which the prisoner deposes is true". And another paper (1960) proposed the use of a different kind of hierarchy for certain purposes. It sought to construct a generalised logic of non-extensionality — which had not previously been attempted. The point was that a merely dichotomous classification misses the richness and variety of non-extensionality, and that we should instead acknowledge the existence of indefinitely many different levels and degrees of resistance to substitution. Standard modal logic could then be generalised so as to formalise these hierarchies of non-extensionality.

In the formal system thus developed the hierarchies of non-extensionality were represented by corresponding hierarchies of modal operators, which ranged, in effect, from "it is physically necessary that..." to, "it is logically necessary that". Now, to prefix such an operator to a proposition implies that the proposition is true. But could there not be analogous modal operators that lacked that implication? If physical necessity is what holds in all physically possible worlds and a physically possible world is one that is subject to the inductively testable uniformities of the actual world, could there not also be lower levels of modality than physical necessity — holding for propositions that are true in worlds subject to only some of the inductively testable uniformities to which the actual one is subject? And would not these lower modalities correspond to what experimental evidence establishes for a scientific hypothesis when the level of justified confidence seems lower than certainty? Would not the controlled variation of an experiment's relevant circumstances, in insulation from other factors, constitute a possible world from which those factors are absent?

It was in this way that in 1964 I first hit on the idea for what has later come to be called a "Baconian" inductive logic. Though I continued to publish occasional articles in the philosophy of language 1964a, 1966f, 1972b, 1973b, 1973e, 1974d, 1975d, 1977a, 1978a, 1979c, 1980c, 1980e, 1981e, 1982f, 1985a, 1986b, forthcoming a), the new idea now formed the basis for two books and several articles. It had to be shown in detail (1966a, 1970e, 1977b) what are the main theorems of such an inductive logic and how well they fit our intuitive ideas about inductive reasoning from controlled experiments. It had to be shown how well the system fits some concrete historical cases of inductive reasoning by well-regarded scientists (1973d). The system had to be shown to advance our understanding also in some new areas of investigation like the psychology of language acquisition (1970b), the study of metaphor (1970d), and the evaluation of medical diagnoses (1980h). The system's relationship to other theories in the modern literature, like Carnap's and Lakatos's, had to be explored (1971, 1976b, 1978c, 1979a, 1980f, 1980k, 1981f, 1985b, 1986c, 1989b, 1989c) along with its debt to the Baconian tradition (1980a). And the independence of Baconian logic and mathematical probability — or "Pascalian probability", as I later came to call it — needed close investigation (1968c, 1970e, 1972a, 1973a, 1977b). But the main inspiration for all this lay in speculation about the logic of modality.

26

A subsidiary inspiration came from another field. When writing the book on the diversity of meaning (1962d, chapter X) I had been particularly struck by one fact about previous writers: philosophers had devoted a lot of attention to how one should assign a truth-value or probability to a proposition, when assuming that one already knows what it means, but had devoted very little attention to the complementary problem of how to determine the exact meaning that a proposition should be supposed to have if its truth-value is already known. And when thinking about contexts in which this latter problem arose it occurred to me that under certain conditions inductive reasoning could be regarded as being concerned with the precisification of hypotheses within appropriate limits. In the face of unfavourable evidence a hypothesis has to be made more complex and restricted in what it claims, if it is to retain tenability, and when a hypothesis survives relevant tests it can be made correspondingly simpler and bolder. Such a non-standard form of inductive reasoning has just the same logical structure as the more familiar form and was explored more fully in §16 of my book about induction (1970e). And I was able to show there how idealisation, as a strategy in the construction of scientific theories, may then be seen as a related form of non-standard inductive reasoning.

This result is achieved, of course, by a redefinition of the term "inductive". Just as we have come to think of a Pascalian probability-function as any function that satisfies the axioms of the classical calculus of chance, rather than as a function that is evaluated by reference to relative frequencies, say, or ratios of chances, so too we can think of a Baconian inductive function as any function that satisfies the axioms of a generalised modal logic, rather than as a function of generalizations that is evaluated by reference to the truth-values of their instances.

3. Some issues in jurisprudence

Preparation for the book on world citizenship, and especially the need to think as clearly as possible about the nature and status of international law, had earlier led me into acquiring some acquaintance with the literature of jurisprudence. This acquaintance had to be expanded considerably when my opposition to "ordinary language philosophy" prompted a detailed criticism (1955d) of the views about definition and theory in jurisprudence which Herbert Hart had published in his 1953 inaugural lecture at Oxford. Further investigations in the same area were prompted by the opportunity to write a critical notice (1962b) of Hart's book *The Concept of Law*. For I felt the book to be flawed by a pervasive incoherence between the "ordinary language" pretensions of its professed methodology and the positivist substance of its arguments. And later I followed Bacon's example in trying to demonstrate (1970e §17), that common law reasoning from judicial precedents to legal principles has the same inductive structure as a scientific reasoning from experimental data to explanatory hypotheses.

From all this a standing interest in jurisprudential problems not unnaturally lingered, and it occurred to me that a satisfactory theory of inductive reasoning ought to be able to contribute towards clarifying the way in which matters of fact can, or should, be proved in the courts. Specifically, so far as the triers of fact are lay persons, like an English or American jury, then within certain limits it ought to be possible to analyse the requisite standards of proof in terms definable within Baconian inductive logic. Such limits would be set by legal requirements about presumptions, about permissible testimony, about the location of the burden of proof, and so on. But the inductive framework ought to be clearly visible, in the sense that the major premiss invoked by any proof of a particular fact — such as of an accused's assault or a tortfeasor's negligence — has to be an inductively supportable generalisation. Moreover, as more relevant facts are put before the court, the more specific becomes the generalisation invoked and therefore the more guarded against falsification, so that the Beconian inductive probability of the conclusion to be proved, on the facts before the court, becomes correspondingly greater (1977b §§58-76, 1980g, 1981c, 1982a, 1982d, 1986a, 1986e, 1986g, 1987a, 1988d).

Some of the classical authorities on the Anglo-American law of evidence, like Wigmore, seem to have had this kind of probability tacitly in mind. Others, like Wills, seem to have assumed that Pascalian probability is at stake here, and that is perhaps the commonest contemporary doctrine (1977b, §15). However, I was encouraged by two academic lawyers at Oxford, Rupert Cross and Tony Honoré, to believe that there was still something to be said on the subject. And in order to be able to say this I was led by my plan for a Baconian analysis of the Anglo-American standards of proof to make a much deeper investigation of Pascalian probability than I had hitherto had occasion to make. In particular the question how both Pascalian and Baconian concepts of probability are possible turned out to be soluble only when one asked also how different probabilistic interpretations of the Pascalian calculus are possible. Just as different types of Pascalian probability — personalist, propensity, relative frequency, etc — are analogous to different types of deductive inferability, so too the additivity or complementationality that distinguishes Pascalian from Baconian probability is analogous to the difference between a complete and an incomplete deductive system (1975b,

28

1977b §§1-3, 1980d, 1986a, 1989c §§12-15). And in the light of these analogies one can replace the mere cataloguing of different types of probability by an adequately principled pluralism.

4. The interface with psychology

My wife — *alias* the cognitive psychologist Gillian Cohen — drew my attention in 1976 to the work of Amos Tversky, Daniel Kahneman and some other like-minded psychologists. Assuming the existence of only one valid mode of probability-judgement, these psychologists were claiming that statistically untutored people are prone to make a variety of systematic errors in their attempts to execute such judgements. So a brief discussion of that claim was included at a late stage in my book about the standards of judicial proof (1977b, §70): what the psychologists described as being mistaken reasoning about Pascalian probabilities looked instead like being correct reasoning about Baconian ones. But as I read more and more of the relevant psychological literature it became clear that the problem was more complex and a much fuller treatment was needed. Not only was there a quite wide variety of experimental data on probability-judgement that needed discussion, but also the discussion of those data could not be kept entirely separate from the discussion of other kinds of alleged fallacy in everyday patterns of reasoning such as the so-called "fallacy of the converse" or the four-card puzzle studied by P. C. Wason and Philip Johnson-Laird. The initial outcome of this was an exploration of how the Pascalian-Baconian duality operates (1979b, 1980m). Then, after an oral exchange with Tversky, and a conversation with Stevan Harnad, during a conference of the Society for Philosophy and Psychology at Ann Arbor in 1980, I published a target article on the subject in *The Behavioral and Brain Sciences* (1981a) which, along with another relevant paper (1982e), led to a good deal of discussion and amplification (1981a, 1983f, 1984d, 1986h §§15-19, 1987d).

No doubt it was primarily through having a pluralist approach to the philosophy of probability that I was led to question some of the conventional assumptions that had prevailed among experimental psychologists in the 1970's about human judgement and rationality. But I had already found the interface between philosophical analysis and experimental psychology an exciting area in which to work. In the mid-60's a Linguistic Circle was formed at Oxford. Its formation was largely due to the initiative of Christopher Ball, but it benefited also — in a small research seminar — from the active participation of two cognitive psychologists, Ann Triesman and John Marshall. Naturally, our attention was captivated by the psycholinguistic turn that had been taken by

Chomsky's revolution in grammatical theory. And I tried to come to terms with some of these developments in the second edition of the book on meaning (1966b, pp. 36-56), in various articles (1965, 1970b, 1973c) and in contributions to a 1966 conference on psycholinguistics at Edinburgh (1966e) and to a 1981 joint symposium of the Royal Society and the British Academy (1981e).

Moreover the computational metaphor came to dominate not only psycholinguistics but all cognitive psychology during the 50's and 60's. My first, and somewhat naive, response to this was an excessively sceptical one (1955e). But invitations to comment on articles by Z. Pylyshyn in my (1978b), and by J. A. Fodor in my (1980l), induced a greater awareness of the metaphor's potential. Another influence in the same direction derived from a desire to investigate what lay behind the apparent analogy (exploited both by Saussure and by Wittgenstein) between chess-games and languages. It was clearly necessary here to explore the literature that had derived from the psychological study of chess-playing and from research into the principles for constructing appropriate expert-systems (1982f). And further investigations in this area were prompted by a felt need to analyse the implications of computational methodology for the theory of meaning and understanding (1984a, 1986h §§20-24). Indeed the very same errors that some philosophers have made in analyses of intellectual history that are not concerned with the computationalist hypothesis have been repeated independently by some computer scientists. Thus the evolutionary theory of thought-development that was mistakenly advocated by Stephen Toulmin, as argued in (1973g), and by Karl Popper as I argued in (1980i), has been echoed by an equally misplaced Darwinianism in regard to software development (1987b).

5. *What is the authority of intuitive judgements about deducibility or probability?*

My challenge to experimental psychologists on the rationality issue was further prompted by an awareness that analytical philosophy often needs to appeal, at crucial points in an argument, to premisses supplied by untutored intuitions. So it was as if in their critique of such intuitions the psychologists were knocking away the props on which analytical epistemologists rest many of their normative conclusions about deducibility or probability. Yet the psychologists were not building up any alternative foundations. Of course, appeals to intuition can be misleading in certain circumstances, especially when they are supposed to support theories about scientific reasoning and the intuitions invoked

30

are philosophers', not scientists (1975c, 1986h §13). But the philosophical analysis of epistemological or ethical norms cannot dispense with a
certain amount of induction from intuitions about particular cases
(1982c, 1986h §§6-14). And in making this point in the 1980's I suppose
that I was at last resuming an early interest in metaphilosophy (1949).
Gilbert Ryle had wisely advised philosophers in the early 1950's to post-
pone further work in metaphilosophy for thirty years or so until a great
deal more analytical philosophy had actually been done. But now that the
structure of the subject could be studied retrospectively from its actual
history rather than just programmatically, it seemed reasonable to write
a book (1986h) about the implications of its methodology.

Another intellectual interest, too, was affected by the rationality issue.
If we cannot in general rely on lay jurors or assessors to be rational in
their judgements about probabilities and deducibilities, then any system
of lay fact-finding is radically defective. A diploma in logic and statistics
is then needed by every trier of fact in the lawcourts, or rules of proof
have to be written into the legal system so as to turn every question of
fact into a question of — adjectival — law. In terms of European legal
history this would be a starkly reactionary move (1983a), and would also
constitute an obvious threat to the moral legitimacy of universal fran-
chise, just as anti-democratic governments in the present century have
often found it helpful to restrict the spread of analytical philosophy
(1986h pp. 33-34 and 61-63).

But, though it was theoretical problems about politics that originally
took me into philosophy, I should not wish to lay too much weight on the
political ramifications of the rationality issue. Intellectual honesty
requires that we treat the outcome of those ramifications as a consequence of what is true about human rationality, not *vice versa*. In that
way a pluralist philosophy of probability comes to support the case for
universal adult suffrage, but the converse does not hold. And in any case
my current interests lie more in the pursuit of epistemological issues than
in a return to political philosophy.

The Queen's College
Oxford
OX1 4AW
U.K.

REFERENCES

(1949) Are philosophical theses relative to language? *Analysis*, **ix**, 72-77.
(1950a) Mr. O'Connor's pragmatic paradoxes. *Mind*, **lix** 85-7.

(1950b) Mr. Strawson's Analysis of Truth. *Analysis*, **x**, 136-40, reprinted in R. P. Ammerman and M. G. Singer (eds.), *Belief, Knowledqe and Truth* (1970), 496-500.

(1951a) Tense-Usage and Propositions. *Analysis*, **xi**, 80-7, reprinted with corrections and additions in M. Macdonald (ed.), *Philosophy and Analysis*, Blackwell, 1954, 180-90.

(1951b) Teleological Explanations. *Proceedings of Aristotelian Society*, **li**, 255-92.

(1951c) Three-valued Ethics. *Philosophy*, **xxvi**, 208-227.

(1952) Survey of the Philosophy of History 1946-50. *Philosophical Quarterly*, **ii**, 172-86.

(1953) Gilbert Ryle's Philosophy of Mind (in Hebrew). *Iyyun*, **iv**, 229-39.

(1954a) On the Project of a Universal Character. *Mind*, **lxii**, 49-63.

(1954b) *The Principles of World Citizenship*. Oxford: Blackwell.

(1955a) A Relation of Counterfactual Conditionals to Statements of What Makes Sense. *Proc. Ar. Soc.*, **lv**, 45-82.

(1955b) On the Use of "the Use of". *Philosophy*, **xxx**.

(1955c) Assertion Statements. *Analysis* **xv**, 66-70 (jointly with A. C Lloyd.)

(1955d) Theory and Definition in Jurisprudence. *Proc. Arist. Soc.* Supp. Vol. **xxix**, 213-38.

(1955e) Can There be Artificial Minds? *Analysis*, **xvi**, 36-41.

(1957a) Has Collingwood been Misinterpreted? *Philosophical Quarterly*, **vii**, 149 ff.

(1957b) Can the Logic of Indirect Discourse be Formalised? *Journal of Symbolic Logic*, **xxii**, 222 ff.

(1958a) Prof. Goodstein's Formalisation of the Policeman. *Jour. Symb. Log.*, **xxiii**, p. 420.

(1958b) Critical Notice of A. N. Prior, *Time and Modality*. *Philosophical Quarterly*, **8**, 266-271.

(1959a) Prof. Prior on "Thanking Goodness that's Over". *Philosophy*, **xxxiv**, p. 360ff.

(1959b) Are Moral Arguments Always Liable to Break Down? *Mind*, **68**, 530-532.

(1960) A Formalisation of Referentially Opaque Contexts. *Jour. Sym. Log.*, **xxv**, 193-202.

(1961) Why Do Cretans Have to Say So Much? *Philosophical Studies*, **xii**, 72-78.

(1962a) Geach on Referring Expressions; and Rejoinder. *Analysis*, **23**, 6-8 and 10-12.

(1962b) Critical Notice of H. L. A. Hart, *The Concept of Law. Mind*, **71**, 395-412.

(1962c) Claims to Knowledge. *Proc. Arist. Soc.* Supp., Vol. **xxxvi**, 7. ff.

(1962d) *The Diversity of Meaning*. London: Methuen (first edition). (Greek translation of §19 in *Deucalion*, **17**, 1977, 92-105).

(1963a) Indirect Speech: A Rejoinder to Prof. A. N. Prior. *Phil. Studies*, **xiv**, 15-18.

(1963b) Crititical Notice of W. & M. Kneale, *The Development of Logic. Philosophical Quarterly*, **13**, 253-262.

(1964a) ,Do Illocutionary Forces exist?' *Phil. Quarterly*, **xiv**, 118-137, reprinted in K. T. Fann (ed.), *Symposium on J. L. Austin* (1969), 420-444, with an additional note on p. 468, and in J. F. Rosenberg and C. Travis (eds.), *Readings in the Philosophy of Language* (1971), 580-598.

(1964b) Indirect Speech: A Further Rejoinder to Prof. Prior. *Phil. Studies,* **xv**, p. 38 ff.

(1965) On a Concept of Degree of Grammaticalness. *Logique et Analyse*, **8**, 141-153.

(1966a) A Logic of Evidential Support. *British Journal for Philosophy of Science*, **21**, 2-43 & 105-126.

(1966b) *The Diversity of Meaning*. London: Methuen (second edition, revised).

(1966c) What has Confirmation to do with Probabilities? *Mind*, **lxxv**, 463-481.

(1966d) More about Knowing and Feeling Sure. *Analysis* **xxvii**, 11-16.

(1966e) Critical discussion of paper by Fodor and Garrett on "Competence and Performance" in J. Lyons and R. J. Wales (eds.), *Psycholinguistics Papers, Proceedings of the Edinburgh Conference 1966*, 163-173.

(1966f) Critical Notice of J. Fodor and J. Katz (eds.), *The Structure of Language. Philosophical Quarterly*, **16**, 165-177.

(1966g) Critical Notice of D. J. O'Connor (ed.). *A Critical History of Western Philosophy*, **41**, 268-273.

(1966h) Articles in *Chambers' Encyclopaedia*, Oxford: Pergamon Press, on Belief; Induction; Knowledge, Theory of; Meaning; Metaphysics; Monism; Pluralism; Philosophy of Common Sense; Rationalism; Semantics; Solipsism; Teleology; Transcendentalism.

(1968a) Criteria of Intensionality. *Proc. Arist. Soc.* Supp. vol. **42**, 123-142.

(1968b) Geach's Problem about Intentional Identity. *Journal of Philosophy*, **lxv**, 329-335.

(1968c) An Argument that Confirmation Functors for Consilience are Empirical Hypotheses. In I. Lakatos (ed.), *The Problem of Inductive Logic*, 247-250.

(1970a) What is the Ability to Refer to Things, as a Constituent of a Language-Speaker's Competence? In *Linguaggi nella Societáe nella Tecnica*. Milan: Edizioni di Communita, 255-268. (Proceedings of the Olivetti Centenary Symposium at Milan, 1968).

(1970b) Some Applications of Inductive Logic to the Theory of Language. *Amer. Phil. Quart.* **7**, 299-310.

(1970c) Critical Notice of J. R. Searle, *Speech-Acts. Philosophical Review* **79**, 545-557.

(1970d) The Role of Inductive Reasoning in the Interpretation of Metaphor (jointly with Avishai Margalit). *Synthese*, **21**, 469-487, reprinted in *Semantics of Natural Language*, ed. D. Davidson and G. Harman, 1972, 722-740.

(1970e) *The Implications of Induction*. London: Methuen.

(1971) The Inductive Logic of Progressive Problem-Shifts. *Revue Internationale de Philosophie*, no. 95-6, 62-77.

(1972a) A Reply to Swinburne. *Mind*, **81**, 249-50.

(1972b) Some remarks on Grice's Views about the Logical Particles of Natural Language. In Y. Bar Hillel (ed.), *Pragmatics of Natural Languages*, 50-689, German trans. in Georg Meggle (ed.) *Analytische Kommunikationstheorie*, Suhrkamp Verlag, 1978, 395-418.

(1973a) The Paradox of Anomaly. In R. J. Bogdan and I. Niiniluoto (eds.), *Logic, Language and Probability*, Dordrecht: Reidel, 78-82.

(1973b) The Non-existence of Illocutionary Forces: A Reply to Mr. Burch. *Ratio*, **15**, 125-131.

(1973c) Is contemporary linguistics value-free? *Social Science Information*, June, 1973, 53-64.

(1973d) A Note on Inductive Logic. *Journal of Philosophy*, **70**, 27-40.

(1973e) Spoken and Unspoken Meanings. *Times Literary Supplement*, October 5, 1973, 1161-1162.

(1973f) Critical Notice of A. N. Prior, *Objects of Thought, Mind*, **82**, 127-142.

(1973g) Critical Notice of S. Toulmin, *Human Understanding*, vol. I, *Brit. Jour. Phil. Sci.* **24**, 41-61.

(1974a) Professor Hull and the Evolution of Science. *Brit. Jour. Phil. Sci.*, **25**, 334-336.

(1974b) Roger Gallie and Substitutional Quantification. *Analysis*, **34**, 69-73.

(1974c) Guessing. *Proceedings of the Aristotelian Society*, **lxxiv**, 189-210.

(1974d) Speech Acts. In T. A. Sebeok (ed.), *Current Issues in Linguistics* vol. **xii**, 173-208.

(1975a) Reply to Wesley Salmon on Statistical Explanation. Reprinted in S. Körner (ed.), *Explanation*, Oxford: Blackwell, p. 152ff.

(1975b) *Probability – the One and the Many*, British Academy Philosophy Lecture, London: Oxford University Press; also published in *Proceedings of the British Academy*, **61**, 1976, 88-108.

(1975c) How Empirical is Contemporary Logical Empiricism? (Bar-Hillel Festschrift). *Philosophia*, **v**, 299-315; reprinted in *Language in Focus: Foundations, Methods and Systems*, ed. A. Kasher, 1976, 359-376.

(1975d) A Reply to T. M. Simpson. *Logique et Analyse*, **69-70**, 209-214.

(1976a) Why should the science of nature be empirical? Royal Institute of Philosophy Lecture, Oct. 25, 1974, Published in *Impressions of Empiricism*, ed. G. Vesey, London: Macmillan, 168-183.

(1976b) The Progress of Science. Read at S. I. L. F. S. Conference, Chiavari, translated in *Il Concetto di Progesso nella Scienza* ed. E. Agazzi, Feltrinelli: Milan, 105-120, under the title "Il progresso nella scienza".

(1976c) A Conspectus of the Neo-Classical Theory of Induction. In R. Bogdan (ed.) *Local Induction*, Dordrecht: Reidel, 235-262. Summarised in *Zentralblatt für Mathematik und ihre Grenzgebiete*, Band **331**, 1977, 16-17: March 23, 1977.

(1976d) How Can One Testimony Corroborate Another? *Essays in Memory of Imre Lakatos*, ed. R. S. Cohen et al., Reidel. 65-78.

(1977a) Can the Conversationalist Hypothesis be Defended? *Philosophical Studies*, **31**, 81-90.

(1977b) *The Probable and the Provable*. Oxford: Clarendon Press. (A Spanish translation of sections of Part II, entitled 'El concepto de probabilidades en pruebas judiciales', appeared in *Teorema* **vii**, 1978, 277-302. A Chinese translation of some sections appeared in *Ziram Kexue Zhexue Wenti Congkan* 3, 1983.).

(1977c) The Causal Theory of Perception. *Proc. Aristot. Soc.* Supp. Vol. **li**, 127-141.

(1978a) The Coherence Theory of Truth. *Philosophical Studies*, **xxxiv**, 351-60

(1978b) Rational reconstruction of inferential processes – a task straddling the A1-C3 boundaries. *The Behavioral and Brain Sciences*, **1**, 101-2.

(1978c) Is Popper more relevant than Bacon for scientists? *Times Higher Education Supplement*, July 14, 1978, 11.

(1978d) What scientists cannot learn from Popper. *Times Higher Edication Supplement*, Sept. 1, 1978, 11

(1979a) Rescher's Theory of Plausible Reasoning. In E. Sosa (ed.), *The Philosophy of Nicholas Rescher*, Reidel, 49-60.

(1979b) On the Psychology of Prediction: Whose is the Fallacy? *Cognition*, **vii**, 4, 385-402.

(1979c) The Semantics of Metaphor. In A. Ortony (ed.) *Metaphor and Thought*, Cambridge: Cambridge University Press, 64-77.

(1979d) Philosophy of Language and Semiotics. In S. Chatman, U. Eco and J. M. Klinkerberg (eds.), *A Semiotic Landscape: Panorama Sémiotique* (Proceedings of the First Congress of the International Association for Semiotics, Milan, 1974), The Hague: Mouton, 19-28.

(1980a) Some Historical Remarks on the Baconian Conception of Probability. *Journal of the History of Ideas*, **lxi**, 219-231.

(1980b) British Philosophy 1945-1977. In *Handbook of World Philosophy*, ed. J. R. Burr, Greenwood, 99-116. Chinese Translation in *Zhexue Yiccong*, **6**, 1983, 77-82.

(1980c) Is a Criterion of Verifiability Possible? *Midwest Studies in Philosophy*, **v**, 347-353.

(1980d) What has inductive logic to do with causality? In M. Hesse, *Applications of Inductive Logic*, ed. L. Jonathan Cohen, Clarendon Press, 156-171.

(1980e) The Individuation of Proper Names. In Z. Van Straaten (ed.) *Philosophical Subjects: Essays on the work of P. E. Strawson*, Clarendon, 140-163.

(1980f) Is the epistemology of science a kind of logic or a kind of history? (in Russian). *Voprosy Filosofii*, **2**, 143-156. (in Chinese) *Ziran Kexue Zhexue Wenti Congkam*, **1**, 1981, 12-18.

(1980g) The Logic of Proof. *Criminal Law Review*, 1980, 91-103, 257-260, 747-752.

(1980h) Bayesianism versus Baconianism in the evaluation of medical diagnoses. *Brit. Jour. Phil. Sci.*, **31**, 45-62. French trans. in Medecine et Probabilites, ed. A. Fagot, Universite Paris Val-de-Marne: Creteuil, 1982. Italian translation in Pegaso-scienza e filosofia 2.

(1980i) Some comments on third world epistemology. *Brit. Jour. Phil. Sci.*, **xxxi**, 175-180. Revised and expanded as Third World Epistemology in G. Currie and A. Musgrave (eds.), *Popper and the Human Sciences*, Dordrecht: Nijhoff, 1985, 1-12.

(1980j) The Problem of Natural Laws. In D. H. Mellor (ed.), *Prospects for Pragmatism*, Cambridge, 211-228.

(1980k) What has Science to do with Truth? *Synthese*, **xlv**, 489-510.

(1980l) Some defects in Fodor's computational theory. *The Behavioral and Brain Sciences*, **3**, 75-6.

(1980m) Whose is the fallacy? A rejoinder to Daniel Kahneman and Amos Tversky. *Cognition*, **8**, 89-92.

(1981a) Can Human Irrationality be Experimentally Demonstrated? *The Behavioral and Brain Sciences*, **4**, 317-331 and 359-370.

(1981b) Induction. In *Dictionary of the History of Science*, ed. W. F. Bynum, E. J. Browne, & R. Porter, London: Macmillan, 203-205. Spanish translation: Barcelona: Editorial Harder, 1986.

(1981c) Subjective probability and the paradox of the gatecrasher. *Arizona State Law Journal*, 27-34.

(1981d) Who is starving whom? *Theoria*, **47**, 65-81.

(1981e) Some remarks on the nature of linguistic theory. *Philosophical transactions of the Royal Society* B 295, 1981, 235-243. Reprinted in *The Psychological Mechanisms of Language*, ed. H. C. Longuet-Higgins et al., London, Royal Society & British Academy, 21-29.

*Poznań Studies in the Philosophy
of the Sciences and the Humanities
1991, Vol. 21, pp. 21-37*

L. Jonathan Cohen

FROM A HISTORICAL POINT OF VIEW

1. The move from political philosophy to the philosophy of language

Sometimes it is easier to understand and evaluate a system of ideas if one knows how the system has evolved. One can grasp more of its inner dynamics that way. So I shall describe here how I think that my own views about probability and induction have developed. But I make no more than the subjective claim that this is how the main part of the development seems to me now to have taken place. The period of time concerned is nearly half a century, during which I have read widely and given talks or lectures in many different countries, and I may well have forgotten some important influences that ought to be mentioned. The original seed or source of an idea may sometimes have been another person's article or book that I no longer remember reading or a conversation in which I no longer remember participating. If so, I can only hope that the original author of that idea, and my readers, will forgive me.

What originally took me into an academic career — in October, 1947 — was a desire for sufficient leisure, and the right critical *ambiance*, to write a book on political philosophy. It was this aspect of philosophy that preoccupied me while I was stationed on war service (in naval intelligence) in Mombasa and Colombo during the years 1943 to 1945. All the available books on the subject seemed to assume that the proper norm for political organisation was a plurality of sovereign states within which each individual had no political obligations beyond the frontiers of his own state. Yet in practice, where it suited them, governments frequently invoked international law and the ties of international association, in ways that implied some fracturing of this long-established norm. So I wanted very much to write a book on the principles of world citizenship.

It was obviously useless to attempt this until I had returned to Oxford, which I had left after only four terms in January, 1941. I needed to spend a lot more time in structured studies. Of course, other exciting avenues also began to open up after hostilities in the Indian Ocean area

(1981f) Inductive Logic 1945-1977. In E. Agazzi (ed.) *Modern Logic – A Survey*. Dordrecht: Reidel, 343-375.

(1982a) The problem of prior probabilities in regard to the standards of forensic proof. *Ratio* **xxiv**, 72-76.

(1982b) Bolzano's theory of induction. *Acta Historiae Rerum Naturalium nec non Technicurum*, special issue 13, Prague: Czechoslovak Academy of Sciences. Reprinted in M. K. Salmon (ed.), *The Philosophy of Logical Mechanism*, Dordrecht: Kluwer, 1981, 29-39.

(1982c) Intuition, Induction, and the Middle Way. *Monist*, **lxv**, 287-301, in honour of W. Sellars.

(1982d) What is necessary for testimonial corroboration?, *British Journal for Philosophy of Science*, **33**, 161-164.

(1982e) Are people programmed to commit fallacies? Further thoughts about the interpretation of experimental data on probability judgment. *Journal for the Theory of Social Behavior* **12**, 251-274. Polish translation in *Poznańskie Studia z Filozofii Nauki* **10**, 1986, 133-165.

(1982f) Chess as a model of language. *Philosophia*, **11**, 51-87.

(1983a) Freedom of Proof. (1982 Austin Lecture to U. K. Association for Legal and Social Philosophy). *Archiv für Rechts- und Sozial Philosophie*, Beiheft **16**, 1-21, (*Facts in Law*, ed. W. Twining, Wiesbaden: Franz Steiner Verlag.).

(1983b) How far is induction rationally justifiable? *Epistemology and Philosophy of Science*, Proceedings of the 7th International Wittgenstein Symposium, Vienna: Verlag Hölder-Pichler-Tempsky, 245-254.

(1983c) The Unity of Reason. *Social Theory and Practice*, **IX**, 245-269.

(1983d) Belief, acceptance and probability. *The Behavioral and Brain Sciences*, **6**, 248-9.

(1983e) The Baconian Tradition: Verisimilitude and Legisimilitude (in Chinese). *Ziran Kexue Zhexue Wenti Congkan* 4.

(1983f) The controversy about rationality. *The Behavioral and Brain Sciences*, **6**, 510-517.

(1984a) Semantics and the Computational Metaphor. In R. B. Marcus et al. (ed.) *Proceedings of the 7th International Congress for Logic, Methodology and Philosophy of Science* North-Holland, 597-621. Polish translation in *Poznańskie Studia z Filozofii Nauki* **11**, 1989, 29-59.

(1984b) On the depth and fit of behaviorist explanation. *The Behavioral and Brain Sciences*, **7**, 591-592.

(1984c) Some Psychological Implications of Tolerance in the Philosophy of Probability. In *Probability, Statistics and Inductive Logic*. Proceedings of the International Conference on the Foundations of Probability and Statistics, Luino, Sep. 17-19, 1981 *Epistemologia* **6**, 1984, 213-222.

(1984d) Can irrationality be discussed accurately? *The Behavioral and Brain Sciences*, **7**, 736-738.

(1985a) A Problem about Ambiguity in Truth-Theoretical Semantics. *Analysis*, **45**, 129-134.

(1985b) The verisimilitude-legisimilitude duality. In *La Nature de La Vérité scientifique*, Archives de l'Institut International des Sciences Théoriques, **27**, 55-66.

(1986a) Twelve Questions about Keynes's Concept of Weight. *Brit. Jour. Phil. Sci.*, **37**, 263-278. Italian translation in R. Simili (ed.) *L'epistemologia di Cambridge 1850 – 1950*, Mulino, 1987.

(1986b) How is conceptual innovation possible?, *Erkenntnis*, **25**, 221-238.

(1986c) Verisimilitude and Legisimilitude. In T. A. F. Kuipers (ed.) *What is Closer-to-the- -Truth?* (*Poznań Studies in the Philosophy of the Sciences and the Humanities*, **10**) Amsterdam: Rodopi, 129-144.

(1986d) An Ambiguity in Egan's Concept of Belief. *Annals of Theoretical Psychology*, **4**, 365-367.

(1986e) The Role of Evidential Weight in Criminal Proof. *Boston University Law Review*, **66**, 635-649. Reprinted in E. Green and P. Tillers (eds.) *Probability and Inference in the Law of Evidence*, Dordrecht: Reidel, 1988, 113-128.

(1986f) What sorts of machines can understand the symbols they use? In *Proceedings of the Aristotelian Society*, Supplementary Volume **60**, 81-95.

(1986g) The Corroboration Theorem: A Reply to Falk. *Mind*, **95**, 510-512.

(1986h) *The Dialogue of Reason: An Analysis of Analytical Philosophy*. Oxford: Clarendon Press.

(1987a) On Analysing the Standards of Forensic Proof: a Reply to Schoeman. *Philosophy of Science*, **54**, 92-97.

(1987b) A Note on the Evolutionary Theory of Software Development. *British Journal for the Philosophy of Science*, **38**, 381-84.

(1987c) Laws, Coincidences and Relations between Universals. In P. Pettit, R. Sylvan and J. Norman (eds.) *Mind, Morality and Metaphysics: Essays in Honour of J. J. C. Smart*, Oxford: Blackwell, 16-34.

(1987d) What are the foundations of normative theories about human reasoning? The *Behavioral and Brain Sciences*, **10**, 312-313.

(1988a) Creativity: Some Ambiguities in the Issue. *Journal of Social and Biological Structures*, **11**, 77-79.

(1988b) Realism Versus Anti-Realism: What is the Issue? In E. Ullmann-Margalit, *Science in Reflection* (The Israel Colloquium: Studies in History, Philosophy and Sociology of Science, vol. **3**, Boston Studies in the Philosophy of Science, vol. **110**), Dordrecht: Kluwer, 81-96.

(1988c) Some Logical Distinctions Exploited by Differing Analyses of Pascalian Probability. In E. Agazzi (ed.) *Probability in the Sciences*, Dordrecht: Kluwer, 67-76.

(1988d) The difficulty about conjunction in forensic proof. *The Statistician*, **37**, 415-416.

(1988e) Who Introduced Mathematical Probability into the Philosophy of Induction? In E. Scheibe (ed.), *The Role of Experience in Science*. Berlin: de Gruyter, 75-81.

(1989a) Belief and Acceptance. *Mind*, **98**, 347-389.

(1989b) Are Inductions Warranted? and What Use are Beliefs that We Do not Take to be Warranted? *Analysis*, **49**, 1-4 and 7-8.

(1989c) *An Introduction to the Philosophy of Induction and Probability*. Oxford: Clarendon Press. (A Chinese translation is forthcoming).

(1990a) What is the Role of Mathematical Models in Psychology. In J. Brzeziński, T. Marek (eds.) *Action and Performance: Models and Tests. (Poznań Studies in the*

Philosophy of the Sciences and the Humanities, **14**). Amsterdam: Rodopi, 5-15, Polish translation in *Poznańskie Studia z Filozofii Nauki*, 12.

(1990b) Idealisation as a Form of Inductive Reasoning. In: J. Brzeziński, F. Coniglione, T. A. F. Kuipers and L. Nowak (eds): *Idealization – I: General Problems. (Poznań Studies in the Philosophy of the Sciences and the Humanities*, **16**). Amsterdam: Rodopi, 153-166.

(Forthcoming a) Ontology and the Philosophy of Language. In *Handbuch der Sprachphilosophie*, Berlin: de Gruyter.

(Forthcoming b) Two problems for the explanatory coherence theory of acceptability, *Behavioral and Brain Sciences*.

(Forthcoming c) *An Essay on Belief and Acceptance*. Oxford: Clarendon Press.

PART II

THE METHOD OF RELEVANT VARIABLES AND OTHER MODES OF SCIENTIFIC REASONING

*Poznań Studies in the Philosophy
of the Sciences and the Humanities*
1991, Vol. 21, pp. 41-63

Leszek Nowak

THE METHOD OF RELEVANT VARIABLES AND IDEALIZATION[*]

I. Introduction

The system of logic of induction elaborated by L. Jonathan Cohen (1966a), (1970), (1977) seems to be fairly close to the research practice. Witness its ability to cope with non-trivial pieces of it and also to conceptualize some methodological concepts of high relevance to understanding the method of science. The concept of idealization is one of them.

It is an explicit Author's aim to manage with idealizations in scientific practice:

> "An adequate analysis (...) must also show how so-called 'fact-correcting' generalizations — idealisations that appear to tell us how nature should behave rather than how nature actually behaves — may be brought within the scope of inductive logic" (Cohen (1970), p. 2). For "in science (...) no comprehensive theory can normally emerge in a given field without a certain element of deliberate idealisation" (ibid., p. 167).

L. J. Cohen also criticizes some widespread methodological approaches for their negligence of the problem of idealization even if taking it into consideration would entail changing their main ideas (e.g., as it is, according to the Author, with the main approaches to *verisimilitude*, Cohen (1987), p. 130). What is to be especially underlined is that L. Jonathan Cohen not only recognizes the importance of idealization in science ((1970), pp. 4, 77, 142ff, 167, (1977), pp. 182ff, 242, (1983), p. 250, and others), or ethics ((1970), p. 74, (1981), p. 321), but also in the philosophy of science itself:

> "... no real understanding, no potentially fruitful explanation, is ever achieved until the description and classification of concrete individuals, populations and substances gives way to study of abstract properties and of the idealised individuals (...). Natural history must give way to theory-construction within scientific enquiry itself, and within philosophy of science an analogous transition has to be made" ((1970), p. 4, cf. also ibid., pp. 157, 167, (1977), p. 166).

Every philosopher of science is aware of the fact that science applies idealization. Not so many of them try to draw from that fact some consequences for their theories. And few pose it close enough to the center of their research fields. Professor Cohen is one of these: he tries to reconstruct idealization in his conceptual apparatus as a special case of what he terms the method of relevant variables and what constitutes for him the central part of scientific logic.

One thing has certainly been proved by L. Jonathan Cohen's theory. That the accusations of the inductivist approach of being incapable of handling idealizations are based on a too simplicistic understanding of what induction consists in (Cohen (1970), p. 77, (1990)). It is in the light of his conception, I must admit, that my own assessments of the kind (cf. my (1971a), p. 71ff, (1980), p. 33) also turned out to be too hasty and unsound.

Although being impressed by the subtlety and innovativeness of L. Jonathan Cohen's conception, I cannot share the main results of his analysis as far as the place of idealization in science is concerned. I am going to argue that what he has reconstructed is not the method of idealization, but a certain similar procedure. None the less, what L. J. Cohen has indeed discovered is of great methodological importance and has not yet been reconstructed in the framework, which I assume here, viz. in one placing idealization in   the center of scientific methods. I shall try to find a place for the method of relevant variables in that framework.

II. The method of relevant variables vs. idealization

1. The main idea that allows L. J. Cohen to maintain that his system includes the method of idealization underlies his method of relevant variables. Let us present the main idea of the Author in a simplified scheme.

Given the hypothesis:

(t) Anything, if it is A, is B.

and the set of relevant variables, say (R, S), with R being more relevant than S, one may find a canonical sequence of more and more thorough tests for (t).

A favorable outcome for canonical test $T1$ is constituted by

(a) an instance of A that is at the same time B — in circumstances that are normal in relation to R and S.

Then the hypothesis (t) gains lst-grade support.

A favorable outcome for the more thorough test $T2$ is constituted by

(a) (as above), and

(b) by an instance of A being B under each variant of the variable R — but still in circumstances that are normal to the variable S.

The favorable result for the test $T2$ witness that, in normal conditions in relation to S, no change of values of the relevant variable R offers a reason to falsify the hypothesis (t). The test $T2$ gives 2nd-grade support for the hypothesis under consideration.

A favorable result for the most thorough — under our simplification limiting the space of relevant variables only to two items — test $T3$ is the following:

(a) (as above), and

(b) (as above), and

(c) by an instance of A being also B under each combination of a variant of R with a variant of S.

This test would give 3rd-grade support for the hypothesis (t) eliminating any possibility of its being false. Obviously, provided that there are no other variables relevant to our hypothesis than R and S. The author frequently stresses the fact that this procedure depends on the truthfulness of the initial decision to enlist, and to order, certain variables as relevant ones (e.g., (1977), p. 180).

If the hypothesis (t) has passed the test $(T1)$ (resp. $(T2)$) but failed to pass the more thorough test $(T2)$ (resp. $(T3)$), it is claimed to gain merely 1st-grade (resp. 2nd-grade) support. If it has passed the test connected with all the relevant variables, i.e. the most thorough test, it is said to be fully supported.

"The reliability of a hypothesis is graded by its ability to survive tests under more and more powerful variation of circumstances, where power is increased by introducing more and more types of relevant circumstances into the variations" ((1980c), p. 166).

2. The Author derives numerous far-reaching consequences from the method of relevant variables. He interprets in its terms not only what can be expected, for instance, the rationality of induction (e.g., (1970), p. 183ff, (1983a)), the structure of confirmation and of probability (e.g., (1970), p. 25ff, #12-13, (1977), pp. 11ff, 188ff) or criteria of rational belief (e.g., (1977), p. 310ff). He also interprets in terms of his ramified theory what one could hardly, if at all, expect: e.g., the notion(s) of necessity (e.g., (1970), p. 219ff, (1977), p. 232ff), the distinction between law-like statements and accidental generalizations (e.g., (1970), p. 114ff, (1980a)), the notion of disposition (e.g., (1977), p. 324ff) or the rationality of acts (e.g., (1983b)). L. J. Cohen applies these results to more concrete topics as well. And not only to reasoning about causes and effects (e.g., (1980c)), but also to reasoning about reliability of a legal rule (e.g., (1970), p. 155ff, (1977), p. 245ff) or of linguistic hypotheses about the meaning of words (e.g., Cohen and Margalit (1970), Cohen (1986)).

44

All these results are of high importance. Here, however, I would like to discuss another trait in L. J. Cohen's approach, namely its ability to grasp the method of idealization as it is normally applied in science.

3. Here it is the typical — in its simplest form — way of use of idealizations in science. Galileo's law of free-fall says that under conditions of the lack of resistance $(R = 0)$, the following holds:

$$(1^*) \quad F = Gm$$

where F is a force, G is the gravitational constant and m is the mass. Formula (1^*) is to be valid for the frictionless free-fall which, obviously, does not occur at all. Therefore, it is corrected in order to take into account the force of resistance:

$$(2^*) \quad F = Gm - R.$$

This can certainly be conceptualized in many ways. One of them, in a simplified form (a more sophisticated reconstruction cf. Such (1974), cf. also Krajewski (1974)), runs as follows (the notions employed here are explained in my (1971b), (1980)). Galileo's law of free fall is an idealizational statement equipped, apart from a realistic condition $ff(x)$ referring equation (1^*) to a free falling body x, in one idealizing condition in its antecedent:

$$(1) \quad \text{if } ff(x) \text{ and } R(x) = 0, \text{ then } F(x) = Gm(x)$$

In order to apply this law to actual bodies falling in conditions of non-zero resistance, it is concretized:

$$(2) \quad \text{if } ff(x) \text{ and } R(x) \neq 0, \text{ then } F(x) = Gm(x) - R(x).$$

The latter statement is supposed here to be a factual one. As a matter of fact, it nevertheless contains additional, silent idealizing conditions (three in the light of Galileo's approach, many more in the light of Newton's theory — cf. Such (1974)). This simplified example suffices in our discussion.

4. There are at least two ways of interpreting idealizations in terms of the method of relevant variables outlined above (cf. above **1**). The Author claims they do not exclude each other ((1970), p. 143/144)[1].

First consists in the experimental approximation:

"Typicaly, a domain of individuals is postulated that are not subject to the operation of whatever variables would otherwise create exceptions to the generalization. For example, if the generalization correlates the velocity of a falling body with the period for which it has been falling, it may be necessary to confine the domain of the generalization to the objects in a frictionless medium" ((1970), p. 144).

In this connection the Author explains:

"But such a situation is not a mere figment of logical imagination. If an experimenter wishes to approximate as closely as he can, under laboratory conditions, to the idealised situations that form the domain of some formed proposed correlational generalization, he may well aim at reports like

(r) (Ex) $(Ax$ and Bx and $T_1x)$

(where) T_1x implies that x is unaffected by any variant of R, S, (the notation adjusted to the convention adopted above, **II.1** $-$ L. N.) (...) Drop your object in as near a vacuum as you can create, and you can measure its speed of fall, after such-or-such an interval, under as nearly ideal, frictionless conditions as you can get" ((1970), p. 152).

The point, however, is that reports of type (r) cannot be obtained in any experimental conditions: it is always so that the resistance R in such conditions is greater than zero. If so, the standard inductive passage from (r) to (t) which the Author refers to ((1970), p. 152), actually fails for idealizational statements. No experiment is able to give a report:

$$(3) \quad ff(a) \text{ and } R(a) = 0 \text{ and } F(a) = Gm(a) = u$$

justifying (1). What the experimenter gets for the free-falling body is merely this:

$$(4) \quad ff(a) \text{ and } R(a) = M \text{ and } F(a) = Gm(a) - R(a) = u - M$$

And (4) does not inductively justify the idealizational statement (1), as is claimed, but, at best, its concretization (2). However small M might be, its size is negligible for physics, not for the philosophy of science. The

46

problem, how an observation of real bodies can be generalized into idealizational statements, still remains open.

Yet, the method of experimental approximation could be modified in this manner. In a certain place ((1977), p. 186) the Author has remarked that the unmodified version of the hypothesis, (t), logically implies the modified one, (t') (cf. below, **5**). Under our reconstruction this does not hold good (e.g., (2) and (1) refer to different levels of reduction — a larger discussion cf. my (1990)) but let us omit this for the sake of discussion. If this is adopted, then one could modify the method of experimental approximation as follows: an idealizational statements is acceptable, if an inductively supported statement, more general than the idealizational hypothesis under consideration, has been found. However, this line of reasoning (e.g., Hempel (1952), Batog (1976), Niiniluoto (1986), Hamminga (1989)) does not do justice to the method of idealization. Such a solution would be susceptible to some objections resulting from the research practice (cf. my (1971b)). If it is possible to introduce inductively the factual statements, what would be the cognitive goal for substituting the limiting cases ("zero-s") in them to get "idealizations"? What could be their role in explaining and predicting phenomena, if all that can be performed by the (factual) generalizations of idealizational laws? Why actually are these, and not any other, special cases of factual statements distinguished and why do they even deserve the dignity of "fundamental laws" or "principles"? It does not seem that the outlined, alternative version of handling the troubles of the experimental approximation method would be in the spirit of the Author's own assumptions, which makes the difficulties mentioned above even sharper.

5. The second method which may be used for dealing with idealizations consists in adding new conditions into the antecedent of the hypothesis.

> "Whatever the evidence, every testable and unmodified proposition has some suitably modified version (or versions) that is (or are) proof against falsification. If the evidence gives some positive, but not full, support, the antecedent of (appropriate hypotheses) may be thought of as being designed to exclude the operation of the obstructive variant or combination of variants" ((1970), p. 148).

Let us assume that the hypothesis (t) has passed the test $(T1)$ but failed to gain a favorable outcome in the test $(T2)$. Now, instead of attaching (t) the 1st-grade support, one may modify it by introducing a certain condition to its antecedent:

(t') Anything, if it is A, and not affected by any variant or combination of variants of R and S, is B.

(t') possesses full support but at the cost of the addition of a new restriction to its antecedent. Similarly, instead of attaching (t) the 2nd-grade support one may present it in an even more restricted form:

(t'') Anything, if it is A and R, and not affected by any variant of S, is B.

Again, (t'') retains full support at the cost of even more restricted formulation of its antecedent. For, in general, there is always some appropriate ith version, U', of the unmodified hypothesis U such that, for any evidence E, U is supported in i/n degree relative to E if and only if U' is supported by evidence E in n/n degree, where n is the number of variables relevant to U (Cohen (1970), p. 150).

Although the procedure described above shows some remarkable similarity to that of concretization of idealizational statements, the main difference is that in the former the formula of (t) remains the same, whereas in the process of explanation through idealization and concretization actually the consequent of an appropriate conditional actually changes. In other words, what changes during the process of inductive testing is the antecedent of (t), not its consequent as is the case while concretizing the idealizational statements; let us add that this difference has been noticed by the Author himself (e.g., (1990)).

As a result, however, the defense against falsification looks quite differently in case of idealizational statements than in case of factual ones. Let us analyze this a little more carefully.

Given is the equation (1^*). Assume that the author of it does not yet recognized the significance of the resistance of the air as the disturbing factor for the free-fall. Given that state of knowledge, a methodological reconstruction of the equation may be expressed in the factual statement:

(1') if $ff(x)$, then $F(x) = Gm(x)$

Let us assume that the researcher applies it to a certain body b obtaining thus the theoretical value of F: $F(b) = Gm(b)$ which differs from the empirical result of measurement by deviation d. Then — quite in spirit of the method of relevant variables (Cohen 1990) — the statement (1') is not falsified but corrected. The researcher namely attempts to identify the source of deviations. Let him find the resistance R as an additional factor significant for the phenomenon of free-fall. Then, under our reconstruction, he does two things. The first is still in the spirit of L. J. Cohen's approach: our researcher adds a new condition to the antecedent of (1'). Namely, he adds the idealizing condition $R(x) = 0$ formulating thus the idealizational statement (1). But the second thing he does is beyond the scope of the method of relevant variables. Namely, he takes into account the working of the resistance of the medium in order to explain the deviations he had faced. As a result, he actually corrects the consequent of (1) obtaining (2). But, if this is so, then the second considered method of handling the peculiarity of idealizations within the framework of L. J. Cohen's theory seems to be descriptively, i.e. from the standpoint of the actual research practice, insufficient.

From the normative standpoint, the second method would bring some unwanted results. Let us consider the rule which seems to be contained in this:

> "if a generalization is falsified by certain variants of a relevant variable but not others, a mention of the latter can be introduced into the antecedent of the generalization as to exclude falsification of the former. In this way a higher grade of inductive reliability can be maintained for a hypothesis, though at the cost of accepting a lower grade of simplicity for it" ((1980a), p. 368).

Now, this rule indeed seems to be acceptable for factual hypotheses. But not for the idealizational ones. There would be no simpler way to make any criticism invalid than by introducing an idealizing condition annihilating the uwanted deviation from the tested statement. Does the thesis that capitalism is being split into two classes, bourgeoisie and the proletariat, fail when contrasted with the historical experience? This presents no problem, let us remove the falsifying circumstances referring the thesis to the idealised world alone! It would obviously be an improper procedure, which, in fact, has sometimes been applied (cf. an analysis of cases of the kind in my (1971a), p. 244ff). A Marxist would instead be obliged to show how these additional circumstances modify this famous claim of the *Communist Manifesto*. Obviously, this is difficult to accomplish in this domain. Let us then refer to an unrealistic but simpler case.

Imagine that our fictitious researcher, who has invented the factual statement (1'), faced with the counter-example decides to neglect the influence of the upon the free-fall and modifies the antecedent of his generalization (1') adopting the idealizing condition $R(x) = 0$. He obtains thus the idealizational statement (1) saving his generalization (1') from falsification. But if he remains on this level of idealization, his way of theorizing could be intuitively inacceptable. He is, instead, obliged to take into account the working of the cause of deviations, i.e. to concretize his idealizational law. In doing this, he follows the rule which is, in fact, adopted at least in the developed science: one may defend his theory against objections by idealization provided that he proves to be able to reject these objections by himself concretizing the theory (cf. my (1977))[2]. In other words, one may defend his thesis against falsification by making it a more idealized statement but only under the condition that he reveals the way of removing this additional simplification and introducing the appropriate correction; briefly, that he concretizes the thesis in relation to the same idealizing condition which allowed him to avoid falsification. Obviously, if he is unable effectively to make an appropriate concretization (this actually happened to the Marxist sociology as far as the above mentioned thesis of the necessary polarization of the bourgeois society is concerned), or if the concretization does not explain the deviation in question, the thesis is falsified.

In sum, in the case of factual statements a modification of the antecedent of a tested generalization, leading to a specialization of it, seems to be quite a legitimate method of preserving the "grain of truth" contained in the initial formula. But in the case of idealizational statements this method is inadmissible. Or, rather, the admission of idealizing conditions is legitimated only if appropriate modifications of the formulae follow.

6. I conjecture that L. Jonathan Cohen is mistaken in claiming that his system of inductive logic includes the problem of idealization and justifies this procedure in the body of scientific methods. The method of idealization is too basic a scientific procedure to be treated as a special case of another procedure, even of the so fruitful as the method of relevant variables. None the less, the Author is entirely right in claiming that idealisation has an important isomorphism with the method of relevant variables ((1990), p. 164). Indeed, the structural similarity goes so far that even quite special problems still preserve it. For instance, the problem of hidden variables in L. J. Cohen's approach (e.g., (1970), p. 38, (1977), p. 132) has a structural analogue in the problem of the epistemic relations between the image of the essential structure and the essential structure itself (cf. Nowakowa (1975a-b), Brzeziński (1978), (1985), my (1980), p. 117ff). This makes the task of comparison even more interesting.

III. The method of relevant variables in terms of idealization

1. Although not handling with idealizations, the method of relevant variables is of great importance for understanding what is actually going on

50

in science. For it fits a large area of the research practice of a more empirical- and/or practical-oriented type[3]. Let us then try to find which of the Author's ideas can be paraphrased in terms of the idealizational approach to science.

To paraphrase a certain conception by another one means much less than to translate it. To paraphrase a fragment f of a (paraphrased) theory T on to a (paraphrasing) theory T' means rather to find in T' such a fragment f' which occupies in T' the same logical position as f in T. If limited simply to this, a paraphrase is trivial. It becomes non-trivial, if f' must be only invented, thus T' being enriched with f', and if due to that some yet unsolvable problems of the paraphrasing theory T' become solved.

In the present part of these remarks I would like to argue that the method of relevant variables is so rich that it might be paraphrased, at least in some of its contents, into the idealizational approach to science allowing the latter much better to handle the complexity of empirical research in science. I shall also try to make this paraphrase not quite trivial.

2. Any paraphrase is an interpretation. And to interpret a conception means, first and foremost, to make decision of what is cognitively crucial in it. As seen from the idealizational standpoint, the essence of the method of relevant variables is hidden in the notion of "normal conditions". And so understood, it permits to an idealization-oriented approach to learn a lot. Indeed, despite what the division of statements into factual and idealizational might suggest, not all the factual statements which describe what happens in the world we live in, are of the same methodological status. Some of them present typical ("normal") cases, some refer rather to specific ("abnormal") ones, both being factual and hence describing what occurs in our world. The straightforward examples are, obviously, medical variables, but the distinction under consideration occurs quite generally. In order to notice this it is sufficient to consider the oppositions: the normal physical conditions (of pressure, of temperature, etc.) vs. abnormal (the very high or the very low) ones, the biological species vs. mixed individuals, the types of personality vs. untypical (outstanding and/or pathological) personalities, the typical embodiments of styles vs. untypical (very creative or eclectic) literary works, etc.

This holds for every variable (factor, magnitude) in the range of which there is an area of normality and the rest constituting the area of abnormality. Assume, for the sake of simplicity, that every variable is composed merely of one area of normality, and of one of abnormality.

The values assigned to the objects from the area of normality constitute the normal interval, while those assigned to the objects from the area of abnormality form the abnormal interval. Leaving aside for the time being the question of explication, let us try to characterize the procedure under consideration in the standard terms of the idealizational approach to science.

I would like to add that the latter implies also some limitations which are not necessary from the standpoint of L. J. Cohen's methodology. Among these there is a limitation to the interval magnitudes, whereas those considered in the method of relevant variables may also be of weaker types, including nominal ones. But, in the idealizational approach, the nature of concretization of qualitative statements is not yet clear enough, and we have to refer to the realm of the quantitative statements.

The initial point is the profound observation of L. J. Cohen that the claim that a (standard) magnitude is composed of the normal area only is a kind of counterfactual assumption. A normalizing assumption might be said one which takes the form:

$$V(x) = N$$

if it is known that V is composed of both the intervals of normality and abnormality. Correspondingly, the set of, linearly ordered, values of V is composed of the two intervals, normal and abnormal; N ranges over the normal interval (or, briefly, the numerical norm) of the magnitude V.

A conditional containing a normalizing assumption is termed a normalizational statement. It is possible to combine normalizing assumptions with other types of deforming conditions — idealizing (cf. my (1971a-b), *ceteris paribus* (Patryas (1975), (1982), protoidealizing (Brzeziński (1978), (1985), quasi-idealizing (cf. my (1980), p. 166ff), abstractive (Zielinska (1981)), stabilizing (Chwalisz (1979)), para-idealizing (Maruszewski (1983), (1985)), etc. Let us limit ourselves, however, to the pure normalizational statements, i.e. ones possessing only realistic (defining the universe of discourse) and normalizing conditions.

Consider, for instance, the expression:

(n) if $G(x)$ and $R(x) = N$ and $S(x) = N'$
  then
$$A(x) = f(B(x)).$$

Of the four magnitudes which are referred to in the above statement, two, A and B, are taken in their full ranges of variability while the remaining two, R and S, are taken only in their areas of normality. If (n) is tested

52

positively, in L. J. Cohen's sense, in relation to the variable R, then it may be transformed into the statement:

(n') if $G(x)$ and $R(x) = r$ and $S(x) = N'$,
    then
$$A(x) = f(B(x))$$

where the symbol r ranges over the whole set of values of the magnitude R. Thus, that (n') is a specification of the normalizational statement (n) means that what has been claimed in (n) as holding good only for the area of normality is now enlarged to the whole range of the variable R, including the area of abnormality. Similarly, the second, more thorough, test allows for a specification of (n) with respect to R and S which is the statement:

(n'') if $G(x)$ and $R(x) = r$ and $S(x) = s$,
    then
$$A(x) = f(B(x)),$$

where s ranges over the whole scope of the variable S, including that of abnormality. Obviously, the conditions $R(x) = r$ and $S(x) = s$ might be omitted in (n'') altogether.

It can now be noticed that the conditional (n) is a factual, and not an idealizational, statement. Its range of application is non-empty: it is composed of those objects which possess property G and properties R and S in their normal size. In contrast to this, the range of application of all the idealizational statements, like (1), is always vacuous. In other words, there are no objects fulfilling an idealizing condition, while there exist some, numerous or not, representing the norm of a given variable (though not all, and possibly even few, variants of the variable belong to its appropriate normal interval).

It can be noticed as well that the specification procedure, i.e. passing from (n) to (n') is not a concretization procedure as the function in the consequent remains constant. Indeed, the procedure of normalization and specification corresponds to the method of relevant variables. One might say that it is a paraphrase, historically correct or not, of the latter in terms of the idealizational approach to science. The problem is, which is its cognitive sense within that approach.

3. The question is which is the cognitive role of the procedure of normalization and specification as conceived of in the terms proposed. It seems to me that the method of relevant variables can be taken as a test for magnitudes which the theoretician considered to be possibly

essential for the phenomenon investigated. Assume that the starting point is the formula:

(f) $A = f(B)$, for members of G.

The test of the hypothesis:

($h0$) if $G(x)$, then $A(x) = f(B(x))$

proves, however, its being false. Against the hypothetist's claim, there is yet no reason to reject (f). This is, instead, a reason to modify it. As to this point both the method of relevant variables and the method of idealization agree perfectly (Cohen (1990), cf. also Nowakowa (1975a-b) and my (1980), p. 207). Let R be suspected of disturbing the working of dependence f of formula (f). From that moment on, R becomes a relevant variable for A, given the initial knowledge of our researcher. He takes R first to be essential in its range of normality hence putting forward the hypothesis:

(5) $G(x)$ and $R(x) = N$, then $A(x) = f(B(x))$.

Assume that the test is positive. This testifies to the effect that R, at least in its range of normality, is not essential for A, as it does not disturb the working of the dependence f. Then, the researcher decides, whether (f) might be generalized over the full scope of variability of R. That is, he tests, still along the line of L. J. Cohen's inductive logic, the hypothesis:

(6) if $G(x)$ and $R(x) = r$, then $A(x) = f(B(x))$.

Assume that the outcome is again positive. This testifies to the fact that the variable R is inessential for A both in its range of normality and in its range of abnormality. Thus condition $R(x) = r$ may be omitted and the researcher comes back to the initial formula (f) with the same question as before: what is it that disturbs the working of the dependence f.

Let him then take S as a relevant variable. At first in the range of normality:

(7) if $G(x)$ and $S(x) = N'$, then $A(x) = f(B(x))$.

Let us assume the outcome will be positive as well. This proves to the effect that S, at least in its normal variability, is inessential for A. And

54

again the researcher tries to generalize this outcome over the whole range of the magnitude S testing the hypothesis:

(8) if $G(x)$ and $S(x) = s$, then $A(x) = f(B(x))$

In this case let the outcome of the test be negative. If so, then it proves that S is disturbing the working of f, if it takes its abnormal values. The result is that instead of accepting the simple image of the essential structure for A admitted so far:

S_A: B,

the researcher introduces a larger one:

S_A: B
$\quad\quad B$, S^{ab},

where S^{ab} is a limitation of the variable S to the range of its abnormality.

Now, according to the idealizational picture of what is done while making a theory, he modifies the antecedent of his initial hypothesis $(h'0)$:

$(h'0)$ if $G(x)$ and $S^{ab}(x) = 0$, then $A(x) = f(B(x))$.

This step could be said to fall under the general scheme of the method of relevant variables. But the second could not. The researcher namely concretizes the modified hypothesis with respect to the second factor, already estabilished as essential though secondary. This requires a correction not in its antecedent but in the consequent:

$(h'1)$ if $G(x)$ and $S^{ab}(x) = s$,
$\quad\quad\quad$ then
$$A(x) = f'(B(x), S^{ab}(x)).$$

If the test is positive, he establishes an (simple, linear, etc.) idealizational theory of A, in our case it is composed only of the two statements $(h'0, h'1)$, limited to the area of abnormality of S.

Assume now that the test for the hypothesis (7) is negative. This proves the fact that S, already in its normal range of variability, is essential for A. Then, an attempt to generalize this result is undertaken by testing the statement (8). And there are two variants possible depending on the outcome of the test. If it is positive, this means that only the limitation of S to its area of normality, i.e. S^n, is esential for A, and hence the image of the essential structure for A is the following:

$$S_A: \begin{array}{l} B \\ B, \; S^n. \end{array}$$

And, accordingly, the researcher at first modifies the antecedent of his initial hypothesis:

$(h''0)$ if $G(x)$ and $S^n(x) = 0$, then $A(x) = f(B(x))$

and then the consequent of the first modification:

$(h'2)$ if $G(x)$ and $S^n(x) = s$,
$\qquad$ then
$$A(x) = f'(B(x),\; S^n(x)).$$

A result will be a (simple, linear, etc.) idealizational theory composed of two other statements: $(h''0, h'2)$.

Let us consider now the second variant, when the test for (8), after (7) has already been negatively tested, brings a negative outcome as well. This implies that S is essential for A not only in the realm of its normality but also in that of abnormality. The image of the essential structure for A turns out then to be the following:

$$S_A: \begin{array}{l} B, \\ B, \; S, \end{array}$$

and the necessary modification of the antecendent of $(h0)$ to be quite general:

$(H0)$ if $G(x)$ and $S(x) = 0$, then $A(x) = f(B(x))$.

Correspondingly, the required concretization of the latter is also quite general:

$(H1)$ if $G(x)$ and $S(x) = s$
$\qquad$ then
$$A(x) = f^*(B(x),\; S(x)).$$

In this case the established (simple, linear, etc.) idealizational theory is composed of two quite general statements $(H0, H1)$.

56

4. The above considerations, although carried out for the extremely simple schemes, lead to several variants which might obscure the main thought. Let us distinguish the uniform result of a test, i.e., the same outcome (either positive or negative) for both the normal and abnormal cases of the relevant variable, from the mixed result, when the outcome for the normal (resp. abnormal) area differs from the one for abnormal (resp. normal). Now, let us compare the standard image of the research, provided the results of tests are uniform, under the method of relevant variables and the method of idealization at the same time.

The initial hypothesis being $(h0)$, the method of relevant variables inclines to accept either this itself, in the case of a positive outcome of a test, or its special case $(H0)$, in the case of negative outcome of it:

$$+: \quad \text{-----------------} > \; h0$$
$$h0$$
$$-: \quad \text{-----------------} > \; H0.$$

In the same conditions the method of idealization simply forces the replacement of $(h0)$ with $(H0)$ and to concretize the latter in order to obtain $(H1)$, the relationsphip between the former two being methodologically not analyzed:

$$h0 \quad H0 \; =============> \; H1$$

If the above observation is correct, then the full image of the procedure in question should instead be the following:

$$+: \quad \text{-----------------} > \; h0$$
$$h0$$
$$-: \quad \text{-----------------} > \; H0 \; =============> \; H1$$

both the relation of specification (-------->) and of concretization (======>) being involved. Notice that the above picture clearly shows why it is so that the falsification of a hypothesis is a more fortunate — for science, not for the poor author of the hypothesis — case. It is not only, and even not so much, a negative criterion of elimination of falsities. First and foremost it is the positive source for a new, improved, hypothesis. This idea, being a development of the pure selectionistic understanding of the principle of falsification (Popper (1959)), is again common both to the method of relevant variables and the method of idealization.

Notice also that the results of the comparison of the two methods would be similar, although graphically more complicated, if the mixed outcomes of the tests were compared. But this is an irrelevant simplification. What matters is that enriching the idealizational picture of the research procedure with the ideas taken from the method of relevant variables leads to a more adequate grasp of scientific practice.

5. One could object to the above considerations because of the lack of any explication of the notion of a normal case. However, although difficult, such an explication seems to be, in principle, possible. The initial idea is that a normal appearance of a phenomenon is one in which the phenomenon undergoes merely the working of its most significant parameters. Or, to put it in a more metaphysical language, the norm of a phenomenon is the manifestation of its pure essence. The simplest way to explain this idea in the conceptual apparatus of the idealizational approach would perhaps be the following.

Given is an arbitrary magnitude F with its scope **F** divided into ordered classes of abstraction to each of which a certain value is assigned. A factor A is essential for F on an object a if there is a subset E of values of F such that the fact that $A(a) = n$ excludes that F takes on the object a any value of the set E (the range of exclusion of F relative to A and a)[4]. The difference of the scope **F** and the range E of exclusion of F relative to A and a is termed the range of admittance of F in regard to A and a. The set of objects from F in relation to any of which A is essential for F will be termed the range of influence of A upon F. Take two factors, A and B, essential for F with their ranges of influence being **A** and **B**, correspondingly. The intersection of **A** and **B** will be termed their common range of influence upon F, whereas the difference between **A** and **B** will be labelled the A-pure range of influence on F. The same in the case of more factors essential for a given one.

Now, A is termed more essential for F on a than B if the A-range of exclusion of F on a is larger than the B-range of exclusion on a. The maximally essential factor excludes the whole set of values of a determined magnitude except for one (and thus the appropriate range of admittance is a singleton); in the case of a nonessential factor, its range of exclusion is empty (and that of admittance is a signleton). Given the magnitude F and an object a from its scope **F**, the essential space of F on a is constituted by all the factors essential for F on a, and this space might be divided into sets of factors being equi-essential for F on a. Ordering these sets respective to the cardinality of an appropriate set of exclusion (of F, on a), one obtains the essential structure of the magnitude F on the object a. The most essential factors of the space (with regard to F, a) will be termed principal factors for F, a. The rest of the space forms the set of secondary factors for F, a.

Now, all the above standard notions are relativized to an object from the scope of a determined variable. Take all the objects from the scope **F** in relation to which a sequence of sets of factors is the essential structure for F on every member of that set. They constitute a F-species. Take all the objects from **F** in relation to which the same factors are the principal for F. They constitute a F-kind. One could say that F-kind is a set of objects possessing the same essence (in relation to F) whereas objects constituting F-species possess not only the same F-essence but also all its forms of manifestation. Assume the magnitude F has n F-species and k $(k \leqslant n)$ F-kinds. Therefore, according to the general rules of the idealizational approach to science, there are n (simple, linear) idealizational theories possible, and there are possible k types of idealizational theories with the same idealizational law.

Let us assume, for simplicity's sake, that the set of objects **F** contains one F-kind and one F-species only, i.e. the two sets are identical with the scope **F**. Now, one may identify the set of F-normal cases (the area of F-normality or, briefly, F-norm) with such subset of the F-kind which is identical to the pure range of influence of the principal factors for F. In other words, the area of F-normality is formed by these objects on which the property F is influenced by its principal determinants and by no secondary determinants. It is a set of such objects on which the F-essence is not disturbed by the working of any additional, secondary factors. Obviously, F-norm is a subset of the (appropriate) F-species. The area of F-complexity is composed of those objects which do not belong to the F-norm and hence on which the property F undergoes not only the working of principal but also secondary factors influencing F. The more factors join their working upon F on a given object, the greater is the degree of its F-complexity. Finally, the area of F-appearance is constituted by these objects which belong to the pure range of influence of secondary factors on F, i.e. which do not undergo the working of principal factor but merely are influenced by parameters of the adventitious nature. As the observation of the F-norm is always fruitful for building a theory of the F-facts and the observation of the F-complexity is usually indifferent for that goal, as the observation of the F-appearance is simply misleading. Take the case of the free-fall again: as far as our knowledge reaches it, the free-fall in the physical vacuum would be close to the norm of that phenomenon, whereas the case of fall of a sheet of paper during a hurricane would be close to the appearance of the free--fall.

Obviously, the above explication of the idea that norm is a non--disturbed transparence of the essence of phenomena is based on numerous simplifications. Some of them are of the ontological, while some other of the epistemological nature[5]. For instance, the F-norm may be, and usually is, represented not by a single value of the magnitude F but by a certain interval of measures of F. Namely, it may be so that more than one class of abstraction from **F** is included into F-norm. Another simplification that has been silenced above is that we refer to single phenomena whereas in fact they appear as members of categorial systems[6]. There are certainly more simplifications hidden above[7], but what we meant was not a multisided analysis of the notion of norm as applied in the empirical research, but simply an outline one of possible constructions adjusted to the framework of the idealizational approach to that research.

IV. Conclusions

The conclusions of the paper are three. First, the method of relevant variables does not reconstruct the procedure of idealization. Second, the method of idealization does not reconstruct the method of relevant variables. Third, if the method of relevant variables is correctly, at least in its main intention, paraphrased as that of normalization and specification, then the idealizational approach can be enlarged sufficiently to do justice to the former. And, if that is correct, this is perhaps a modest proof of the richness of the method of relevant variables.

Institute of Philosophy UAM
Szamarzewskiego 89
60-568 Poznań

FOOTNOTES

* I would like to thank to the Netherlands Institute for Advanced Study (Wassenaar) for making the present writing possible. I am indebted to Anne Simpson (NIAS) for correcting my English.

[1] The Author enlists also the third method of "adjusting the meaning of a hypothesis" in case of the threat of falsification ((1970), pp. 142/143), namely a redefinition of a term, or terms, in it. This will be omitted here, because we do not possess sufficiently elaborated a comparative background. The methodological conception underlying the present analysis is still based on the simplicistic assumption neglecting the problems of meaning of scientific theories. Certain attempts to cope with that problem are contained in Tuchańska (1980), chap. I and Łastowski (1987), chap. IV but they are too limited for the possible use in this context.

60

[2] Nowakowa (1975a-b), (1982) proves that this rule works in the development of scientific theories. Cf. also some generalizations of that conception in Patryas (1976), Nowak and Nowakowa (1978), Łastowski (1987) and Paprzycka (1989).

[3] This problem is not often conceptualized in terms of idealization (cf. only Tuchańska (1977)). There is, to be sure, a conception of proto-idealization eleborated for empirical disciplines (Brzeziński (1978), (1980), (1985)), but its intended scope of application are procedures of theory-building in behavioral sciences, not empirical research as such.

[4] This definition is modified, in comparison to one given in (1990), under the influence of criticism of Paprzycki & Paprzycka (1991). Obviously, there are many other possibilities to define the notion of essential factor. Cf. Nowakowa et al. (1978) and Machowski (1988).

[5] Cf. the distinction between the ontological and epistemological models of the idealizational conception of science in Gaul (1988).

[6] E.g., in the sense of Brzeziński et al. (1976).

[7] On multiplicity of intuitions connected with the notion of norm in the empirical science cf. Sanocki (1983). Apart from the intuition explicated in the main text (norm as a pure manifestation of the essence of a phenomenon) there are two other main intuitions: norm as a statistical regularity and norm as a value. It may be argued that the three are not so separated from one another as it might seem. And even that they are closely connected with the essentialist intuition being the main one (cf. my (1974), (1984)).

REFERENCES

Batóg, T. (1976). Concretization and generalization. In: *Idealizational Conception of Science* (*Poznań Studies in the Philosophy of the Sciences and the Humanities*, **2**, 3), Amsterdam: Grüner.

Brzeziński, J. (1978). *Metodologiczne i psychologiczne wyznaczniki procesu badawczego w psychologii* [The Methodological and Psychological Determinants of the Research Process in Psychology]. Poznań: Adam Mickiewicz University Press.

Brzeziński, J. (1980). The methodological awareness of a researcher in psychology. *Polish Psychological Bulletin*, 4, 252-264.

Brzeziński, J. (1985). The protoidealizational model of the investigative process in psychology. In: (ed.) J. Brzeziński, *Consciousness: Methodological and Psychological Approaches* (*Poznań Studies in the Philosophy of the Sciences and the Humanities*, **8**) Amsterdam: Rodopi.

Brzeziński, J. and I. Burbelka, K. Łastowski, A. Klawiter, L. Nowak (1976). Law and theory. A contribution to the idealizational interpretation of the Marxist methodology. In: *Idealizational Conception of Science* (*Poznań Studies in the Philosophy of the Sciences and the Humanities*, **2**, 3). Amsterdam: Grüner.

Chwalisz, P. (1979). Stałe w idealizacyjnej koncepcji nauki (Constants in the idealizational conception of science). In: A. Klawiter, L. Nowak, *Odkrycie, Abstrakcja, Prawda, Empiria, Historia − a Idealizacja* [Discovery, Abstraction, Truth, Experience, History − and Idealization]. Warszawa-Poznań: PWN.

Cohen, L. J. (1966a). The logic of evidential support. *British Journal for the Philosophy of Science*, **XVII**, 1, 21-43 (part I) and **XVII**, 2, 105-126 (part II).

Cohen, L. J. (1966b). What has confirmation to do with probabilities?. *Mind*, **LXXV** (October), 463-481.

Cohen, L. J., (1970). *The Implications of Induction*. London: Methuen.

Cohen, L. J. (1977). *The Probable and the Provable*. Oxford: Clarendon Press.

Cohen, L. J. (1979). On the psychology of prediction: Whose is the fallacy?. *Cognition*, **7**, 385-408.

Cohen, L. J., (1980a). The problem of natural laws. In: (ed.) D. H. Mellor, *Prospects for Pragmatism*, Cambridge: Cambridge Univ. Press.

Cohen, L. J. (1980b). Inductive logic. 1945-1977. In: (ed.) E. Agazzi, *Modern Logic – A Survey*. Dordrecht: Reidel.

Cohen, L. J. (1980c). What has inductive logic to do with causality? In: (eds.) L. J. Cohen and M. Hesse, *Applications of Inductive Logic*. Oxford: Clarendon Press.

Cohen. L. J. (1981). Can human irrationality be experimentally demonstrated?. *The Behavioral and Brain Sciences*, **4**, 317-331.

Cohen, L. J. (1983a). How far is induction rationally justifiable?. In: *Epistemology and Philosophy of Science* (*Proceedings of the 7th International Wittgenstein Symposium*), Wien: Hoelder-Pichler-Tempsky.

Cohen, L. J. (1983b). The unity of reason. *Social Theory and Practice*, **9**, 2-3, 245-268.

Cohen, L. J. (1987). Verisimilitude and legisimilitude. In: (ed.) T. A. F. Kuipers, *What Is Closer-To-The-Truth?* (*Poznań Studies in the Philosophy of the Sciences and the Humanities*, **10**). Amsterdam: Rodopi.

Cohen, L. J. (1990). Idealisation as a form of inductive reasoning. In: J. Brzeziński, F. Coniglione T. A. F. Kuipers and L. Nowak, (eds.) *Idealization-I* (*Poznań Studies in the Philosophy of the Sciences and the Humanities*, **16**), Amsterdam: Rodopi, 153-166.

Cohen, L. J. and Margalit, A. (1970). The role of inductive reasoning in the interpretation of metaphor. *Synthese*, **21**, 469-487.

Gaul, M. (1988). Modele rzeczywistości czy modele poznania (The models of reality or the models of cognition). In: J. Brzeziński, K. Łastowski (eds.) *Filozoficzne i metodologiczne podstawy teorii naukowych* [Philosophical and Methodological Foundations of Scientific Theories] (*Poznańskie Studia z Filozofii Nauki*, **11**). Warszawa-Poznań: PWN.

Hamminga, B. (1989). Sneed versus Nowak: an illustration in economics. In: (eds.) W. Balzer, B. Hamminga, *Philosophy of Economics – II*, *Erkenntnis*, **30**, 1-2, 247-265.

Hempel, C. G. (1952). Problems of concept and theory formation in the social sciences. In: *Science, Language and Human Rights*, Philadelphia.

Krajewski, W. (1974). Kopernik i Galileusz vs. Arystoteles [Copernicus and Galileo vs. Aristotle]. *Studia Metodologiczne*, **12**, 3-22.

Łastowski, K. (1987). *Rozwój teorii ewolucji. Studium metodologiczne* [The Development of the Theory of Evolution: A. Methodological Study]. Poznań: Adam Mickiewicz University Press.

Machowski, A. (1988). Pojęcie istotności (wpływu). Próba interpretacji wariancyjnej (The notion of essentiality (influence). An attempt at the variance interpretation). J. Brzeziński, K. Łastowski, (eds.) *Filozoficzne i metodologiczne podstawy teorii naukowych* [Philosophical and Methodological Foundations of Scientific Theories]. (*Poznańskie Studia z Filozofii Nauki*, **11**). Warszawa-Poznań: PWN.

Maruszewski, T. (1983). *Analiza procesów poznawczych jednostki w swietle idealizacyjnej teorii nauki* (An Analysis of the Cognitive Processes of Individual in the Light of the Idealizational Theory of Science). Poznań: Adam Mickiewicz University Press.

Maruszewski, T. (1985). Are the idealizational procedures used within the scope of common-sense knowledge? In: J. Brzeziński (ed.), *Consciousness: Methodological and Psychological Approaches. (Poznań Studies in the Philosophy of the Sciences and the Humanities*, **8**), Amsterdam: Rodopi.

Niiniluoto, I. (1986). Theories, idealizations and approximations. In: R. B. Barkus, G. J. W. Dorn and P. Weingartner (eds.), *Logic, Methodology, and Philosophy of Science*, VII. Amsterdam: North Holland.

Nowak, L. (1971a). *U podstaw Marksowskiej metodologii nauk* [Foundations of Marxian methodology of science]. Warszawa: PWN.

Nowak, L. (1971b). The problem of explanation in Marx's *Capital. Quality and Quantity*, **V**, 2, 311-338.

Nowak, L. (1974). Value, idealization, valuation. *Quality & Quantity*, **VII**, 2, 107-119.

Nowak, L. (1977). *Wstęp do idealizacyjnej teorii nauki* [An Introduction to the Idealizational Theory of Science]. Warszawa: PWN.

Nowak, L. (1980). *The Structure of Idealization. Towards a Systematic Interpretation of the Marxian Idea of Science*, Dordrecht: Reidel.

Nowak, L. (1984). The review of Sanocki (1983). *Przegląd Psychologiczny*, **XXVII**, 4, 1029-1034.

Nowak, L. (1990). Abstracts are not our constructs. The mental constructs are abstracts. In: J. Brzeziński, F. Coniglione, T. A. F. Kuipers, L. Nowak, (eds.) *Idealization-I, (Poznań Studies in the Philosophy of the Sciences and the Humanities*, **16**). Amsterdam: Rodopi.

Nowak, L. and I. Nowakowa (1978). Pewne metodologiczne osobliwości rozwoju teorii ekonomicznych (Some methodological peculiarities of the development of economic theories). In: P. Buczkowski and L. Nowak (eds.), *Teoria ekonomiczna: metodologia i rekonstrukcje* [The Economic Theory: The Methodology and Reconstructions]. Poznań: The Academy of Economic Sciences Press.

Nowakowa, I. (1975a). *Dialektyczna korespondencja a rozwój nauki* [The Dialectical Correspondence and the Development of Science]. Warszawa-Poznan: PWN.

Nowakowa, I. (1975b). Idealization and the problem of correspondence. *Poznań Studies in the Philosophy of the Sciences and the Humanities*, **1**, 1, Amsterdam: Grüner, 65-70.

Nowakowa, I. (1982). Dialectical correspondence and essential truth. In: W. Krajewski (ed.), *Polish Essays in the Philosophy of the Natural Sciences (Boston Studies in the Philosophy of Science*, **68**). Dordrecht: Reidel.

Nowakowa, I and W. Patryas, W. Szaban (1978). O pojęciu istotności (On the notion of essentiality). In: L. Nowak (ed.), *Założenia materializmu historycznego* [Assumptions of Historical Materialism]. (*Poznańskie Studia z Filozofii Nauki*, **3**). Warszawa-Poznań: PWN.

Paprzycka, K. (1990). Reduction and correspondence in the idealizational conception of science. In: J. Brzeziński J., and F. Coniglione, T. A. F. Kuipers, L. Nowak (eds.) *Idealization-I. (Poznań Studies in the Philosophy of the Sciences and the Humanities*, **16**) Amsterdam: Rodopi.

Paprzycki M. and K. Paprzycka (1991). A Note on the Unitarian Explication of Idealization. In: (eds.) J. Brzeziński, L. Nowak, *Idealization* – III (*Poznań Studies in the Philosophy of the Sciences and the Humanities*, **25**). Amsterdam: Rodopi.

Patryas, W. (1975). An analysis of the caeteris paribus clause. *Poznań Studies in the Philosophy of the Sciences and the Humanities*, **1**, 1, Amsterdam: Grüner, 59-64.

Patryas, W. (1976). *Eksperyment a idealizacja* [Experiment and Idealization]. Poznań- -Warszawa: PWN.

Patryas, W. (1982). The pluralistic approach to empirical testing and the special forms of experiment. In: W. Krajewski (ed.), *Polish Essays in the Philosophy of the Natural Sciences* (*Boston Studies in the Philosophy of Science*, **68**). Dordrecht: Reidel.

Popper, K. R. (1959). *The Logic of Scientific Discovery*. London: Hutchinson.

Such, J. (1974). O zasadzie korespondencji (On the principle of correspondence). In: J. Kmita (ed.), *Metodologiczne implikacje epistemologii marksistowskiej* [The methodological implications of the Marxist methodology]. Warszawa: PWN.

Sanocki, W. (1983). *Pojęcie normy w psychologii* [The Notion of Norm in Psychology]. Gdańsk: Gdańsk University Press.

Tuchańska, B. (1977). An idealizational view on measurement and indicator-based reasoning. *Poznań Studies in the Philosophy of the Sciences and the Humanities*, **3**, Amsterdam: Grüner, 180-198.

Tuchańska, B. (1980). *Czynnik – wielkość – związek – zależność. Przyczynek do idealizacyjnej koncepcji nauki* [Factor – Magnitude – Regularity – Correlation. A Contribution to the Idealizational Conception of Science]. Poznań-Warszawa: PWN.

Zielińska, R. (1981). *Abstrakcja, Idealizacja, Uogólnienie* [Abstraction, Idealization, and Generalization]. Poznań: Adam Mickiewicz University Press.

*Poznań Studies in the Philosophy
of the Sciences and the Humanities
1991, Vol. 21, pp. 65-79*

Menachem Fisch

LEARNING FROM EXPERIENCE

1. We learn from experience by prudently testing the hypotheses that we form. To test for example a simple first-order generalisation of the form (x) $(Ax \supset Vx)$ is, as the case may be, to prudently seek, occasion or attempt to construct A's that or not B's. Subsequently the quality and standing of our initial conjecture will depend on how prudent, on the one hand, and unsuccessful, on the other, we are in our efforts. This is the essence both of Popper's falsificationism and, as L. J. Cohen has pointed out, of Bacon's eliminative induction. However, even with respect to the most elementary first order generalisations such a formulation requires considerable fleshing out both methodological and logical. Such has been L. J. Cohen's undertaking in his superb work on induction.

Related as they undoubtedly are, Cohen has been careful to keep his methodology and logic well apart — and rightly so. It is one thing to ask how, under given circumstances, are non-B-ish A's best sought for, yet quite another to assess the support bestowed upon an hypothesis in the event of success or failure. Indeed as Cohen has convincingly shown (late of Quine[1]) Hempel's notorius ravens paradoxes are resolved when the methodological and logical denotations of the term confirmation are appropriately teased apart. (Requiring only of the latter to conform to the Equivalence Principle[2]). With respect to empirical testing at least, methodology not only differs from, but actually precedes logic. For if, as Cohen observes, "the scale of inductive support runs, not from logical contradiction to logical implication, but from no support by appropriate experimental tests to full support"[3], a methodological analysis of appropriate test procedures is a prerequisite for a meaningful explication of the notion of inductive support forthcoming in the light of their possible outcomes.

Cohen's work on induction subsequently falls under three major and ordered headings. First he developed an account of empirical test procedure entitled the Method of Relevant Variables. Second, and based upon the first, he devised a non-probabilistic syntax for inductive support mapped on to a class of generalisations of the Lewis-Barcan modal system S4. Finally, especially in his later work, he has suggested an interpretation of inductive syntax viewing it as grading what he terms "legisimilitude". Cohen has not limited himself to the philosophy of science. In a series of insightful and knowledgeable studies he has made the Method of Relevant Variables and its accompanying "logic of reliability ranking" the basis for a convincing analysis of reasoning about legal, ethical, grammatical and even philosophical hypotheses. The latter culminating in a suggestive and reflexive analysis of analytical philosophy itself. His latest work — *An Introduction to the Philosophy of Induction and Probability* — is a critical and historical survey of the attempts to articulate ampliative modes of learning from experience in which his own contribution justly takes its place as the most adequate account of inductive reasoning to date. In what follows I shall be less concerned with the formal aspects of his work *per se*. The inductive syntax eventually adopted, Cohen submits, should be regarded "as a kind of first-stage analytical conclusion, rather than as criteria of adequacy for the analysis as a whole"[4]. My remarks will therefore be directed in the main towards assessing the methodological and semantic aspects of his analysis. I shall also limit my comments to scientific investigations. Cohen no doubt regards the understanding of scientific test procedures as basic to his entire system. But whether or not my criticisms apply equally to the other areas of discourse studied by Cohen I shall leave for others to judge.

2. To borrow Kuhn's phrase, Cohen's Method of Relevant Variables is a methodology of normal science. Yet, at the same time, it is basically Popperian (or, as Cohen has it: Baconian) in the sense that hypotheses are viewed as proving their mettle in the face of attempted refutation. These two basic aspects of Cohen's methodology, I shall argue, eventually run against each other in a manner unaccounted for in his writings.

According to Cohen to prudently test an hypothesis H_1 of the form: $(x)\,(Ax \supset Bx)$ is to seek, occasion or attempt to construct non-B-ish A's in circumstances in which to the best knowledge of the scientific community they are most likely to be found. In Cohen's view, positive reports of B-ish A's, numerous and varied as they may be, do not lend H_1 any support at all unless they are found, occasioned or constructed under conditions prudently contrived with an explicit view to yield their contrary. Empirical testing is thus contextualized by Cohen with respect to background knowledge. But unlike standard Bayesian accounts of confirmation it is done so in a very specified manner. It is here that the Kuhnian aspect of his work becomes apparent.

Within reasonably mature fields of enquiry, Cohen argues, viably testable hypotheses form classes of what he calls "materially similar" propositions. A likeness implying that members of the same class of materially similar hypotheses be falsified in similar circumstances. If propositions H_1 and H_2 are regarded materially similar by the scientific community, it follows that according to the present state of knowledge potential falsifiers of the former are most likely to be found (occasioned or constructed) in circumstances similar to those that prevailed when in the past the latter was in fact found not to obtain. Testable hypotheses, in other words, come accompanied by a *set of relevant circumstantial variables* by virtue of their material similarity to the other members of their class. However, Cohen observes, the notion of material similarity between hypotheses cannot itself depend on the outcomes of testing those hypotheses. "Otherwise we should not be able to say what variables are relevant for a hypothesis without having tested it already, and so, since designing tests for a hypothesis requires us to presume what variables are relevant for it, no test would ever take place"[5]. From this Cohen concludes that "material similarity must always be linguistically defined — by reference to a set of non-logical terms". In other words, two simple first-order generalisations such as: (x) $(Ax \supset Bx)$ and (x) $(Cx \supset Dx)$ will be regarded as materially similar if and only if A and C, and B and D belong to "the same semantic categories", and that "the domain of discourse is a specified set of elements to which these terms may be significantly applied". This may be a necessary condition for material similarity but it is undoubtedly far from being a sufficient one. There are any number of semantic categories two sets of predicates may belong to without implying that generalized conditionals formed around them be tested under similar conditions. The mere fact that our linguistic practices regularly associate two pairs of attributes (e.g. kinds of organism and kinds of behavioral traits, or kinds of chemical compounds and kinds of reactions with other chemical compounds) does not in itself imply that conditional conjectured correlations between members of different pairs will also be similarly associated. Cohen seems aware of this for he is quick to replace the more or less formal semantic category thesis by a more sensitive and less purely linguistic approach.

> It is for scientists themselves ... not philosophers, to determine from time to time the exact criteria of categorical identity for this purpose, each in regard to his own field of research. If they do not do so, at least implicitly, they have no rational basis for selecting some sets of circumstances in which to test a new hypothesis in their field rather than others — out of the indefinitely wide variety of circumstances that have been observed to affect hypotheses of different kinds[6].

However, if it is for philosophy to analyse the nature and grading of inductive support, it is for philosophers to attempt to analyse the kind of criteria by which scientists (tacitly) judge hypotheses to be comparable with respect to test procedure. And as I have tried to show elsewhere[7] these arguably involve more than mere semantic categorization. If as a matter of fact scientists elect to test (x) $(Ax \supset Bx)$ in similar circumstances to which (x) $(Cx \supset Dx)$ had been formerly falsified, it is not simply because they tacitly affiliate predicates A and C, and B and D to the same semantic categories. What seems to be tacitly assumed is *that whatever normally causes C's to exhibit or sustain their D-ness* (or bars the possibility of them exhibiting non-D-ness) is what normally causes A's to exhibit or sustain their B-ness (or bars the possibility of them exhibiting non-B-ness). Otherwise it would be meaningless for anyone to conclude simpliciter, as scientists apparently do, that similar conditions to those that evidently occasion the existence of non-D-ish C's should as a matter of course be expected to similarly occasion the existence of non-B-ish A's. The fact that scientists elect to test "All ravens are black" in circumstances in which otherwise white swans were formerly found to lose their whiteness indicates, I suggest, that a higher level hypothesis is implicitly at work; namely, that whatever causes or sustains the whiteness of swans is akin to whatever causes or sustains the usual colour of ravens. Only then is it reasonable to conclude that conditions material to the collapse of the former should act similarly on the latter. Such a conclusion is in no way implied merely by semantic categorization. However, if that is the case, Cohen's methodology is rendered problematic.

The test procedure appropriate to an hypothesis H_1 is determined, according to Cohen, by the class of materially similar propositions M to which it belongs. M, I suggest, consists of generalized conditionals held by the scientific community to normally obtain by virtue of *the same type of causal mechanism*. To state all this somewhat more formally M, I submit, is such that H_i and $H_j \in M$ if H_i and H_j are suspected to hold true *because of* C_i and C_j respectively, where C_i and C_j are causal mechanism of a kind C. Not all members of M, however, hold true in all circumstances. There exists a subset M_i of M known to have been successfully falsified in certain circumstances. Let $\{V_i\}$ $i = 1...n$ be the joint set of circumstantial variables, variants of which are held to have been directly responsible for the falsification of the members of M_i. Together $\{V_i\}$ form the set of all known *obliterators* of C-type mechanisms. Accordingly they are held by the scientific community to be the maximal set of circumstantial variables relevant to the testing of M-type hypotheses — H_1 included.

If this is an acceptable account of Cohen's notion of material similarity we are in a position to take stock with respect to background knowledge. For an hypothesis H_1 to be viably testable it is required by Cohen to belong within the framework of a mature field of enquiry about which the following is asserted:

(a) H_1 is a member of a class of materially similar propositions M.

(b) Members of M are held to normally obtain by virtue of causal mechanisms of a kind C. (C itself, note, need not be specified in any way. All that need be assumed is that all members of M are expected to pertain to the same type of causal explanation.)

(c) There exists a specified and ordered list $V_1,...,V_n$ of all known obliterators of C-type causes.

(Cohen's notion of normal science, we may note in passing, is thus in important ways very different from Kuhn's. New hypotheses are not investigated according to Cohen with reference to unquestioned paradigms of success as in Kuhn. What is assumed by Cohen is that well-formed fields of enquiry characteristically make reference to entrenched categorizations of *hypotheses* (formed with respect to tacit causal hunches) and paradigmatic examples of successful falsifications of their like.)

Given the set $\{V_i\}$ of circumstantial variables relevant to the testing of H_1 a telescopic hierarchy of $n+1$ canonical trials are theoretically constructed varying progressively in terms of their thoroughness. In the first and most basic test to all possible effects of $V_1, V_2,...,V_n$ are screened off and H_1 is instantiated. (This is to ensure that A-ness is in principle coexistent with B-ness — i.e. is not itself an obliterator of the causal mechanism assumed to prevail.) In t_1 $V_2,...,V_n$ are screened off and H_1 is instantiated in all known variants of V_1. In t_2 $V_3, V_4,...,V_n$ are screened off and H_1 is instantiated in all possible combinations of all known variants of V_1 and V_2. And so on until t_n in which H_1 is instantiated in all possible combinations of all the known variants of the entire set of V_i's — t_n thus being the most thorough test of H_1 possible with respect to the community's present state of knowledge.[8]

The inductive support lent to an unsuccessfully tested hypothesis, Cohen submits, cannot be measured since no scale exists against which the outcomes of tests of different hypotheses can be meaningfully compared. But then there is no sense in comparing, and indeed no conceivable need to compare, the support assigned to hypotheses in different fields of enquiry in the face of tests of varied thoroughness. However, Cohen argues, the *reliability* of an hypothesis may be *graded* with

70

reference to the relative thoroughness of the trials it survives. In the event that an hypothesis H_1 is reported to have survived the i-th of n possible tests, the support function $s(H_1, E)$ will be assigned the value i/n (where E is the report that despite the combined variation of $V_1,...,V_i$ all A's observed were nevertheless found to retain their B-ness). The reliability of an hypothesis is hence associated with its proven resistance to the combined effects of what are thought to be its potential obliterators. If the hypothesis is found to have survived t_n, the most thorough test conceivable, it is assigned maximum reliability.

The first point to notice is that in the course of formalizing and subsequently interpreting the grading of inductive support thus construed, Cohen seems to impose upon his methodology of test procedures a finality his analysis is incapable of supporting.

Taking for the moment as unproblematic Cohen's basic assimilation of inductive support with the reliability of an hypothesis in the light of empirical evidence, let us look at the manner in which he proposes to formulate what he terms "the logic of reliability-ranking". He does so by a generalisation of modal logic which makes use of a series of square-operators $\square^1$, $\square^2$, ..., $\square^d$, where $\square^d$ represents logical necessity, $\square^{d-1}$ represents natural law and $\square^1$, $\square^2$, ..., $\square^{d-2}$ represent thresholds of increasing reliability that fall short of natural law. From which follows that $\square^1 H \supset H$ is provable only for $i \geqslant d-1$. The formulation itself is achieved by associating $s(H, E) = i/n$, the reliability ranking of H, with the truth of $E \supset \square^1 H$. However, rather than leave it at that Cohen goes one step further and by equating n (the number of canonical tests) with $d-1$ he obtains $s(H, E) = n/n$ for $E \supset \square^{d-1} H$. In *The Implications of Induction*, n, the number of canonical tests, is equated more generally with "the weakest modalisation of A that still implies the truth of A". Either way it follows that if E is the report that H has withstood the most thorough test conceivable H is provably *true!* If initially the ranking of inductive support was meant to run "from no support by appropriate experimental tests to full support" by mapping it on to a scale of legisimilitude it is found to run between "the vanishing point, as it were, of A-assertion" at the lowest boundary, to "the level of establishment" of the truth of A at the highest. Hence unlike standard accounts of verisimilitude, particularly that of Popper, Cohen seems to maintain (as did Bacon and Whewell) that it is in the power of science to not merely approximate, but to actually prove the truth of a universal correlational generalisation. But given that on Cohen's own showing the construction of t_n is at any given time grounded upon a revisable *empirical* conjecture concerning the material similarity of the hypothesis under consideration, such a conclusion is blatantly inadmissible.[9]

3. *Prima facie* there seem to be three diferent ways in which one could counter this objection. One obvious way out would be to modify Cohen's formalization and set $n \leqslant d-2$ by definition. However, while the introduction of such a formal limitation certainly ensures the basic openness of empirical testing it seriously handicaps Cohen's proposed logical syntax in which $\vdash \Box^n H \supset H$ is set as a criteria of theoremhood[10].

The two remaining options leave Cohen's inductive logic intact. One way would be to ascribe to him some type of theory of truth by convention by rendering the entire logic of inductive support relative to background knowledge. Accordingly "$\vdash \Box^n H \supset H$" would be taken as shorthand for "if it is true that H is reported to have survived the most thorough test implied by what we now know, then (for the time being) H is justifiably acceptable". But Cohen seems to aiming much higher than that. "Judgements of inductive support", he writes explicitly[11],

> unlike judgements of acceptability, are not relative to time, place or circumstances, because they are independent of evidential availability. A man may feel he knows just how much a certain experimental result would support his hypothesis, even if the actual performance of the experiment is quite outside his power. Admittedly, his views about how much E supports H may change, and rightly change, as he comes to see that new tests on hypotheses of a given type are superior to the old ones. ... But that is no reason for supposing that inductive support itself is relative to a time, let alone to a place or to other circumstances. Rather, the discovery that one is ever wrong about the extent of E's support for H is going to be very difficult indeed if this extent itself can alter.

How then are we to interpret n, the number of canonical tests, in Cohen's formulae? If n is taken as the number of canonical tests at a given time without relativising the relation of inductive support it would be absurd to regard $\Box^n H \supset H$ as provable since, as Cohen admits, tests superior to those currently practiced may well come to light at a later date, some even that H may be found to fail. The only option left is to regard n as entirely independent of current knowledge — namely, as the *real* number of all potential obliterators of M-type hypotheses. Upon such a reading Cohen's logic of support, as opposed his Method of Relevant Variables, is to be viewed as formulated with respect to total omnipotent and omniscient knowledge of the full range of circumstantial variables relevant to the testing of an hypothesis[12]. Such a move might also seem to resolve a second difficulty dealt with by Cohen — namely that of infinite regress. Since it might be objected that forming sound judgements with respect to the support of an hypothesis requires us first to form a sound judgement with regard to the higher-level hypothesis concerning its material similarity to the class of hypotheses it is related to, which in turn requires us to form a sound judgment with regards to the even higher-level hypothesis concerning the ways in which *that* judgement should be formed etc. However, if the logic of judgement formation is envisaged as turning on total knowledge of the variables relevant to the hypothesis in question no such backing is called for.

In short, upon this view Cohen's logic of support is not to be taken as a formalization *simpliciter* of his methodology, but rather as an idealization of it with respect to the background knowledge it presupposes. Legisimilitude should then perhaps be construed not as a measure of the thoroughness of a test with respect to the highest degree of thoroughness known at a given time, but rather as a measure of the extent to which an actual set of thought-to-be circumstantial variables approximates the true full list.

Curiously, though, and plausible as all this might sound, Cohen's Method of Relevant Variables in principle does not admit of the kind of idealization his logic was perhaps meant to capture. As we shall see, insofar as his notion of material similarity is grounded upon hunches about causal mechanisms and is attested to by former falsifications, the Method of Relevant Variables cannot be taken even as an approximation of testing against omniscient background knowledge. Not only that, but if my argument is sound, Cohen's very notion of reliability ranking is seriously undermined.

4. The first thing to notice is that as a hypothesis is subjected to tests of greater thoroughness two rather than one induction are called into question. In seeking for non-B-ish A's we are testing more than the hypothesis that all A's are B's. Whether we like it or not we shall also be testing the higher level assumption concerning the material similarity of the hypothesis in question — an assumption on which the entire test procedure rests.

The second and more important thing to notice is that the problem then is not one of infinite regres. It is not simply the case that in order to confidently test a hypothesis we must first fully confirm the higher level hypothesis on which its test procedures are founded. Assumptions of omniscient background knowledge, as we have seen, seem to allow Cohen to ignore that kind of objection, at least with respect to his logic. The trouble with (or rather the novelty of) Cohen's induction is that it is eliminative. By testing ravens for blackness in circumstances known to have caused swans to lose their natural white, ornithologists are viewed by Cohen as seeking to eliminate one of two contrasting hypotheses; namely that all ravens (quite unlike swans) are black, or that ravens, though normally black, tend (exactly like swans) to lose their natural blackness in certain circumstances. In other words, what ornithologists are seeking to learn from experience is whether or not ravens and swans are of a feather with respect to whatever is responsible for their plumage hue. What severely complicates matters is the fact that the tests themselves are set up *on the assumption that they are*! The circumstantial variables selected as appropriate for testing the ravens hypothesis are exactly those believed by the ornithological community to be the potential obliterators of whatever it is that normally causes their blackness. Hence, as it gradually becomes clear that ravens evidently retain their normal blackness even in circumstances in which they were expected to lose it, the supposition that all ravens are indeed black is in some sense supported. But at the same time, and by the very same token, we are gradually led to suspect that as far as their colour is concerned, ravens are rather different from swans, exactly because it is now becoming evident that the mechanism responsible for the ravens blackness is indifferent to the known obliterators of the mechanism thought to be responsible for the normal whiteness of swans. Yet that formed the "rational basis" for selecting the circumstantial variables manipulated in testing the ravens in the first place. In other words passing the most thorough test conceivable in itself strongly indicates that perhaps it was not such a thorough test at all! Alternatively, if, in the course of testing, a hypothesis is found not to hold in circumstances it was initially assumed not to hold, the higher-level hypothesis regarding its material similarity to similarly falsified hypotheses gains support while it itself is disconfirmed. This reciprocal feedback effect of the possible outcomes of empirical tests upon the higher-level inductions on which they were intially based, is a general and necessary feature of the Method of Relevant Variables. Cohen himself has not dealt with this problematic aspect of his methodology, and proceeds throughout on the assumption that in the course of empirical investigations relevant background knowledge concerning the material similarity of the hypotheses involved remains intact. Here is where the incompatible static "Kuhnian" and dynamic "Popperian" (Baconian) aspect of his work are seen to run against each other. If the "rational basis" for testing the reliablility of one hypothesis depends in principle on the reliability of another *which necessarily implies its negation,* I cannot see how Cohen's logic of reliability ranking can be preserved in its original form. Neither does there seem to be any point in idealizing background knowledge for the sake of formalization, since in the event of truly omniscient knowledge of the *real* obliterators of a hypothesis no testing would be called for in the first place.

We are left with the following: In experimental science The Method of Relevant Variables will typically apply within the context of a preliminary and as of yet loosely formed causal explanatory theory. Since only within such a context can the notion of material similarity be ascribed the cogency to uphold Cohen's methodology. Such an assumption, however, necessarily implies that as a hypothesis is found to survive tests of greater thoroughness, the reliablility of the test procedure itself is seriously called into question. In which case the inductive support lent to a hypothesis in the event of such testing can no longer be construed as ranking its reliability *simpliciter*. Nor can the logical syntax for inductive support be attributed the conclusiveness ascribed to it by Cohen, since, by the same token, the idea of perfect background knowledge is rendered inapplicable in principle.

Are we to conclude therefore that when experimental science is denied the stable setting ascribed to it by Cohen, his notion of support as reliability ranking and its underlying modal logic are rendered inadequate? Despite all that has been said I think not. In fact I shall try to show in the remainder of this paper that quite the opposite is the case, namely, that when reassessed in the context of a truly dynamic research programme Cohen's entire system can be retained unproblematically.

5. What I think renders Cohen's analysis basically problematic as it stands is not the Method of Relevant Variables or his logic of induction themselves, so much as what is taken by him to be the *subject matter* of evidential support. If it is sincerely assumed, prior to testing, that the ravens' hypothesis is materially similar to other hypotheses such as the swans', it will not be asserted *simpliciter*. Background knowledge is relevant to more than test procedure, for it is within the context of background knowledge that hypotheses are put forward in the first place. Within the type of relatively mature sciences Cohen has elected to deal with we will never simply be testing whether a low-level correlational generalisation is true.

As stated above the Method of Relevant Variables applies within the context of partly articulated explanatory theories. Such a theory, T, asserts that a set M of correlational generalisations H_1, H_2, ..., H_m are materially similar. That is to say that they normally hold true by virtue of the same kind of causal mechanism C. In its most primitive form such a theory will be no more than a hunch. In its most advanced form each and every member of M will be shown to follow rigorously from a detailed description of C (the existence of which would have to be warranted independently by experience). The maturation of such a theory will typically involve two quite different yet connected research efforts: One, to determine the precise *applicandum* of T — i.e. the precise list of generalisations to be eventually explained by C. The other, to define, locate and fully explicate C itself.

Not untill C is fully articulated and sufficiently warranted will the set of materially similar lower-level hypotheses explicable in T be finalized. And even then we may still encounter what William Whewell termed consiliences of inductions — i.e. explanatory surprises — as a result of which M may be expanded even further. In any event, prior to the full articulation of C the appropriate set of materially similar generalisations will not be fully defined. Viewed thus, it is reasonable to assume, as does Cohen, that in the first instance the boundaries of M will indeed be decided tentatively on the grounds of semantic categorization, which will probably be all there is to go on. However, linguistic practices alone will not suffice to warrant the existence of causal mechanisms. Lacking as of yet any positive knowledge of C itself, the most important indicators of both the possible *explanans* and probable *explanandum* of T will at this stage be what seem to be C's potential obliterators — i.e. circumstantial variables evidently responsible for the obliteration of the type of causal mechanism suspected to be at work. Gradually, within M, the initial loosely defined list of candidate *explananda*, a core-set M^1 will begin to take shape. The core-set will consist of generalisations found *not to obtain* in similar circumstances. Accordingly, a list V of thought to be circumstantial obliterators of C will take form with reference to which preliminary conjectures as to the precise nature of C will be put forward (a stage corresponding to Bacon's original Table of Absence), and other candidate generalisations belonging to M will be tested empirically for membership in the core-set. The idea being to expand M^1 and V as far as possible. It is here that I believe Cohen's Method of Relevant Variables should be made to apply.

What then is learnt from experience by the Method of Relevant Variables? The final goal of such an investigation is to fully determine both the *explanans* and *explanandum* of T — i.e. both the precise nature of C and the precise boundaries of M. As already noted the two objectives are largely independent. Knowledge of C, detailed and precise as it may be, will not suffice to establish exactly which lower-level correlations are in fact governed by such a mechanism — especially if C is not directly observable. Conversely, knowing exactly which correlations are governed by C-type mechanisms will not suffice to disclose the nature of C itself. Yet knowledge of the obliterators of such mechanisms will be material to both lines of enquiry.

With respect to the latter, the early stages of such an investigation will be characterized by two working hypotheses constantly put to test:

(a) That M constitutes the entire range of application of T and

(b) That V exhausts all potential obliterators of C.

One by one all members of M — i.e. all hypotheses initially believed to be governed by C-type mechanisms — will be tested with reference to V, the known circumstantial obliterators of the core-set. Ideally each hypothesis will be subjected to the most thorough test possible — t_n — one in which all members of V are manipulated. Let H be such a hypothesis; i.e. a member of M but not of M^1.

What researchers will typically seek to determine empirically is whether or not H holds true in circumstances in which other members of M^1 are already known to fail. Accordingly H will be subjected to a series of trials, corresponding exactly to those analysed by Cohen, in which, ideally, a prudent effort will be made to instantiate H in all possible variants of V. The main objective of such testing, and this is the point to stress, will not be to determine the truth value of H *simpliciter*, but to determine to what extent H belongs in M^1. Naturally, the outcomes of such tests, both positive and negative, will bear to a certain degree on the reliability of H itself. But that will always be a matter largely incidental to the main thrust of the enquiry. Even in the ideal case hypotheses will not be put to test with firm prior knowledge of their obliterators as in Cohen's account. In other words, known members of M^1 will not be retested in the light of V. Rather, candidate hypotheses, members of M but not of M^1, will be tested for membership in the core-set.

Let us suppose that the investigation has reached a relatively advanced stage. For simplicity we shall assume that there is yet no working hypothesis available with regards to the precise nature of C, but that there is wide agreement with respect to a considerable membership of M^1 all thought to have been falsified by virtue of variants of a joint list of circumstantial variables V. When a new hypothesis H, not yet known to belong in M^1, is subjected to t_n we can expect one of three possible outcomes. If H survives t_n convincingly then (depending of course on the reliability of what is already known of M^1 and V) researchers will confidently conclude that H is inexplicable by T. Similarly, if H totally fails t_n, it will be added to the core-set M^1 without hesitation. Both extreme cases lead to conclusive rulings with respect to the initial conjecture; namely that $H \in M^1$. In the first case it will be flatly rejected, and in the second it will be adopted without further ado. The third and most interesting case is, of course, that in which H ends up somewhere in the middle — successfully falsified by variants of only some subset of V. In such cases researchers will normally be required to rethink the situation. Again the problem will not be to determine the truth value of H so much as to clarify the relation between H and M^1. Generally speaking it will be a question of deciding between two competing hypotheses. It could be the case that the causal mechanism responsible for H is not a C-type mechanism at all despite the evident overlap of obliterators. Certain circumstances may be pertinent to the obliteration of more than one type of causal mechanism. One the other hand, it could very well be that C-type mechanisms come in a variety of shapes and sizes, and that H is governed by one more stable than those responsible for the other known members of the core-set. In which case, despite its anomalous resilience, H should nevertheless be annexed to M^1. The data in hand will not suffice to decide the issue. New tests will need to be devised; tests pertaining perhaps to other classes of hypotheses whose lists of known obliterators overlap those of M^1 in similar fashion. Subsequently preliminary suggestions as to the nature and structure of C itself will also be put forward.

The Method of Relevant Variables is hence by no means an exhaustive description of the methodology of experimental science, nor was it ever intended by Cohen as such. Viewed thus, however, within the context of a truly dynamic research programme, its purely methodological value, both descriptive and prescriptive, is, I believe, beyond dispute. In exposing the centrality of eliminative test procedures based on the known obliterators of familiar correlational generalisations, Cohen has contributed decisively to our understanding of the ampliative modes of reasoning necessarily involved in learning from experience.

6. But what of the formal implications of denying Cohen's methodology its original stable setting? Here, as indicated above, one major modification of his original logical syntax of evidential support is called for. Since a clear distinction needs to be made between the hypothesis actually tested — namely, the lower-level correlational generalisation H — and that assessed in the light of the evidence — namely, the higher-level working hypothesis $H \in M^1$. The support function itself will be of a form similar to Cohen's $s[U, E]$, though E and U will now be interpreted differently. The evidential argument E will be shorthand for "H was *successfully falsified* in the i-th of the $n+1$ canonical tests determined by V", and the hypothesis argument will not denote H, as in Cohen's original account, but $H \in M^1$. As in Cohen's account s will be assigned values of the form i/n ($i = 0, 1, 2, ..., n$) according to the degree of thoroughness of the test reported in E. In this manner it is easy to see that much of Cohen's original system is preserved. In particular the values of s may be viewed unproblematically as ranking the reliability of U — i.e. of

78

annexing H to the core-set — in the light of H's proven falsification by members of V. And the notion of reliability ranking is no longer complicated by the reciprocal feedback effect that, as we have seen, plagues Cohen's original syntax. Since the background knowledge relevant to testing (namely, that V are obliterators of members of the core-set) is unaffected by the confirmation or disconfirmation of the hypothesis assessed (namely, that H be annexed to M^1). If H is successfully falsified in t_n researchers can rationally and confidently conclude that H indeed belongs in the core-set. Conversely, prior to systematic testing against V, $H \in M^1$ may be regarded viably assertable (though no more) by virtue of H's initial membership in M. From which follows that under ideal conditions of omniscient knowledge as to the entire list of obliterators of all possible *explananda* of T, the logic of reliability ranking thus construed may be mapped unproblematically on to Cohen's original version of the Lewis-Barcan modal system S_4. Since then it indeed follows logically that if a correlational generalization is successfully falsified when subjected to the most thorough test possible, it is demonstratively true that it belongs in M^1.

The Cohn Institute for the History and Philosophy of Science and Ideas
Tel Aviv University,
Ramat Aviv
69978 Israel

NOTES AND COMMENTS

[1] Quine (1969), Ch. 5.

[2] For a comparison and critical assessment of Quine and Cohen's solutions of Hempel's paradoxes see Fisch (1984), pp. 45-62.

[3] Cohen, L. J. (1970), p. 61.

[4] *Ibid*, p. 1.

[5] *Ibid*, p. 39.

[6] *Ibid*, p. 40.

[7] Fisch (1984), pp. 55ff. In that paper, however, I failed to appreciate the problematic consequences of my account for Cohen's system as taken up in what follows.

[8] As result of criticism, especially that of Isaac Levi, Cohen had modified his methodology in *The Probable and the Provable* requiring that rather than screen off variables not manipulated in intermediate tests, to allow normal variants of them to prevail. Such a move, I have argued, is uncalled for: cf. Fisch (1985), pp. 254-5, ftn. 17. For as Mary Hesse has shown – (1980), pp. 202-17 – allowing normal variants of non manipulated variables to prevail renders Cohen's logic of inductive support expressable in Bayesian terms.

[9] Cf. Cohen (1970) p. 92, ftn. 1.

[10] *Ibid*, p. 221, formula 108.

[11] *Ibid*, pp. 9-10.

[12] This is a matter quite different from the problem of the requirement for total evidence dealt with by Cohen at length. (See in particular Cohen (1970), pp. 197f.) For the formation of judgements about inductive support in the face of eliminative trials, there is no need, Cohen argues correctly, for total evidence as in non-eliminative inductive logics such as Carnap's. There is, however, a need for total knowledge of the *obliterators* of any given hypothesis in order for tests on them to yield conclusive judgments.

REFERENCES

Cohen, L. J. (1970). *Implications of Induction*. London: Methuen and Co.

Cohen, L. J. (1977). *The Probable and the Provable*. Oxford: Clarendon Press.

Cohen, L. J. (1989). *An Introduction to the Philosophy of Induction and Probability*. Oxford: Clarendon Press.

Fisch, M. (1984). Hempel's Ravens, the Natural Classification of Hypotheses and the Growth of Knowledge. *Erkenntnis* **21**, 45-62.

Fisch, M. (1985). Whewell's Consilience of Inductions – An Evaluation. *Philosophy of Science*, **52**, 239-255.

Hesse, M. B. (1980). Inductive Appraisal of Scientific Theories. In Cohen, L. J. and Hesse, M. B. (Eds.), *Applications of Inductive Logic*. Oxford: Clarendon Press.

Quine, W.V.O. (1969). Natural Kinds. In his *Ontological Relativity and Other Essays*. New York and London: Columbia University Press.

*Poznań Studies in the Philosophy
of the Sciences and the Humanities
1991, Vol. 21, pp. 81-95*

Maurice A. Finocchiaro

INDUCTION AND INTUITION IN THE NORMATIVE STUDY OF REASONING

1. Introduction

In the contemporary philosophical scene, the work of L. Jonathan Cohen[1] stands out in several ways. One stems from the significance, range, and combination of topics on which he focuses, which are primarily reasoning and rationality, induction and probability, metaphilosophy, philosophy of science, and philosophy of law. Another relates to the character of the approach he follows in his investigations, and which I would describe as both theoretically deep and practically relevant, both insightful and well-argued, rigorous without being pedantic, and original without being cranky. Cohen's work also deserves admiration and emulation because of the particular stand he takes on the issues and the particular conclusions he arrives at in regard to the topics studied; and here I have in mind his defense of human rationality from the attacks of experimental psychologists, his defense of democratic values and principles in regard to the use of lay juries in jurisprudence, and his defense of pluralism, open-mindedness, and dialogue in analytical philosophy.

In the light of all this, it is obvious that my remarks here cannot do full justice to the depth and variety of Cohen's work. I shall limit myself to some issues in his latest book (1986), and even in regard to them I feel that all I can do is to provide what may serve as the beginning of a potential dialogue, rather than a full analysis and resolution of them.

2. Reasoning in Analytical Philosophy and the Inductive-Intuitive Method

The cluster of issues I want to focus on involve the nature of one type of argument in analytical philosophy, its connection with inductive reasoning, and the question of its viability and correctness in general. The topic is, I feel, especially promising in two ways. First, it is simply a sociohistorical fact that the amount of work in the field of analytical philosophy is considerable: this is the work of philosophers like Ayer, Carnap, Davidson, Hempel, Hintikka, Quine, Rawls, Ryle, and Scriven, and their followers. Given the sheer quantity of writings, and the fact that they obviously contain reasoning and argument, the field demands serious attention by any serious theorist of reasoning. One could say that its presence is so robust that reasoning in analytical philosophy deserves to be studied simply because it is there, if for no other reason.

The second reason for the special relevance of the topic stems from the fact that Cohen claims that a greal deal of analytical-philosophical argument is (1) about reasoning, specifically about (2) norms of reasoning, and consists of (3) inductive reasoning (4) based on intuitions. If he is right, then we may say that there exists a distincitive approach to the study of reasoning, an approach which may be labeled the intuitive-inductive approach. His account would therefore deserve the attention of anyone concerned with the methodology of the study of reasoning, whether he is an advocate or a critic[2] of the experimental approach prevalent in cognitive psychology, or an advocate or critic of the apriorist approach prevalent in formal logic. In fact, Cohen may be taken to have identified, elaborated, and defended a different approach.

Therefore, in his latest work, Cohen's contribution is at least two-sided, metaphilosophical and methodological; metaphilosophical, insofar as it advances a thesis about the nature of reasoning in that particular domain of human inquiry which is analytical philosophy, and methodological insofar as it articulates an approach to the study of reasoning which is worthy of serious consideration in general.

If from the viewpoint of significance Cohen's contribution has this double dimension, from the viewpoint of content it is striking in three ways.

First, there is his emphasis on the importance of induction in philosophical thought. Now, although this sort of recognition goes back as least to Aristotle's interpretation of Socrates's definitions as the result of inductive arguments,[3] such an inductivist viewpoint goes against the prevailing view that all philosophical argumentation is deductive[4]. And here it should be added, by way of clarification, that Cohen attaches his own special meaning to the notion of induction:[5] on the one hand he does retain the contrast with deduction, and the idea that inductive reasoning proceeds from the particular to the general; however, he emphasizes that what is particular and what is general is something that depends on the context (p. 82)[6] and more importantly that what he has in mind is not the enumerative but the eliminative type of induction, according to which generalizations are tested by reference to relevant variations in kinds of instances rather than merely being inferred from a multiplicity of instances (p. 69).

Second, there is Cohen's recognition of the importance of singular intuitions as the basis for philosophical conclusions, and here it is important to understand what he means by intuition. He does not mean the infallible apprehension of essences, or a heuristic faculty that is part of the creative imagination, but rather a fallible inclination of judgment originating from a system of tacit rules rather than from any conscious or explicit inferential or causal process (pp. 73-82).

Third, Cohen's central thesis has normative content in the sense that he is attributing a normative aim to analytical philosophy. His critique of naturalized epistemology and his plea for a normative epistemology and for a normative study of reasoning, make it clear that he is concerned with the normative principles of reasoning.

So far then we might attempt to formulate Cohen's central thesis as saying that analytical philosophers support norms of reasoning by means of inductive reasoning based on singular intuitions. However, there is another dimension of Cohen's thesis that needs elaboration for a fuller appreciation. To see this let us ask whether he is claiming that this is what analytical philosophers are in fact doing, or what they ought to be doing, or both; in other words, are analytical philosophers essentially right in engaging in this type of reasoning?

It is obvious that Cohen's thesis has a second normative dimension, at the metalevel; that is, his own thesis is primarily a normative principle of philosophical reasoning, a normative principle about the intuitive and inductive basis of norms of reasoning. This is obvious partly from the parts of the book concerned with vindicating the essential correctness of this inductive-intuitive method, and also from the self-referential requirement that his own metaphilosophical exercise is meant to be a piece of analytical philosophy and therefore subject to the same principles being attributed to the latter.

The normative status of Cohen's own metaphilosophical thesis does not, however, undo its factual, descriptive, and explanatory content. So what I am saying is that a full appreciation of it needs to recognize both its normative, evaluative force and its factual, descriptive content. In other words, for Cohen a chief involvement of analytical philosophers *is and ought to be* inductive reasoning based on singular intuitions and supporting norms of reasoning.

These two aspects of the thesis could also be stated as generalizations about norms of reasoning: for example, that norms of reasoning may be properly supported by inductive reasoning based on singular intuitions, and that norms of reasoning are supported in this manner in analytical philosophy. When so stated these two aspects seem to correspond to the above mentioned methodological and metaphilosophical aspects.

3. Reasoning Versus Nonreasoning

Let us now turn into a more critical direction. The first difficulty I should like to raise involves the distinction between reasoning about reasoning and reasoning about nonreasoning. The question is whether the subject matter of analytical-philosophical reasoning is reasoning or something else. This question is different from the one concerning whether or not analytical philosophers are engaged in reasoning when they study whatever it is they are studying. My point is that it is not immediately obvious that, for example, when Hempel (1965) is engaged in analyzing the concept of explanation or when Gettier (1963) is engaged in analyzing the concept of knowledge, they are dealing with the same subject matter as Harman (1986) when he is analyzing the notion of reasoning. Or to give an example from Cohen's own work, when he is engaged in clarifying the concept of intuition (pp. 73-83) or the concepts of belief and acceptance (pp. 91-97), he seems to be dealing with a subject matter significantly different from the case when he is engaged in reconstructing how subject solve the taxi-cab problem (pp. 159-64).

I believe Cohen is addressing this issue when he points out that in the analysis of explanation a central concern is whether or not "the inclusion of statements in such-and-such a relationship to one another or to the world at large counts as a reason for calling some given string of sentences an 'explanation' " (p. 73). So in general we may say that in the analysis of concepts philosophers are studying how ordinary people, or the relevant group of practitioners, as the case may be, reason when they apply the concept. Some clarifications are in order here.

First, if we distinguish between the word and the concept, between lexicography and philosophy as it were[7], then the paraphrase of Cohen's quotation I have just given may be incorrect. I should have said that analysts are studying how people reason when they apply the word, how they would justify the application of a word. To be more exact, we should speak of how people reason when they apply the word correctly. Moreover, a concept may be in someone's mind even without the proper word, and so we would have to say that analysts also study how people would reason if they were to apply the word correctly.

However, it may be objected that the degree or intensity of reasoning here is too weak, that the ratiocination is not sufficiently explicit to be called "reasoning". Moreover, all such reasoning would have a metalinguistic or metaconceptual character that would render suspect its claim to being typical or representative of human reasoning. Finally, in the analysis of the concept of reasoning as such, a philosopher would be examining not only the reasoning involved in the proper application of the notion of "reasoning", but also the more explicit reasoning constituting the content of the argument in question.

In other words, the investigation of what X is (where X stands for a term like explanation, belief, knowledge, intuition, argument, and reasoning) is also an investigation of the reasoning whereby a person judges that this is an X. In fact, for the justification of the judgment that this is an X one would require some such reasoning as: this is an X because this is A, B, C, D, etc., and X is anything which is A, B, C, D, etc. Therefore, the explication of X (produced by the analysis of X) serves as the major premise of such arguments; it is in a sense the principle of reasoning connecting the minor premise and the conclusion. It follows then that the analysis of concepts practiced by analytical philosophers is indeed the investigation of the principles of reasoning in such arguments. Therefore, I would conclude by saying that there is a sense in which one may analyze analysis as reasoning about reasoning.

However, I have a lingering dissatisfaction with this. Since any statement may become part of reasoning when it becomes a premise of an appropriate argument, it appears that in Cohen's account the study of anything would become the study of reasoning, namely of the reasoning involving that something. And then we would want to introduce a distinction between explicit and actual reasoning and merely implicit and potential reasoning.

4. Norms

Another point of difficulty concerns the part of Cohen's thesis claiming that the focus of analytical-philosophical reasoning is the *norms* of reasoning.[9] We may agree with him that "if the normative study of reasoning did not already exist, it would have to be invented" (p. 63).

However, this thesis would not be correct if it meant that whenever reasoning is the subject matter of philosophical reasoning, the relevant aspect of that subject matter is normative principles of reasoning. A common concern of analytical philosophers is the nature of reasoning as reflected in its structure and as expressed by descriptive generalizations about reasoning.

For example, in the analysis of explanation, it seems to me there is a great difference between the question of the structure of scientific explanation and the question of the conditions under which a scientific explanation is acceptable, or preferable to another. Part of Hempel's account claims that scientific explanations have a deductive and nomological structure, and that their logical structure is identical to that of predictions. But then, within the class of scientific explanations so defined, Hempel tries to distinguish various subclasses depending on various virtues or degrees of merit they may or may not possess, involving such questions as whether or not such explanations are true, or highly confirmed, or sketchy, or *ad hoc*, or self-evidencing, etc. And he also formulates what he calls a general *condition of adequacy for any rationally acceptable explanation of a particular event.* That condition is the following: any rationally acceptable answer to the question 'Why did event X occur?' must offer information which shows that X was to be expected — if not definitely, as in the case of a *D-N* explanation, then at least with reasonable probability" (Hempel 1965, pp. 367-68). I believe it might be useful to think of generalizations about the logical structure of scientific explanations as principles of *interpretation*, and of generalizations about their various degrees of merit as principles of *evaluation*.

One question I have is how Cohen's norms of reasoning fit into this scheme. The principles of interpretation just mentioned correspond, I believe, to the explications of concepts mentioned earlier. That discussion above was limited to identifying a role in reasoning for such principles, the context being that of the reconstruction of the rationale for a judgment such as that this is a scientific explanation. Now we can add that in certain other contexts those major premises would have normative force, and thus might be regarded as normative principles of reasoning; these would be contexts where one is engaged in the construction of the original argument, namely in the original reasoning arriving at the original judgment.

My initial comment here would be that this normative force of interpretative principles is rather meager in comparison with the normative force of the evaluative principles. In the latter the normative force is full-blown and explicit. But although this distinction between interpretation and evaluation is thusly analogous to our earlier one between reasoning and nonreasoning, the effect on Cohen's discussion is here, I believe, the reverse of the earlier one. What I mean is that here Cohen does properly focus on explicitly normative questions; or to be more exact, when he shows how analytical philosophers engage in the normative study of reasoning, the norms of reasoning in question are of the explicit full-blown variety. This is as it should be if one wants to call attention to the normative dimension of epistemology.

However, this leads me to the criticism that Cohen is not paying sufficient attention to the descriptive, interpretative dimension of analytical philosophy; to the fact that an equally important aim of analytical philosophers is the study of the structure of reasoning and of interpretative principles, independently of their normative dimension or their normative connections. To repeat, one does not have to deny the connection between descriptive and prescriptive principles, nor the implicitly normative dimension of descriptive principles of reasoning. The point is merely to admit the distinct status and philosophical significance of descriptive principles of interpretation.

To conclude this section, it is best to add an example from Cohen's own practice to the one from the theory of explanation already given. His work is in a sense, as he subtitles it, "an analysis of analytical philosophy." And a main result of this analysis is that "the unifying force in analytical philosophy is its engagement with the reasoned investigation of reasons at that level of generality (varying in accordance with subject-matter) where no conclusions can be taken as universally granted" (p.57). Another is that "a great deal of reasoning in modern analytical philosophy is inductive, not deductive" (p. 67). These are important generalization about analytical-philosophical reasoning, but their primary import is descriptive and interpretative rather than normative; and I believe their very significance would be diluted to the extent that we were to construe them in normative terms.

5. Induction

Next I should like to focus on that part of Cohen's thesis which is in many ways the most striking and the most exciting, namely the part which attributes an inductive character to analytical-philosophical reasoning.[10]

The difficulties I feel in this regard concern the relationship between inductive reasoning and (1) deduction, (2) the empirical approach, and (3) what Cohen himself calls "inductivism".

To begin with, it is useful to stress that Cohen is not denying that a great deal of philosophical reasoning is deductive (pp. 66-67), and that he himself points out (p. 144) that the writings of analytical philosophers normally contain a mixture of both deductive and inductive reasoning. How does the metaphilosopher distinguish the two?

Cohen states that "in deductive reasoning our conclusion must be true if the premises are true. Through inductive reasoning — of the kind relevant here — we may acquire some level of justified confidence in a generalization if we accept reasons for supposing that the generalization holds good in certain varieties of instance" (p. 67). These look to me like definitions of deductively *valid* reasoning and of inductively *correct* reasoning, for they seem to exclude the possibility of deductive arguments which happen to be invalid and of inductive reasoning which happens to be inductively faulty. And I am sure Cohen does not want to allow this consequence. That is, when as examples of deductive reasoning he gives "Descartes's proof of an external world ... Spinoza's *Ethics* ... transcendental argument" (p. 67), I am sure he does not mean that all these arguments actually confer the requisite level of justified confidence in the generalizations being supported. However, if the definitions given really make a distinction between two *types of evaluation* of an argument, how is the distinction between *argument-types* to be made?[11]

In regard to the connection with empiricism, I should like to raise the question whether the inductive approach is a type of empirical approach, and I should like to explore the possiblility of an affirmative answer. To be sure, I have no problem grasping the point, which Cohen wisely never tires of reiterating, that inductive reasoning cannot be equated with reasoning based on observation or experiment, and that the latter in only one species of the genus (pp. 67-73). But in my view this does make the inductive approach nonempirical; in fact I would not want to equate the empirical with the observational or experimental, partly because experimentation often contains *a priori* elements, partly because I am inclined to regard the historical approach as a type of empirical approach,[12] and partly because I would want to analyze the empirical in terms of open-minded fallibilism.[13]

It may also look like Cohen is driving a wedge between induction and empiricism when he says that "the choice between deductive and inductive patterns of argument can be quite independent of any preference for rationalist or empiricist epistemology" (p. 135). However, his meaning turns out to be such as to dispel this initial impression. What he means is that philosophers sometimes use deductive argumentation in the service of an empiricist epistemology, a classic example being Hume's attitude toward his epistemological principle that every simple idea is a copy of a previous impression.[14] Similarly, philosophers sometimes use inductive arguments in support of rationalist epistemologies, Cohen's example being Whewell's historical arguments to justify his Kantian epistemology (pp. 135-36). However, I think this merely means that many philosophers are not completely consistent, and that their own philosophical practice often belies the general conclusions they preach, that their general principles often run into self-referential difficulties.

Thirdly, it looks like Cohen is advocating at least the *compatibility* between inductive reasoning and apriorism in view of the fact that some of his most frequent examples of analytical philosophy come from the work of Carnap and Hempel, and yet their theories of confirmation and of explanation are clearly cases of excessive apriorism in the procedure they followed in their own work,[15] a fact which Cohen himself recognizes (p. 126).

However, perhaps all that Cohen is saying is that apriorism is a species of *faulty* inductive reasoning. For on the one hand he regards these philosophers as part of "the curious tendency that existed for a while in modern analytical philosophy to treat certain kinds of scientific problem as philosophical rather than scientific issues" (p. 118), and he criticizes this as a "mistakenly a priorist tendency" (p. 121). On the other hand, he later diagnoses their problem by saying that they "invoked intuitions in relation to certain problems about confirmation, explanation, etc. that are really part of the domain of scientific inquiry ... in order to achieve an a priori global resolution for issues that in fact require empirical determination in local contexts" (p. 126); and he goes on to suggest that "what is necessary instead is to discern the appropriate domain for philosophical generalization about scientific reasoning — i.e. to discern the level of abstraction or generality at which the analysis of this reasoning is possible" (p. 126). Thus, Cohen's point may be that, their reasoning did indeed have an inductive character, but that it was unsound in the sense of being excessively apriorist. Therefore, their work would be instantiating both induction and apriorism, but in a harmless manner.

However, a new difficulty arises at this point. Why should one describe their theories of confirmation and of explanation as examples of faulty inductive reasoning (faulty because apriorist or "hasty") rather than as an example of objectionable *deductivism?* For that is how Cohen describes Chisholm's theory of inductive evidence (p. 140), which seems to me to be an exactly parallel case since Cohen's complaint is its lack of (sufficient) empirical content.[16]

Finally, I am not exactly clear about the connection between inductive reasoning and what he calls "inductivism" in the context of a discussion of "deductivism and inductivism as alternative strategies in analytical philosophy" (pp. 129-47). Obviously, inductivism is partly the strategy of employing inductive arguments, and deductivism the strategy of employing deductive arguments. However, he also draws the contrast in terms of, respectively, a critical[17] versus a conservative orientation, a localized versus a globalist orientation, and an interest in traditional philosophical problems versus an interest in problems stemming from contemporary nonphilosophical areas.

In this section the most striking feature of the discussion is Cohen's judiciousness. For he admits that "within analytical philosophy, both deductivism and inductivism may run into characteristic perils if allowed

90

go dominate excessively" (p. 133); and he recognizes that "in practice, of course, inductivism and deductivism often complement one another in a single text" (p. 144); and he argues that "the thesis that both inductive and deductive reasoning are admissible in any branch of analytical philosophy is a categorical and inductive one, in that intuitively recognizable examples of both kind of reasoning are cited in support of the thesis" (p. 147).

One comment I have here is that it seems implausible to include all these features under the label of "inductivism." I do see their family resemblance to one another and to the employment of inductive arguments, in the more or less technical sense of induction intended by Cohen. However, I am not sure I see how all of them could be reduced to the technique of employing inductive reasoning, whose possibility is suggested by his choice of the label "inductivism." Another question would be why the empirical orientation is not included in the list.

6. Intuition

Appeal to intuition is certainly one of Cohen's central themes. As mentioned above, it is important to understand that he is talking about fallible inclinations of judgment. It should be mentioned that he argues conclusively, in my estimation, in favor of the primacy of singular intuitions. It is also clear that singular intuitions are seen as providing the data on which inductive reasoning in analytical philosophy is based.

The only issue I should like to raise here is a question about the relative importance of intuitions about imagined artificial examples and intuitions about actual historical cases, which relates to the question whether such intuitions are a priori or empirical.

Let us begin by asking what such intuitions are about. For Cohen they are "about what should be inferred, judged or meant in such-and-such a context" (p. 73), as distinct from what is in fact inferred; they are about "what *is* a reason for what, not ... what is taken by scientists, lawyers, or others to be a reason for what" (p. 77). This is in accordance with the rest of his account which makes these intuitions the basis of inductive reasoning in support of *normative* principles. In short, the relevant intuitions are intuitions about normative issues in particular cases. We might speak of normative singular intuitions.

Are the particular cases merely imagined or empirically real? Are they artificially contrived or historically reconstructed? I feel that this is an important question because its answer determines whether or not the intuitive-inductive method advocated by Cohen is empirical. I believe he is somewhat ambiguous on the matter.

In fact, on the one hand we can find explicit and clear statements of the superiority of real or historical examples over imagined and artificial ones. For example, while discussing this issue as it emerges in the different fields of moral philosophy, grammatical theory, and epistemology he says:

> No doubt it is better to rely on real-life decisions of conscience if one can, just as a grammarian may prefer to rely on real-life utterances if he can: there is less risk that the intuitions may be biased, or otherwise influenced, by the preferred theory. But some recourse to intuitions about imaginary situations or utterances may be unavoidable. Similarly a philosopher of science needs to take his examples from the history of science whenever this is possible. But he may not be able to dispense altogether with appeals to avowed intuitions about imaginary cases or his own thought-experiments [p. 88].

Moreover, Cohen's own examples from the writings of analytical philosophers are almost always actual, historical, and "real-life"; therefore, in his own metaphilosophical theorizing he is certainly practicing what I would call an empirical, historically oriented method.[18]

On the other hand, most of these examples are from analytical philosophers who tend to give imagined and artificial examples, thus shunning the empirical method in their own practice and engaging in some form or other of apriorism. Further, at one point Cohen declares explicitly that an intuition that *p*, in his sense of intuition, is *a priori*, while admittedly not analytic.[19]

There is no problem with the denial of analyticity, which I take to be another way of stressing the fallibility of the kind of intuitions he is talking about. But the atrribution of an *a priori* status is puzzling. The reason Cohen gives for it is that such intuitions are "not checkable by sensory perception" (p. 75), and do not depend on a "form of introspection" (p. 75). However, it seems to me that the crucial issue would be whether they depend on some form of experience. Now, while I would agree with Cohen that such intuitions do not depend *consciously* or *deliberately* on education, it seems to me that they cannot fail to depend at least *unconsciously* on education, as Cohen admits that they may (p. 77).

In conclusion, while his claim about the *a priori* status of intuitions is likely to be untenable, there can be no question that he is ambiguous in regard to this whole issue.

It may be that Cohen's partial inclinations toward apriorism stem, at least in part, from his rejection of the experimental-psychological approach to the study of reasoning. The latter is certainly a species of empiricism, but I share Cohen's deep concerns about it, and have advanced some objections of my own.[20] Therefore, it should be clear that I for one am not advocating the substitution of his inductive-intuitive approach by the experimental method of cognitive psychology.

7. Conclusion

In summary, I began this paper by calling attention to the significance of Cohen's work in general. Then I focused on his latest book, which contains two interrelated themes. One is a metaphilosophical, descriptive, and interpretative account of the nature of reasoning in analytical philosophy; here his thesis is that a chief aim of analytical philosophers is to elaborate normative principles of reasoning in the various object domains by means of inductive reasoning based on normative singular intuitions. The other theme is the methodological, normative, and evaluative one that this inductive-intuitive-normative method of studying reasoning is a viable one, being preferable, for example, to the experimental method of cognitive psychology.

I am impressed by the originality and fruitfulness of both theses, and my primary aim has been to provide, or ask for, a number of needed clarifications. For example, I have suggested that the study of explicit instances of reasoning has priority over the study of implicit instances, and that much analytical philosophy focuses on nonexplicit reasoning. I have also argued that the study of reasoning ought not to be restricted to the study of the norms of reasoning, and that the work of Cohen himself and of many other analytical philosophers is in fact not so restricted, but rather advances important and insightful descriptive interpretations about the nature and structure of reasoning as such. Thirdly, I have attempted to add to the value of the inductive approach to the study of reasoning by relating it to an empirical anti-apriorist approach, and I have suggested the the connection needs further elaboration. Finally, I have stressed the importance of distinguishing between two types of appeals to singular intuitions, and I have suggested that if we want to avoid objectionable apriorism the intuitions in question should be about actual or historical cases rather than imagined or contrived ones.

Department of Philosophy
University of Nevada − Las Vegas
Las Vegas, Nevada 89154
USA

NOTES

[1] See, for example, Cohen (1966, 1970, 1977, 1981, 1986).

[2] The present author has advanced his own criticism of the experimental method as practiced by cognitive psychologists; see, for example, Finocchiaro (1980a, pp. 256-72, 1987a; 1987b; and 1989).

[3] Aristotle, *Metaphysics* 1078b, 29-32, cf. Cohen (1986), pp. 136-37.

[4] This view is so common that it hardly needs documentation; Cohen (1986, p. 67) mentions Passmore (1961, p. 6).

[5] This, of course, corresponds to ideas he has elaborated in Cohen (1970, 1977).

[6] From here on, references to Cohen (1986) will be given in parenthesis in the text by mentioning just the page numbers.

[7] In view of his critique of the linguistic interpretation of analytical philosophy, Cohen would be the first one to do so.

[8] For a nontrivial example of this type of reasoning, see Finocchiaro (1988b, pp. 197-211) where I reconstruct Hegel's Preface to the *Phenomenology of Spirit* as the argument that philosophy is dialectical because it is pluralistic, conceptual, concrete, self-reflective, spiritual, systematic, negative, and self-referential, and because it sublimates the oppositions between truth and falsity, method and result, change and permanence, form and content, and subject and predicate.

[9] See, for example, his assertion that "if we examine seriatim the problems that actually puzzle analytical philosophers we shall find that the problems of analytical philosophy are all normative problems connected in various ways with rationality of judgment, rationality of attitude, rationality of procedure, or rationality of action" (Cohen 1986, p. 49).

[10] I should also say that my hesitation is also greatest here, for in view of Cohen's long-standing and brilliant contributions to the study of induction I cannot help but feel that he must have thought about these questions and must have ready answers for them. Now, although this makes any criticism on my part rather risky, perhaps I can adopt the attitude that, if I can motivate him to provide the answers, the general gain will be considerable.

[11] I should mention that I have a hunch that the answer may lie in Cohen's theory of probability as the gradation of provability, which he develops in the Chapter 2 of Cohen (1977), but I am not sure.

[12] See, for example, Finocchiaro (1987a, 1987b).

[13] For example, along the lines of Shapere (1984).

[14] Hume (1911), pp. 13-14; cf. Cohen (1986), pp. 132-33, 135; cf. Finocchiaro (1988a), p. 314.

[15] For such a critique of Hempel's theory of explanation, see for example, Finocchiaro (1973), pp. 17-56.

[16] Cohen is referring to Chisholm (1970).

[17] This is my own word; Cohen (1986, pp. 129-35) speaks of "sceptical" conclusions.

[18] If it should be thought that this is unavoidably so in metaphilosophy, I would give the counterexample of Johnstone (1959, 1978). His metaphilosophical theory focuses on the process of *ad hominem* argumentation, and I have elsewhere pointed out its many merits (Finocchiaro 1980b); however, without suggesting that this is a major criticism, I would say that his examples *tend* to be imagined and artificial.

94

[19] "Nothing positive is implied thereby about the specific nature of p's content, where p is the intuited proposition. Certainly p need not be concerned with non-natural facts or essences. But also, though the judgment that p is obviously classifiable as a priori (because not checkable by sensory perception), it need not be analytic" (Cohen, 1986, p. 75).

[20] See, for example, Finocchiaro (1980a, 1987a, 1987b, and 1989).

REFERENCES

Chisholm, R. M. (1970). On the Nature of Empirical Evidence. In L. Foster and J. L. Swanson, (eds.), *Experience and Theory*, pp. 103-34. London: Duckworth.

Cohen, J. J. (1966). *The Diversity of Meaning.* 2nd edition. London: Methuen.

Cohen, L. J. (1970). *The Implications of Induction.* London: Methuen.

Cohen, L. J. (1977). *The Probable and the Provable.* Oxford: Clarendon Press.

Cohen, L. J. (1981). Can Human Irrationality Be Experimentally Demonstrated? *The Behavioral and Brain Sciences*, **4**, 317-70.

Cohen, L. J. (1986). *The Dialogue of Reason: An Analysis of Analytical Philosophy.* Oxford: Clarendon Press.

Finocchiaro, M. A. (1973). *History of Science as Explanation.* Detroit: Wayne State University Press.

Finocchiaro, M. A. (1980a) Galileo and the Art of Reasoning. *Boston Studies in the Philosophy of Science*, vol. **61**. Boston: Kluwer.

Finocchiaro, M. A. (1980b). Review of H. W. Johnstone, Jr., *Validity and Rhetoric in Philosophical Argument. Review of Metaphysics*, **34**, 143-44.

Finocchiaro, M. A. (1987a). An Historical Approach to the study of Argumentation. In F. H. van Eemeren et al., (eds.), *Argumentation: Across the Lines of Discipline*, pp. 81-91. Dordrecht: Foris.

Finocchiaro, M. A. (1987b). Six Types of Fallaciousness: Toward a Realistic Theory of Logical Criticism. *Argumentation*, **1**, 263-82.

Finocchiaro, M. A. (1988a). Empiricism, Judgment, and Argument: Toward an Informal Logic of Science. *Argumentation*, **2**, 313-35.

Finocchiaro, M. A. (1988b). *Gramsci and the History of Dialectical Thought.* Cambridge: Cambridge University Press.

Finocchiaro, M. A. (1989). Methodological Problems in Empirical Logic. *Communication and Cognition*, **22**, 313-335.

Gettier, E., Jr. (1963). Is Justified True Belief Knowledge? *Analysis*, **23**, 121-23.

Harman, G. (1986). *Change in View: Principles of Reasoning.* Cambridge, Mass.: M. I. T. Press.

Hempel, C. G. (1965). *Aspects of Scientific Explanation and Other Essays in the Philosophy of Science.* New York: Free Press.

Hume, D. (1911). *Treatise of Human Nature.* London: J. M. Dent & Sons Ltd.

Johnstone, H. W., Jr., (1959). *Philosophy and Argument.* University Park, Penna.: Pennsylvania State University Press.

Johnstone, H. W., Jr. (1978). *Validity and Rhetoric in Philosophical Argument*. University Park, Penna.: The Dialogue Press of Man & World, Publishers.
Passmore, J. (1961). *Philosophical Reasoning*. London: Duckworth.
Shapere, D. (1984). *Reason and the Search for Knowledge*. Boston: Kluwer.

PART III

SUBJECTIVE PROBABILITY:
PASCALIAN AND BACONIAN CONCEPTIONS

*Poznań Studies in the Philosophy
of the Sciences and the Humanities
1991, Vol. 21, pp. 99-145*

David A. Schum

JONATHAN COHEN AND THOMAS BAYES ON THE ANALYSIS OF CHAINS OF REASONING[*]

1.0. Setting The Stage

My task is to provide an account of what Baconian and Pascalian views of probabilistic reasoning have to say about various issues that arise when reasoning from observable evidence to hypotheses is indirect and involves some number of intermediate stages. As a vehicle for illustrating the workings of these two probabilistic systems on chains of inferential reasoning, I have chosen a commonly-encountered inference task that can be construed in terms of a chain of reasoning; this task involves assessing the credibility of a human source of evidence. The relationship between the credibility of a human source and the inferential value of what this source tells us has been of interest to probabilists ever since the first scholarly writing on probability over three hundred years ago. However, both the multiattribute nature of human source credibility and the chain of reasoning which serves to identify credibility attributes have gone unrecognized in most probabilistic accounts of credibility-testimony issues. The task of credibility assessment is very rich in terms of the inferential issues it involves. Thus, it provides a suitable vehicle for illustrating the workings of the Baconian and Pascalian systems for probabilistic reasoning, each of which has many things to say about the analysis of chains of reasoning.

1. 1. Finding Common Ground

I believe this to be a most exciting time to have an interest in the study of probabilistic reasoning. In the past dozen years or so we have witnessed the emergence of several well-articulated formal systems of probabilistic reasoning that do indeed capture various attributes of probabilistic reasoning that have been recognized, for quite some time, as being elusive under the "conventional" system of probability we all first learn about. Jonathan Cohen refers to this conventional system as the "Pascalian" system, to acknowledge the debt we owe to Blaise Pascal, from whose work stems the extensive and very useful calculus of conventional probability. In this conventional calculus, Bayes' rule emerges as a canon for the task of revising probabilistically-expressed opinion about hypotheses or possibilities on the basis of evidence. This rule follows from the axioms of conventional probability and the *definition* of a conditional probability. The posterior probabilities calculated using Bayes' rule are numbers between zero and one and behave in accordance with these conventional axioms.

But Cohen argues that there is another conception of probability that has an equally respectable intellectual lineage, one tracable to the works of, among others, Francis Bacon and John Stuart Mill; this alternative conception he refers to as "Baconian" (Cohen, 1977). The reader encountering Cohen's most extensive work on Baconian probability (1977) for the very first time must be prepared for a syntax of properties quite unlike the ones that Bayes' rule obeys. For example, Baconian probabilities are *ordinal* in nature and, as such, cannot be meaningfully combined in any algebraic way; we can compare two or more Baconian probabilities but we cannot add, subtract, multiply, or divide them. Such algebraic operations are necessary in calculations using Bayes' rule; as I shall later mention, such operations are often only arguably justifiable given the nature, under certain cirumstances, of the probabilistic ingredients Bayes' rule requires. In the Pascalian system we expect that the probability of the joint occurrence of events A and B is always less than the probability of A or the probability of B. But the Baconian probability of the joint occurrence of A and B is never less than the smaller of the probability of A or the probability of B. The Baconian probability of the disjunction of events A and B is, unlike a Pascalian probability, never greater than the larger of the probability of A or the probability of B. In Pascalian terms the probability of E^c [read E^c as the complement or negation of E, i.e. "not-E"] is always one minus the probability of E. The Baconian negation rule is, as I shall illustrate, quite different from the Pascalian negation rule.

In fact, if one examines Cohen's syntax of properties of Baconian probabilities with the syntax of properties Bayes' rule obeys, about the only similarity one observes is that the probabilities in both systems are numbers greater than or equal to zero. A fair question is: given such a degree of formal dissimilarity, how is a comparison between these systems ever possible or meaningful? I shall now begin the construction of a situation in which such comparison is certainly possible and, I believe, quite meaningful. In this situation a "Baconian" and a "Bayesian" both use exactly the same evidence; but they use it in different ways. A body of evidence has very many interesting characteristics; the Baconian and the Bayesian simply "resonate" to different, but equally important, characteristics of this evidence. Each usage will tell us something different about the inferential value of this evidence. I shall argue that Baconian and Bayesian analyses are not incompatible and that one form of analysis can, indeed, enhance the value of the other.

1.2. Chains Of Reasoning In "Cascaded" Inference

For reasons not entirely clear to me the study and analysis of chains of reasoning has been of greater interest to evidence scholars in jurisprudence and to psychologists than it has been to logicians, probabilists, and statisticians. The earliest systematic study of chains of reasoning seems to have been that of the American jurist John Henry Wigmore. Wigmore distinguished between what he called "simple" and "catenate" inferences (Wigmore, 1937, p. 13). In "simple" inference we proceed in a single step from evidence to some conclusion. A friend comes into my office carrying a dripping umbrella and raincoat and so I conclude, directly, that it is very probably raining outside. In catenate inference, however, there are reasoning steps that may be interposed between evidence and possible conclusions; this suggests Wigmore's metaphoric chain consisting of some number of links, each corresponding to a reasoning step. Consider the following: from believable evidence that victim insulted defendant's wife two weeks before the crime we reason that defendant probably harbored animosity toward victim at this time; from the probable existence of such animosity two weeks before the crime we reason that it probably still existed at the time of the crime; from the probable existence of this animosity at the time of the crime we reason that defendant was probably the one who committed it. Such a chain of probable reasoning might be formed in an effort to defend the relevance of evidence of the insult on the ultimate matter at issue, namely, whether or not defendant was the one who committed the crime. Wigmore lamented the fact that logicians of his time focused their attention on simple inference and had very little to say about catenate inference. He argued that "simple" inferences, those involving just a single reasoning step, are rare in trials at law (Wigmore, 1937, p. 13).

Whether an inference is simple or catenate, to use Wigmore's terms, we may have some doubt or uncertainty about our conclusions and so we hedge on them, as I have done in the examples above. In catenate inferences there are simply more sources of uncertainty or doubt; each identified stage of reasoning interposed between evidence and possible final conclusions opens up a potential source of uncertainty or doubt. Wigmore made it clear throughout his analyses that different persons might form different reasoning routes from the same evidence to possible conclusions. A similar point was made by Venn who noted that the interposition of a certain reasoning step depends upon the "minuteness" to which we decide to work and upon the existence of appropriate names for events at this step (Venn, 1907, p. 506). For example, we might have inserted any number of additional reasoning stages in the catenate example above, each identifying defendant's probable animosity toward the victim at some additional temporal point during the two-week interval between the insult and the alleged crime. So, the uniqueness of any particular chain of reasoning or argument we might construct from evidence to hypotheses can rarely be defended. Wigmore suggested that an advocate preparing for trial would be well-advised to lay bare specific stages in his/her arguments from evidence to matters at issue, in part to identify sources of uncertainty or potential weaknesses that might be exploited by an opponent. Such a task becomes very complex when there is a mass of evidence and many, possibly interrelated, chains of reasoning to be considered; Wigmore developed an evidence charting scheme he believed would facilitate this process (Wigmore, 1937, p. 858-919)[1]. Jurists also use the term "inference upon inference" in discussing chains of reasoning. Peter Tillers has recently updated Wigmore's discussion of the many difficult and interesting forensic issues that arise in the analysis of chains of reasoning (Wigmore, 1983, vol. 1-A, p. 1106-1138).

In the early 1960s a number of psychologists (including D. Schum) became interested in both formal and behavioral issues in research on probabilistic reasoning. At this time, as far as probability was concerned, we knew about Bayes' rule and not much else. However, some of us quickly observed that, as a canon for probabilistic reasoning, the pristine form of Bayes' rule doesn't capture much of the behavioral richness so apparent in the probabilistic reasoning tasks people actually perform in so many important contexts. In an effort to make Bayes' rule responsive to elements of this richness, we began to study how it might be expanded to account for various evidential subtleties we thought commonly observable in probabilistic inference. During the past twenty years I have provided Bayeasian accounts of a wide variety of evidential subtleties including those associated with: (i) the several species of evidential redundancy, (ii) conflicting and contradictory evidence from multiple sources, (iii) corroborative and convergent evidence, (iv) hearsay evidence, (v) negative, missing, and equivocal evidence, (vi) temporal factors on the value of evidence, and (vii) an assortment of credibility-related factors (for reviews see Schum & Martin, 1982; von Winterfeldt & Edwards, 1986, 163-204; Schum, 1987).

I don't know who the person was who first described inference based on a chain of reasoning as "cascaded" inference; among behavioral scientists this term has persisted since the 1960s although some persons use the terms "hierarchical" or "multistage" inference when it involves chains of reasoning.[2] Whatever we call them, inferences based on chains of reasoning of varying length and complexity seem to be a universal feature of the probabilistic reasoning tasks people perform in so many different contexts. It seems fair to ask why there should be any special attention paid to chains of reasoning; it might be argued that a chain of reasoning involves just a sequence of those simple inferences that have been the subject of inquiry among logicians for many centuries. One answer is that, upon close examination, all of the evidential subtleties mentioned above, as well as others I will mention, rest upon identifiable chains of reasoning; they do not involve just "simple" or single-stage inferences. As an example, suppose we follow the lead of both Wigmore and Keynes in distinguishing between a person's testimony E^*, that event E occurred, and event E itself (Wigmore, 1937, p. 4; Keynes, 1957, p. 181). Many probabilistic analyses of credibility-testimony issues have been muddled by treating an inference from testimony E^* to the occurrence of event E as a single-stage inference. In fact, an inference from testimony E^* to event E can be decomposed into a chain of reasoning that takes account of specific attributes of the credibility of the person providing this testimony. This particular chain of reasoning will be a major part of my present story about the usefulness of both Baconian and Bayesian accounts of "cascaded" inference.

1.3. Cohen And Bayes: Not Natural Adversaries

An ecumenical posture with respect to the concept of probability invites criticism as Jonathan Cohen and I have both discovered. In *The Probable And The Provable* and in an earlier paper Cohen discussed how different criteria for inductive proof lead to different conceptions of probability, each of which has merit under certain circumstances (Cohen, 1975; 1977). In a recent paper I attempted to show how different conceptions of probability *seem necessary* at different points during the processes of discovery, proof, deliberation, and choice that are played out in litigation (Schum, 1986). One critic said that Cohen had "let a hundred notions of probability blossom in the garden of philosophy" (Cohen & Hesse, 1980, p. 194). At a recent symposium on probability and inference in the law of evidence (see Green & Tillers, 1986) a commentor on my paper said that I had "let a hundred notions of probability blossom in the garden of evidence law". The only difference between these two instances is that my critic simply wondered how the alternative views I mentioned could be brought together in any sensible way; Cohen's critic was less charitable in his claim that only one view (the Pascalian) was necessary. More than any other person, Jonathan Cohen has helped to convince me that all of the richness of human probabilistic reasoning is not to be captured within the confines of any single axiom-based system. This conclusion may not be as obvious as it sounds; in many disciplines reference is still being made to the Pascalian system as "the" system of probability.

So, in my juxtaposition of Baconian and Bayesian accounts of cascaded inference I will not threaten Cohen with a "dutch book" for his certain refusal to accept my Bayesian account of chains of reasoning in credibility assessment as *the* rational account; I already know what his response would be to such a threat (Cohen & Hesse, 1980, p. 196). My account of inferential issues that arise in chains of reasoning will be similar in spirit to the account Cohen has given of the complementary nature of Baconian and Bayesian approaches to medical diagnosis (Cohen, 1980). Study of the cascaded nature of credibility assessment invites examples involving testimony given by witnesses at trial and so I will draw upon the very rich legacy of experience and scholarship on credibility-testimony issues as recorded in the literature in jurisprudence. The legal context exposes many interesting evidential issues that should be noticed in others. Witness credibility seems to have become an important preoccupation in the Anglo-American legal system when the roles of witnesses and jurors began to be separated and a petty jury was conceived as a body of disinterested persons who would, presumably, provide an impartial or unbiased evaluation of the evidence in a case. In medieval courts jurors were often themselves witnesses to matters at issue; there is an informative and entertaining account of the difficult position in which this placed the defendant in a case (Wells, 1914).

Cohen has argued that a Baconian rather than a Pascalian view of probability is necessary in order to capture the patterns of reasoning that occur in the Anglo-American system of law. I have already had my chance to react to this aspect of Cohen's work (Schum, 1979a). Debate on this matter continues and was a major focal point in the recent symposium I mentioned above. In *The Probable And The Provable* Cohen's discussion of inference upon inference is one of the very few I have ever seen outside of literature in law and psychology. From his discussion I first learned about the importance of transitivity issues in the examination of reasoning chains as well as other issues concerning the location of possible "weak links" and "rare" events in a reasoning chain (Schum, 1979b). In this work I argued that there are ways to construe cascaded inferences in Bayesian terms that are quite different from the one Cohen offered and I shall employ one of them in my present analysis. In addition, Cohen makes frequent reference to credibility-related matters throughout *The Probable And The Provable*. So, my juxtaposition of Cohen and Bayes on chains of reasoning and credibility assessment has an already-existing basis. Though I shall put both Cohen and Bayes on stage at the same time, I cannot combine their roles in any way; to do so would ignore the fundamental differences between Baconian and Pascalian views that Cohen has explained (Cohen, 1977, pp. 188-198, 219-229). What makes the following analysis interesting is that the two views being presented are so different.

2.0 Credibility Assessment And Cascaded Inference

2.1. Inference From Testimony Based On Direct Observations

Consider the following generalization: "if a person asserts that an event happened, then this event did happen". On a certain occasion a witness W_i gives us testimony E_i^*, that a particular event E happened, and so we conclude in accordance with this covering generalization that event E did happen. Apparently, this generalization gives us license to infer E on testimonial evidence E_i^* from W_i in this particular case. The trouble, of course, is that invoking this rather strong generalization puts us out on a very long and slender limb when we begin to consider the variety of ways in which our faith in it might be undermined in any particular situation, for any particular witness, and for any particular reported event. In the first place, we might well inquire about the basis for W_i's testimony. Perhaps W_i made a direct observation in a situation in which event E either occurred or did not. But W_i might instead have simply heard about the occurrence of event E from another person or source and is simply passing this "second hand" information along to us. Another possibility is that W_i might have inferred the occurrence of event E from direct observations and/or second-hand information about other events, say C and D. In evidence law these three situations are distinguished. In the first case we say that W_i's testimony E_i^* rests upon direct observation or "personal [or first-hand] knowledge"; in the second E_i^* rests upon "hearsay"; and in the third E_i^* is simply "opinion" evidence. Each of these three possible bases for testimony E_i^* raises interesting inferential issues, not all of which I can cover in my present analysis. In what follows I shall ask the reader to assume that we have some evidential basis for believing that W_i's testimony E_i^* rests upon a direct observation W_i made relevant to the occurrence or non-occurrence of event E.

Assuming that W_i made a direct observation, we begin to consider what common experience teaches about human sources of evidence. The first thing we note is that people do not always tell us what they believe and so we must consider whether or not W_i believes that event E occurred. This, of course, is a *veracity* issue; we would say that W_i is untruthful only if W_i reported against his/her beliefs. But we also know from experience that people do not always believe what their senses record and might, in fact, harbor different beliefs about a sensory impression at subsequently different points in time. Here we have a wide assortment of behavioral matters to consider including W_i's expectancies, wishes, and memory capacity. Perhaps, for example, W_i so strongly expected or desired event E to happen that W_i would believe E happened regardless of what his/her senses recorded. So, we have to consider matters relevant to a person's *objectivity in the formation of beliefs* and the stability of these beliefs over time. What any person believes immediately following an observation may be different from what this person believes much later while giving testimony. Finally, our senses do not always give us a faithful recording of events that happen. As we know, the discriminative power or resolution of our senses varies from person to person. In addition, the conditions of observation are all-important; even the most acute auditory sensitivity is disturbed by a high level of background noise and the most acute visual discrimination reduced under a very low level of illumination. So, we have to consider a variety of factors associated with W_i's *observational sensitivity or accuracy* in light of the conditions under which an abservation is made.

It seems then that we have at least three atrributes of the credibillity of W_i to consider, those associated with veracity, objectivity in belief formation, and observational sensitivity. These attributes suggest that our inference from testimony E_i^* to event E is not "simple" involving just a single stage. In many situations it is necessary for us to regard testimony E_i^* as just inconclusive evidence about what W_i believes; W_i's beliefs are just inconclusive evidence about what W_i "sensed"; and, what W_i "sensed" is just inconclusive evidence about whether or not event E happened. The trouble, of course, is that all we have to go on so far is testimony E_i^*; we have uncertainty about what W_i believes as well as what W_i "sensed". Thus, our reasoning chain from E_i^* to E has intermediate stages involving at the very least a person's beliefs and sensory impressions. We might assert generalizations covering inferences at each of these separate stages[3]:

(i) "If a person testifies that an event happened, then he/she believes this event happened (i.e., "people report what they believe to have happened").

(ii) "If a person making a direct observation believes an event to have happened, then this event is the one recorded by that person's senses" (i.e., "people believe what their senses record"), and

(iii) "If a person's senses record an event, then this event actually occurred".

So, in using testimonial evidence E_i^* as a basis for inferring event E, our inferential "limb" has at least three segments. Without difficulty, we can identify many ways in which our faith in any one of these three credibility-related generalizations could be undermined in any particular situation and for any particular witness.

2.2. *"Knowledge" And Credibility Attributes*

In my identification of veracity, objectivity in belief formation, and observational sensitivity as attributes of human sources of evidence, I made appeal to "common experience". However, in a recent paper (Schum, 1989), I argued that the identification of these three atrributes does have an epistemological basis. For courts in the United States, Federal Rule of Evidence FRE 602 asserts: "A witness may not testify to a matter unless evidence is introduced sufficient to support a finding that the witness has *personal knowledge* of the matter" (there is another rule, FRE 703, providing exceptions for the opinions of "expert" witnesses)[4]. FRE 602 is quite interesting since it supposes that we have considered what it means to say that a witness has "personal knowledge" of the matter he/she asserts. On the view of knowledge as "justified true belief" our witness W_i has "knowledge" of event E if: (i) E happened, (ii) W_i believes that E happened, and (iii) W_i is justified in believing that E happened. Shown in Figure 1 below is the manner in which this view of knowledge suggests at least two of the three credibility attributes I have identified. The sequence of events in Figure 1-A corresponds to knowledge as justified true belief: E happened, W_i "sensed" E [E_s], and W_i believes E [E_b]. My justification condition here will bring to mind the one offered by Chisholm (1957, p. 16); suppose we say that W_i is justified

108

in believing E if he/she received good sensory evidence of E. But, as Figure 1-B illustrates, we must add another event in the sequence, one representing W_i's testimony. From our perspective, as far as our inference about E is concerned, all we have to go on is E_i^* and what we manage to learn about W_i.

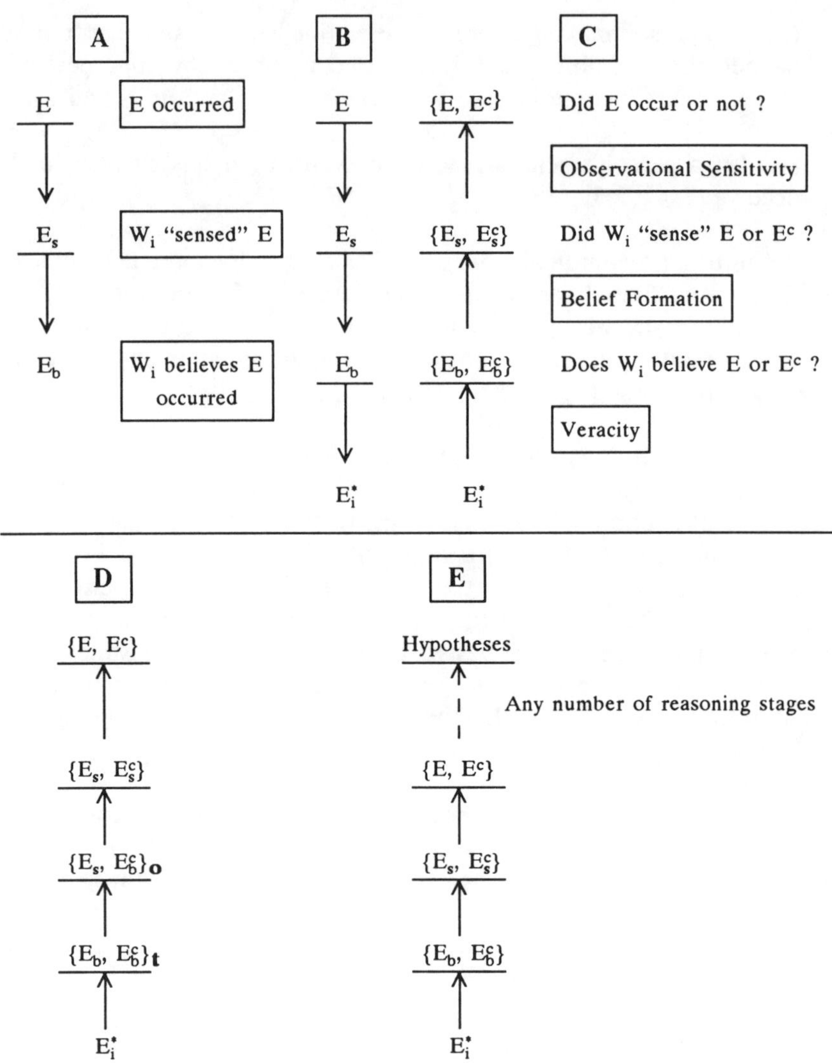

Figure 1: Chains Of Reasoning In Credibility Assessment.

FRE 602 has an awkward feature when it is considered on the view of knowledge as justified true belief or on any other view of knowledge that has as a requisite the "actual" occurrence of event E^5. Suppose we are jurors and W_i is a trial witness who gives testimony E_i^*, that some past event E happened. No one in court, including the judge, the advocates, or the jurors, "knows" whether or not E did happen; that is one reason why W_i is presently being interrogated. In fact, W_i might well be the only witness judged to be competent to give testimony about event E. We might, of course, have ancillary evidence that W_i made a relevant observation could have made this observation. For the knowledge requirement in FRE 602 to make sense regarding W_i, we would have to have *other* sources or witnesses whose testimony would make us certain that E happened. If we had such other evidence, then the testimony of W_i could be objected to on grounds that it is at least partially redundant. I dwell upon this matter only because there is a view of probability that asserts that testimony from a witness can have value only if the witness "knows what he/she is talking about" (Freeling & Sahlin, 1983). How shall the court "know" that W_i has knowledge of E except through reliance upon other witnesses or sources whose credibility is also at issue? Courts must proceed in the face of uncertainty about whether or not witnesses "know what they are talking about". There would never be any litigation possible under the requirement that witnesses *know* what they are talking about, if such knowledge requires positive verification from other sources. Pehaps the intent of FRE 602 is simply to require that the information a witness gives the court is based upon personal experience or personal observation. The requirement of personal "knowledge" raises issues that seem to have escaped the awareness of the drafters of FRE 602 as well as that of Freeling and Sahlin.

Figure 1-C shows both the direction of and the major stages in our inference from testimony E_i^* to event E. Each stage in this inference corresponds to one of the credibility attributes I have identified. The notation in two of the event classes in this chain of reasoning bears some explanation. As Cohen notes (1977, p. 321), even though they are often used synonymously, there is a distinction to be made between the statements: "W_i believes that E did not occur" and "W_i does not believe that E occured". In the second statement we allow for the possibility that W_i might not have a belief one way or the other. Similarly, we should distinguish between the statements: "W_i sensed the nonoccurrence of E" and "W_i did not sense E". In Figure 1-C I have used E_b^c, which means" W_i believes that E did not occur" , and E_s^c, which means "W_i sensed the nonoccurrence of E". I had to make a choice in my present discussion; I might instead have used the symbols

110

$(E_b)^C$, which says "Wi does not believe E", and $(E_s)^c$, which says" W_i did not sense the occurrence of E". However, the formal processes I shall discuss works essentially the same for both kinds of statements.

Before we take the chain of reasoning in Figure 1-C too seriously as representing the exact nature of our cascaded inference from E_i^* to events $[E, E^C]$, we should heed the advice given by Venn and Wigmore regarding the element of arbitrariness in constructing a chain of reasoning. Here are just two reasons why I make no claim that the chain of reasoning in Figure 1-C is uniquely suitable as a representation for credibility assessment as a cascaded inference. In the first place we have to recognize what Nozick (1981, p. 284) describes as the "suppleness" of our beliefs. One behavioral fact is that our beliefs change over time in light of new experiences we have. So, what W_i now believes while giving us testimony E_i^* might be different from what he/she believed immediately following the observation W_i made; perhaps, as is common in legal affairs, considerable time has elapsed between the observations made by a witness and his/her subsequent testimony. In Figure 1-D I simply distinguish between W_i's possible beliefs at the time of testimony (t) and at the time of observation (o). But this does not completely cover the matter; as Venn noted, time is continous and so we might reasonably insert any number of other intermediate belief related stage between $[E_b, E_b^c]_t$ and $[E_b, E_b^c]_o$.

The second matter of concern in Figure 1-C involves my representing either E_b or E_b^c as direct evidence on sensory events $[E_s, E_s^c]$. The precise connection between our beliefs and our sensory impressions has, to my knowledge, not been settled upon. Who among us can tell with any precision what chain of neurophysiological or other events is interposed between the responses of our sensory end-organs to physical stimuli and the subsequent cognitive states we call "beliefs"? In the absence of such insight, I have little choice but to represent possible belief states as direct evidence on possible sensory states. The identification of other intermediate stages between beliefs and sensory events would simply expose other credibility-related attributes. This is why I will not claim that the three attributes I have listed form an exhaustive class.

2.3. Evidential Bases For Cascaded Inference In Credibility Assessment

In examining the basis we might have for reasoning up the chains shown in Figures 1-C or 1-D we begin discussion of matters that are quite important to both Baconian and Bayesian views. To frame the matters of interest I now pose two quite different situations.

2.3.1. Relative Frequencies And The Theory Of Signal Detactability

Psychologists have long been aware of the fact that our sensory/perceptual apparatus is not a blank tablet upon which external stimuli or signals make their mark; many other things are already written on this tablet when any signal arrives. Some of the things already written concern our expectancies and motivations. The trouble is that, in many forms of behavioral research, reliance has to be placed upon the testimony of human observers about what they have seen, heard, and so on. So, for example, in experimental evaluations of our sensory capabilities researchers have common awareness that the event a person reports and the event a person "sensed" may not be the same. For example, when a person reports having heard some low intensity auditory signal against a backgound of internal and external noise, how do we tell whether this signal was actually detected by this person's auditory apparatus or whether this person simply expected or wished this signal to occur, or perhaps thought that this was the response expected by the experimenter?

Modern study in the field of sensory psychophysics has been greatly advanced by the application of various elements of signal detection theory [SDT]. This theory supposes that a person, asked to make an observation and then to report the results of this observation, is not simply a passive collector and transmitter of sensory information. Human observers are thought of as active decision-makers who, following an observation, are faced with a choice about what to believe and to report. Thus, SDT presumes a distinction between the sensing of event E and a belief that E happened (i.e. between E_s and E_b in Figure 1-C). In sensory psychophysics another common presumption is that an experimental subject will always report what he/she believes to have occurred; this amounts to an acceptance of the veracity related generalization in 2. 1 above. Thus, in SDT the bottom stage in the chain of reasoning shown in Figure 1-C is "pruned" under the presumption that E_i^* entails E_b. This presumption is, of course, taken as rebuttable in other contexts we shall examine. So, as far as SDT is concerned there are two attributes of interest: observational sensitivity and what I have called "objectivity in belief formation"; in SDT the latter is usually referred to as a "decisional" attribute to acknowledge that an observer always has a choice about what to believe based upon fallible sensory evidence.

In the conduct of an SDT experiment (see Swets, 1964; Green & Swets, 1966; and Egan, 1975) often long sequences of events are presented; the experimental subject makes a response to each event. Typically, these events involve either the occurrence or nonoccurrence of some information-carrying signal. Over a long sequence of trials relative frequency estimates of $P(E^*|E) = h$ and $P(E^*|E^c) = f$ are recorded, where h stands for "hit rate" and f stands for "false positive" rate. Under various testable assumptions about the effects of stimulation upon our sensory systems, SDT allows inferences to be made about a person's observational sensitivity and belief formation on the basis of estimated values of h and f obtained in a long series of trials. The crucial fact here is that such inferences are based upon the repeated presentation of events and the recording of relative frequencies.

2.3.2 Singular Events: Variety Is The Spice Of Credibility Assessment

Suppose W_i's testimony is as follows: "I saw defendant running away from the house (where the murder was committed) shortly after 10:10 (on the day in question)". The trouble here is that we cannot play the world over again *n* times to see what proportion of these instances will involve defendant's actually running away from the scene, what proportion will involve W_i's "sensing" that he did, what proportion will involve W_i's believing that he did, and what proportion will involve W_i's testifying that he did. The event reported as well as the events representing W_i's sensory images and beliefs are singular or unique and pertain to a specific *nonreplicable* situation. What is at issue in court is what this witness observed and believed about an event that either occurred or did not on exactly one occasion. Nonreplicable situations such as the one just described occur in many other contexts besides court trials; in such situations how does one proceed with credibility assessment in the absence of the enumerative luxuries afforded in SDT experiments? The answer, well-attested by centuries of experience in jurisprudence, is that (i) we seek answers to a *variety* of questions about the credibility-relevant behavior of the witness, and (ii) we examine the witness' testimony within the context of other evidence. What I shall illustrate is how credibility-related ancillary evidence can be sorted out in terms of the three attributes I have identified.

Centuries of experience in our Anglo-American judicial system have produced a now standard collection of grounds for impeaching or supporting the credibility of a witness. Some of these grounds suggest categories of *ancillary evidence* that are relevant in tests of the credibility of a witness. This evidence is ancillary in the sense that it is indirectly relevant to the matters at issue. For example, suppose there is evidence that W_i was not, as usual, wearing his strongly corrective lenses at the time he alleges seeing defendant fleeing from the house in question. Such evidence, by itself, has no bearing on whether or not the defendant actually committed the murder. It acquires relevance, however, since it does bear upon the behavior of a witness who makes an assertion that is relevant to defendant's guilt or innocence. So, in a sense, ancillary evidence is evidence about evidence. Another ground for credibility impeachment involves directly relevant evidence from other witnesses or sources and concerns the existence of conflicting and/or cotradictory evidence. The existence of such evidence can actually bear upon any of the three credibility attributes I have discussed. There is one other ground that bears upon all three credibility attributes; it involves *bias*, of which there are three different species I shall mention. The following kinds of credibility-relevant evidence are all mentioned in legal treaties (e.g., Wigmore, 1937, 1983 Vol 1-A, Tillers Revision; Cleary, et al, 1972; and Lempert & Saltzburg, 1977). However, to my present knowledge, jurists have never categorized them in the following way. But we need to distinguish these grounds very carefully; we should not like to confuse evidence regarding the veracity of a witness with evidence that actually concerns his/her belief formation or observational sensitivity.

(i) Veracity Related Evidence

As a test of self-consistency, whether or not a witness has given prior inconsistent testimony is of interest. Another test of consistency involves collateral details in a witness' testimony. Evidence of mendacity regarding some collateral detail in testimony brings into question a person's truthfulness regarding the "interesting" part of this testimony. Evidence of the past behavior, conduct, or character is deemed relevant in credibility testing and bears upon witness veracity as does evidence of any mutual influence among witnesses. One species of bias might be labeled "testimonial bias"; we may encounter a witness who, on ancillary evidence, appears to have a distinct bias for or against testifying to a certain event. Finally, the demeanor and bearing of a witness while giving testimony are commonly thought to be credibility-relevant however inconclusive such evidence might be on any occasion.

(ii) Objectivity In Belief Formation

Evidence about what a witness expected to observe or wished to occur would be relevant to this attribute of credibility assessment since it bears upon a person's readiness to believe the results of this observation. This evidence might be said to concern the objectivity of the witness in forming his/her beliefs on the basis of sensory evidence. Acknowledging the compliance of memory with time, it seems fair to inquire about the

114

memory capacity of a witness and the extent to which beliefs held by a witness following an alleged observation might have been altered by events that have occured in the meantime. Another species of bias concerns our beliefs; we might encounter evidence that a witness was biased for or against believing that a certain event had occurred.

(iii) Observational Sensitivity

Questions about the capabilities and limitations of any sensory modality involved in an observation are quite relevant. Evidence of inaccuracies about collateral details may be suggestive of similar inaccuracy about "interesting" details in testimony. Any evidence concerning the conditions under which an observation was taken are, of course, relevant. In addition to obvious interest in the ambient conditions of observation such as illumination, background noise level, etc., it is necessary to assess the quality of the observation itself. Here we might be concerned about such matters as the duration of observation. Finally, we have to consider various ways in which a person's sensory apparatus might be biased in one direction or another. A clever pattern of deception might "set up" a witness to "sense" event E, whether or not it occurred.

2.4. Credibility Assessment And The Value Of Testimony

In law and in other contexts the behavior of a witness is not subjected to careful scrutiny simply because the witness may seem to be an interesting person. Credibility assessment is a process embedded in the further task of determining the inferential value of what the witness tells us. Consider the chain of reasoning shown in Figure 1-E above. Suppose we could construct an argument having any number of discernible stage from either the occurrence or nonoccurrence of event E to one or more of the hypotheses we are considering. For example, E might refer to the insult defendant's wife allegedly suffered at the hand of the victim in the example in 1.2. above. In this case E is separated from hypotheses involving defendant's guilt or innocence by at least three reasoning stages. The three-stage credibility assessment process I have just described can be thought of as providing the foundation for any argument to be based on the event(s) reported in testimony. Thus, in the insult example, a witness' testimony about this insult [E_i^*] seems to be at least six reasoning stages removed from major or ultimate facts in issue. This is an example of what two recent reviewers of my work on cascaded inference referred to as "ungodly" inference (von Winterfeldt & Edwards, 1986, p. 170). I shall now subject Baconian and Bayesian views to inferences which, however ungodly, seem to be commonplace in the reasoning task we all perform.

There is one final matter I must attend to before I proceed and it involves the concept of the "weight" of evidence. Readers of Cohen's works (e.g., 1977, 1985, 1986) will know that his conception of the weight of evidence differs rather sharply from one usually provided in Bayesian accounts of inference. In the Bayesian analyses to be described I make use of likelihood ratios which, for quite some time, have been the standard Bayesian metric for evidential "weight" (e.g. Good, 1983, p. 36f). Since I make no claim that my Bayesian analysis is "the" normative account of credibility assessment, I will avoid saying that my likelihood ratios grade "the weight" of evidence; I have other terms that will suffice. I have no wish to let terminological matters obscure what I believe to be the special contributions of both Cohen and Bayes to the analysis of these "ungodly" inferences.

3.0 A Baconian Analysis Of Credibility Assessment

Let us see how a person reasoning in accordance with Cohen's inductive or Baconian probabilities might draw a conclusion about whether or not event E occurred based upon witness W_i's assertion E_i^* that it did occur. I cannot, of course, presume that Cohen will necessarily endorse the cascaded inference from E_i^* to E that I have presented. It makes sense to me and I will suppose that it makes sense to the Baconian whose belief revision process I will represent. In both my Baconian and Bayesian analyses I shall assume the *nonreplicability* of the events identified in the cascaded inference shown above in Figure 1-C; in doing so, I bring my analyses into contact with the credibility assessment tasks people routinely face in court trials and in so many other contexts. The Baconian I shall represent tests credibility-related generalizations on the basis of evidence and uses these tested generalizations as a basis for reasoning from one stage to another. As Toulmin noted (1958, p. 97f) the manner in which we can provide evidential "backing" for a generalization depends upon the field of inquiry and that, in some cases, this backing might be statistical in nature. All I am assuming is that there are no appropriate statistics of veracity or objectivity in belief formation available for witness W_i, as there would rarely be for any human witness. As I will note, there might be some statistical evidence relevant to one form of backing evidence regarding observational sensitivity.

To illustrate how my Baconian reasons from one stage to another I shall employ inference "trees" of the sort commonly used in cascaded inference analysis. Cohen himself uses an arboreal simile in discussing single-stage Baconian reasoning (1977, 205-207); my Baconian trees are just more elaborate versions of the one he discusses. The arrangement of the branches on my Baconian tree may seem to defy gravity on occasion; but we should not be troubled since this defiance of gravity has an explanation within Cohen's system. I shall also mention instances in which Bayesian analyses seem to defy gravity but offer less satisfactory explanations.

3.1. The Baconian "Weight" Of Credibility-Relevant Evidence

I shall use the symbol $I(-, -)$ with reference to *inductive probabilities'* for example, $I(E, E_i^*)$ is the inductive probability of event E on evidence E_i^* from W_i. As we know, inductive probabilities grade the *amount* of uncounteracted evidence favorably relevant to generalizations we invoke in particular situations. The very global generalization covering a direct inference from E_i^* to E is: "if a person testifies that an event happened, then this event did happen". If all we had was E_i^* as $I(E, E_i^*)$ asserts, we would indeed be out on a very long and slender limb in applying this generalization to W_j's testimony on this particular occasion; the reason is that we now know of at least three large classes of evidentiary tests, the results of which might serve to invalidate this global generalization as far as W_i is concerned. So, the weight of evidence E_i^*, by itself, cannot be very large; in fact, in Cohen's terms it has a weight of just $I(E, E_i^*) \geqslant 1/(N + 1)$, where N refers to all of the credibility-relevant tests anyone could think of that apply in the present instance involving W_i. The "1" in the numerator and denominator of $I(E, E_i^*)$ comes from giving this global generalization an initial benefit of the doubt. So, what $I(E, E_i^*) \geqslant 1/(N + 1)$ indicates is a *shortage* of evidence about E and not a preponderance of evidence favoring E^c. In other words, if we took an "inductive leap" from E_i^* [by itself] to E, invoking this global generalization, the leap might be prodigiously long depending upon how large N truly is. The reason is that we have not ruled out any specific reason why this generalization may fail to hold as far as W^i and E_i^* are concerned.

So, we decide to ask some question about the behavior of W_i and about other matters affecting his/her observation. In thinking about these questions, we decide that they seem to fall into three categories, each corresponding to an attribute of the credibility of W_i, the only issue as far as our inference about event E is concerned. If our thinking about these attributes corresponds to the representation in Figure 1-C, it is apparent that we have a chain of inferences at each stage of which there appears a relevant covering generalization; perhaps we can make a sequence of shorther inferential leaps. Take the first stage involving the veracity of W_i; here our inference concerns the events $[E_b, E_b^c]$; i.e., does W_i *believe* that E happened or did not happen. Suppose, as in a court trial, W_i is testifying under oath and so we assert the following generalization: "people testifying under oath tell us what they believe". However, we recall a line from Shakespeare: "weigh oath with oath and you will nothing weigh"[6]; and so we decide to ask some specific questions about the veracity of W_i. In doing so we first consider $I(E_b, E_i^*)$ and note that it has grade $I(E_b, E_i^* \geqslant 1/(n_v + 1)$, where n_v is a list of all of the veracity-related questions that could be thought of in the present situation. Here we have given an initial benefit of the doubt to the above veracity generalization as we did earlier to the "global" generalization. So, we have an initial shortage of evidence regarding an inference about W_i's beliefs; but we note that this shortage is not as great as it is in our inference from E_i^* to E. Thus, our inductive leap from E_i^* to E_b, based on the generalization about witnesses testifying under oath, is not as long as the one from E_i^* to E, based on the very global generalization intially asserted. The reason is that n_v is necessarily smaller than N. Before we can discuss the use of evidence in reasoning from E_i^* to E_b and then further up the inference "tree" to E or E^c, we must consider carefully Cohen's concept of *relevant variables*.

3.2. The Partitioning And Use Of Credibility-Relevant Variables

I have let N be the total number of questions anyone could ask about W_i that are relevant in testing the present applicability of the global credibility generalization. What we seek is license to infer E based on E_i^*. But I have argued that this inference has at least three stages, each associated with an attribute of the credibility of W_i, and that there are specific covering generalizations appropriate to each one of these stages (see 2.1. above). Let n_v be the number of questions anyone could ask in testing the present applicability of the veracity generalization, n_b be the total number of questions anyone could ask in testing the generalization about objectivity in belief formation, and n_s be the total number of questions anyone could ask in testing the observational sensitivity generlization. If I had been able to argue that my three-stage inference from E_i^* to E is uniquely adequate and that the collection of credibility attributes so identified is exhaustive, then I could have said that $N = n_v + n_b + n_s$ and that I have partitioned all credibility-related questions into three categories. But, for reasons stated in 2.2, I cannot make such an argument and so I simply have to suppose that $(n_v + n_b + n_s) < N$. I could, of course, define a "catch-all" category C having n_c questions that might be exposed in future studies in which other credibility attributes are identified; at least this lets me account for "all possible" credibility-relevant tests since $(n_v + n_b + n_s + n_c) = N$. In short,, there may be credibility-relevant questions my present analysis does not expose and so, without this catch-all, I cannot claim to have completely partitioned N. What I will claim is that my analysis exposes a more complete set of credibility-related questions than do any previous probabilistic accounts of credibility-testimony matters.

Each question identified exposes an eviden tiary test of one of the three credibility-related generalizations and so corresponds to what Cohen calls a *relevant variable*. A variable is *relevant* if one of its possible states, levels, or conditions can serve to invalidate a generalization. Thus, for example, ancillary evidence of W_i's giving prior inconsistent testimony E_i^{c*}, that E did not occur, is a state of a "consistency" relevant variable that can serve to invalidate, in this particular instance involving W_i, the veracity generalization that witnesses under oath testify in accordance with their beliefs. Cohen allows for the invalidating effects of evidence to be counteracted in various ways that I shall mention. Thus, further evidence of a threat or other form of coercion might explain why W_i asserted, on a prior occasion, that event E did not happen. Cohen tells us that the identification of relevant variables is itself an empirical process (1977, p. 141). The credibility-relevant variables suggested in my account in 2.3.2. arise as a result of an empirical process that has been taking place over several centuries in our Anglo-American legal system. Cohen further tells us (1977, p. 137) that knowledge of the actual state of some relevant variable can be gained by a variety of means including observation, inference, introspection, or even hearsay. In trial proceedings there are rules about the admissibility of hearsay and opinion evidence that are not enforced in other contexts.

Let V represent the class of veracity-relevant variables of which there are n_v, B the class of belief-formation relevant variables of which there are n_b, and S the class of observational sensitivity relevant variables of which there are n_s. I will consider a bit later what possible values n might have in these three classes. Corresponding lower-case letters carrying a subscript will indicate a particular relevant variable in one of these classes; i.e., v_j indicates the jth veracity variable in Class V. I shall here suppose that, for any relevant variable in any of these classes, there are either "favorable" or "unfavorable" states of this variant on one of the three veracity-related generalizations. Thus, v_{jF} will indicate an evidentiary test of variant v_j that has a "favorable" result as far as E_b is concerned; i.e., it is favorable to applying to W_i the generalization that people under oath report what they believe. The symbol v_{jU} means that test v_j has an "unfavorable" result as far as E_b is concerned but a "favorable" result as far as E_s^c is concerned. I must assume that an item of veracity-relevant ancillary evidence cannot simultaneously favor W_i's believing E happened and believing E did not happen; the same assumption will apply to evidence on the other attributes. As Cohen notes (1977, p. 275) jurors might legitimately disagree about whether any particular evidential result is favorable or unfavorable; in some cases they may not even be able to tell whether an item is favorable or unfavorable. The terms "favorable" and "unfavorable" have one unfortunate connotation. As I shall discuss, we may end up concluding that event E did happen when we also conclude that W_i *believes* that event E did not happen, in spite of his/her testimony E_i^*. Thus, the favorableness or not of a credibility-relevant variant refers only to the covering generalizations as stated in 2.2. and not to the particular reasoning route we may take from testimony E_i^* to events $[E, E^c]$. As I will note in 3.4. below, reasoning routes from E_i^* to event E need not necessarily pass through E_b and E_s. Of course, our reasoning route from E_i^* might terminate at E^c instead of E.

In discussing relevant variables and their role in determining inductive probabilities Cohen attends to their independence, their relative importance, and to the order in which they are to be considered. On independence, I believe there are two species that have to be discerned as far as my present analysis is concerned. One species concerns the "manipulability" of random variables and is discussed by Cohen (1977, p. 132). In any scientific study the manipulation of two variates is independent if changes in one do not force changes in the other. Thus, for example, we can vary the spectral composition of a light source quite independently of the physical intensity of this source. But there is another species of independence that Cohen treats somewhat obliquely and this one concerns the joint effects of relevant variables on the inductive conclusions we draw. As he notes in his example involving British weather (1977, pp. 202-207), a falling barometer counteracts the unfavorable effects of an offshore wind on the generalization: "if there are dark clouds, then rain is imminent". This brings to mind a species of nonindependence called "interaction" that is familar to students of multivariate statistical inference. Two independently manipulable variables are said to interact if they operate nonindependently in their joint effects on some dependent variable. I believe it is this species of independence Cohen has in mind when he discusses the manner in which the result of one evidential test can counteract the effect of a result of another. Atmospheric pressure, wind direction, and cloud condition "interact" in influencing the likeliness of rain.

Cohen (1977, 138f) places emphasis on the ordering of relevant variables in terms of their importance; the most important should be considered first in order to preserve their potential for invalidating generalizations applied in particular cases. But he also acknowledges (1977, 253) that there are situations, court trials for example, in which relevant variables may have to be considered in any order; at trial the order of evidence presentation depends upon the whims of the advocates as they offer evidence in accordance with the antiphonal procedures characteristic of Anglo-American courts. In my construal of credibility assessment as a cascaded inference, there is a natural ordering of credibility relevant variables in terms of the attributes on which they bear as shown in Figure 1-C. Other orderings of the event classes shown in this figure make no sense. For example, no psychologist would believe that E_i^* is direct evidence on events $[E_s, E_s^c]$ for all the reasons made so clear by SDT theorists and others. In reasoning from E_i^* to $[E, E^c]$ we consider the veracity, then the belief formation, and finally the observational sensitivity relevant variables. But the ordering here is not in terms of importance but in terms of evidential relationships that occur in reasoning from one stage to another. Settling on the importance-ordering of relevant variables *within* a credibility attribute category is, however, of importance. As Cohen notes (1977, p. 277) a jury would be quite at home arguing about the relative importance of, say, witness demeanor and witness reputation. We can expect great variability from juror to juror in settling the precedence of credibility-relevant evidence within any attribute category. I know a person who once said that President John F. Kennedy could never be trusted because his eyes were too close together.

3.3. *Distributing Evidential Incompleteness Among Credibility Attributes*

The incompleteness of credibility-relevant evidence and the manner in which this incompleteness is distributed among the three credibility attributes I have mentioned have a most important influence on Baconian reasoning up a tree from E_i^* to events $[E, E^c]$. In fact, evidential incompleteness is the basis for the "gravity-defying" characteristics of Baconian inference trees I will illustrate in 3.4. below. A distinguishing feature of Baconian probabilities is the manner in which they can grade the extent to which we have covered relevant matters in an inference from one stage to another; indeed, this is the basis for Cohen's conception of the weoght of evidence. Evidence has greater weight as it takes more things into account. In the credibility assessment task I have been using as an instance of catenated or cascaded inference we have, at least, some idea of the categories of relevant evidentiary tests to which we might subject a witness. Within each of these three categories there are certain classes of evidence that centuries of experience in court trials allow us to identify; but we must be a bit careful here. Take "character" evidence and its relevance to witness veracity. The "character" of a human being is certainly multiattribute and so there are many possible character-related questions we might ask about W_i that may seem particularly relevant to the situation in which he/she is now giving testimony. The same can be said of any other credibility-related behavioral characteristic appearing in the three categories of credibility-related evidence in 2.3.2. above. The point here is that my jurisprudential account in 2.3.2. only serves to identify *categories* of relevant tests for each attribute; we may never know how many specific evidential tests we could consider in any one of these categories for any particular witness.

One thing seems fairly certain, however, and it is that we may have more evidence on one credibility attribute than we do on another. As a result of direct and cross-examination we may discover quite a bit about the veracity of W_i but very little about W_i's objectivity in forming beliefs or about the conditions under which W_i made an observation. In some extreme cases one or more of the three credibility attributes might not even be put to any test. So, what we have to suppose is that the incompleteness of ancillary and other credibility-relevant evidence is distributed among the three attributes in any conceivable manner. Stated another way, we may have more evidence bearing on one stage of the reasoning in Figure 1-C then we do on another.

Finally, the objective of obtaining *more* evidence on some credibility attribute is not served by simply obtaining repeated instances of some form of ancillary evidence. At trial there is a limit to the number of witnesses who might be heard if all of them are to testify to W_i's past honesty; at some point an objection would be raised and very likely sustained. This is an instance of Cohen's discussion of replicability and its relation to the weight of evidence. Repeated evidence about some event may make us more certain that this event occurred but it usuallly tells us nothing new[7]. It seems, however, that some replication is both allowable and necessary. Suppose the credibility of W_i is challenged by W_k who tells us some unpleasant thing about the past behavior of W_i. Of course, the credibility of W_k is also of interest and so evidence corroborating W_k's story might be sought from other witnesses. The regress of evidence about evidence about evidence... and infinitum... is finally halted when people run out of time, resources, or patience.

122

3.4. Baconian Cascaded inference

In describing the behavior of my Baconian, who seeks to draw a conclusion about events $[E, E^c]$ based on testimonial evidence E_i^*, I must, of course, present this person's behavior in a manner consistent with what Cohen has told us about Baconian probabilities. Here follows a brief account of what I take to be the essentials of Cohen's arguments as they concern cascaded inference.

3.4.1. Baconian Reasoning Mechanisms

Cohen makes it quite clear (1977, pp. 267-269) that we cannot combine inductive probabilities across the three credibility-related stages of my argument from E_i^* to events $[E, E^c]$. More than one stage in this argument is necessary because there are different kinds of connections that occur. The connections (i) between a person's testimony and his/her beliefs, (ii) between a person's beliefs and his/her sensory images, and (iii) between a person's sensory images and the actual occurrence of an event are all different since they involve different behavioral processes. In Cohen's terms they each involve particular covering generalizations of the sort I asserted in 2.1. above. The result is that inductive probabilities at each stage of reasoning have to be evaluated separately. The inductive probabilities so evaluated at each reasoning stage have only ordinal properties and so they cannot be meaningfully combined anyway. So, the question arises: if we cannot combine inductive probabilities in any meaningful way, how do we make the transition from one stage of reasoning to another? The answer Cohen gives involves what he has to say about the "detachment" of beliefs.

Consider the argument from E_i^* to events $[E_b, E_b^c]$ at the veracity stage of the chain of reasoning in Figure 1-C. First suppose there are n_v relevant tests of W_i's veracity and that the j tests that have been made at trial have all produced results favorable to W_i's believing what he/she testifies. So, at this first stage of reasoning, I (E_b, E_i^* & v_{1F} & v_{2F} & ... & v_{jF}) $\geqslant$ (j + 1)/(n_v + 1). Suppose further that these tests have involved important and not trivial matters. Now, Cohen allows a Baconian to "detach" a belief in E_b or, essentially, to believe "as if" E_b were true (i.e., to believe that W_i believes what he/she is telling us). This facility for "detaching" beliefs cannot come as a mystery to anyone who has ever performed behavioral research involving cascaded inference. In fact, the suggestion that experimental subjects do quite a bit of "as-iffing" was commonly made in even the earliest empirical studies of cascaded inference (several such studies appear in Peterson, 1973). But on the

Bayesian views of the authors of these studies this "as-iffing" was usually tossed-off as the simple suppression of uncertainty; Cohen's explanation is quite different since it involves consideration of those evidential test W_i's veracity passed, those it failed, and those that were not made at all.

Now suppose that not all of the first j test were favorable to the conclusion that W_i believes what he/she is telling us; i.e., they do not all favor the covering generalization that witnesses under oath tell us what they believe. Here we have to acknowledge the asymmetry of favorable and unfavorable evidence as it arises in accordance with the Baconian negation rule (Cohen, 1977, p. 207-211). Suppose veracity test k results in evidence v_{kU} which is unfavorable to this generalization in the case of W_i. With reference to veracity testing, this rule states that, if $I(E_b^c, E_i^* \& v_{kU}) > 0$, then $I(E_b, E_i^* \& v_{kU}) = 0$, *provided that we believe* v_{kU}. In other words, believable evidence item v_{kU} cannot simultaneously favor E_b and E_b^c. Evidence v_{kU} could, of course, be counteracted by other evidence, the effect of which would be to resuscitate the inductive probability of E_b. But, in addition, credibility assessment is a very good example of situations in which we may be given grounds for not believing v_{kU}. As I noted, the credibility of the source of this ancillary evidence is also subject to scrutiny. The result here is that we do not have to suppose that Baconian reasoning mechanisms always have the violent asymmetry that the negation rule may sometimes suggest.

Returning to the detachment of beliefs, at any evidential test level j, we are entitled to detach a belief in E_b [or in E_b^c]. But, in doing so, we have to take into account the $(n_v - j)$ veracity matters that we have not considered. At test level j the evidence will, on balance, favor either E_b or E_b^c, and so one of these events has inductive probability at least $(j + 1)(n_v + 1)$, assuming an initial benefit of the doubt for the veracity generalization; what the other one has will depend upon the extent to which we believe the evidential results of these j tests. In any case, Baconian probabilities grade the extent to which we have so far covered the n_v credibility-relevant matters. A Baconian who detaches a belief, say in E_b, at test level j makes the following statement: "Other things equal [i.e. assuming favorable results on the remaining $(n_v - j)$ tests] I conclude that W_j believes what he/she is telling us and is, therefore, being truthful". As Cohen notes (1977, p. 212) Baconian probabilities grade the content of this *ceteris paribus* assumption[8]; the larger j is, the *weaker* is this assumption since more veracity-relevant matters have been taken into account. This holds, by the way, whether or not we are ever able to specify n_v. I cannot help noticing that the reasoning here bears some resemblance to the "default" reasoning schemes currently being discussed in the field of artificial intelligence (e.g. Reiter, 1980, Charniak & McDermott, 1985).

It is very natural to suppose that a person might adopt thresholds for belief about events at any stage of reasoning. At some reasoning stage a person might say: "If at least k relevant matters have been covered, then I will decide one way or the other depending upon the evidential test results". Of course, we expect that such thresholds, if they exist, are quite variable across people and situation. Cohen argues (1977, p. 319) that Baconian probabilities are more adequate than Pascalian ones in setting acceptance rules. A jury's verdict is more soundly based if it takes more relevant considerations into account and this is precisely what Baconian probabilities grade. But now suppose a person whose threshold is at k but at trial only j < k evidential matters have been considered? Must this person suspend belief? I believe Cohen's answer here would be that we can *always* detach a belief that is qualified by some *ceteris paribus* assumption. Indeed, as noted earlier, we could infer E, based on just E_i^* and nothing else, proveded that we were willing to live with the enormously rich *ceteris paribus* assumption involved. In short, we have to go out on inferential limbs all the time; Baconian probabilities have the interesting characteristic that they say something about how long and strong are the limbs upon which we so often find ourselves.

3.4.2. Reasoning Up A Baconian Tree

I will describe a Baconian inference tree in two stages because of its complexity; the first stage is illustrated in Figure 2. The trunk of this tree is the testimony W_i provides. Branches stemming from this trunk refer either to sequences of evidential tests (the heavy bold branches) or to *ceteris paribus* assumptions (the thin branches). Note that there are three levels of heavy bold branches, each one corresponding to a credibility attribute. Also note that we must consider all possible routes from E_i^* to events [E, E^c] and that some of these routes will not correspond to any statement about W_i's having "knowledge" of E or E^c. For example, we might well believe, on ancillary evidence, that E occurred., W_i "sensed" its non-occurrence [E_s^c], disbelieved the evidence of his/her senses and believed that E did happen [E_b], and then gave faithful testimony E_j^*. Here is another possiblity; ancillary evidence may suggest this sequence of events: E happened, W_i sensed E [E_s], believed E did not occur [E_b^c], and then lied to us by telling us that E did occur. So, there is no way in which a Baconian analysis requires W_i to "know what he/she is talking about" in order for us to draw a conclusion about whether or not E happened, as W_i testifies.

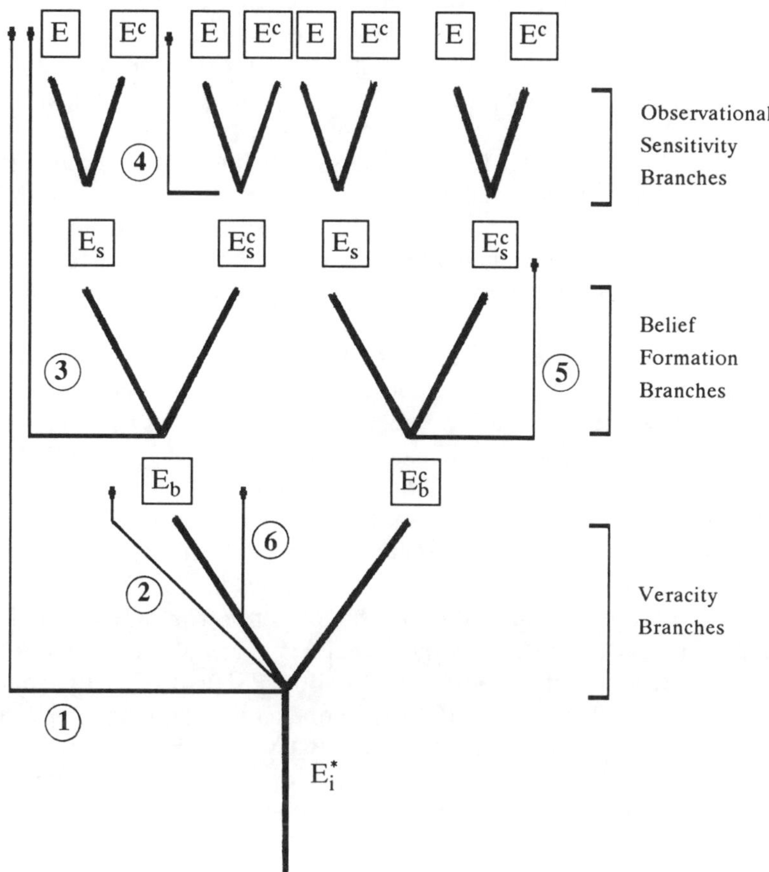

Figure 2: A Baconian Inference Tree: Basic Reasoning Routes.

Each one of the heavy bold branches in Figure 2 actually represents a much richer arboration as I will illustrate later using Figure 3; for example, there will be many possible branches that terminate in either E_b or E_b^c. Consider the thin *"ceteris paribus limbs"* in Figure 2. The limb labeled (1) is the one we have already discussed that involves the simple invocation of the global generalization that, when a person tells us that an event has happened, then this event has indeed happened. This is, of course, the longest, thinnest limb we could possibly atempt to climb; but this is precisely the limb we would have to climb in a direct inference, without supporting evidence, from E_i^* to event E. No credibility-related evidence at all has been considered and so the content of the *certeris paribus* assumption on limb (1) is as rich as it could possibly be. Other *"ceteris paribus* limbs" are not this long; consider the one labeled (2) in Figure 2. If no veracity-releted evidence were considered, we might have to rely on the generalization: "people testifying under oath report what they believe", and so infer E_b. This is a shorter limb than (1) but is is no stronger; it has minimal weight since it involves no evidence. Branches (3), (4), and (5) are illustrations of other possibilities involving *"ceteris paribus* limbs". The one labeled (6) in Figure 2 illustrates a deficiency in Figure 2 that will be removed in Figure 3. We cannot supoose that all evidence-based branches, for example from E_i^* to E_b, will be based upon complete evidentiary tests; suppose that only j of the n_v possible veracity tests are reflected by the evidence we have. On a *ceteris paribus* assumption, we could detach a belief in E_b at test level j provided that the j items of evidence, on balance, favored E_b. The content of this assumption is inversely related to the difference between j and n_j.

Figure 3 comes closer to representing the true complexity of Baconian cascaded inference. To illustrate what it tells us I shall ask the reader to suppose that there are just three relevant variables at each of the three credibility-related stage of the inference from E_j^* to [E, E^c]; my interpretations do not depend upon what n is at any of these levels. First, please follow the route indicated by the dollar signs ($). The payoff for being on this evidential route corresponds to what could be called: *beyond a reasonable doubt.* At the veracity level, the result of test 1 was favorable to E^b, the one at test 2 was unfavorable, but was counteracted by the result at test 3. The three results at the belief-formation and observational sensitivity levels were all favorable to E_s and E respectively. Since we have covered "all possible" relevant matters, we have no *ceteris paribus* assumptions to trouble about and the weight of our evidence is maximal. On this evidence we conclude, beyond a reasonable doubt, that event E did occur as W_i testifies; i.e. we have removed all the reasons for doubt that this chain of reasoning exposes. Remember from the discussion of Figure 2 that there are other routes we could take to this same conclusion with the same strength if the evidence justified; some of these routes will involve E_b^c and E_s^c. However, there are other possibilities we have to consider in which we do have reasonable doubt and our inductive probabilities are not maximal.

It is time to consider what I have referred to as the "gravity-defying" aspects of a Baconian tree. Consider in Figure 3 the thin *ceteris paribus limb* labeled (1). Suppose we have just one of three possible evidential tests of W_i's veracity; it is a favorable result. *Assuming* similarly favorable

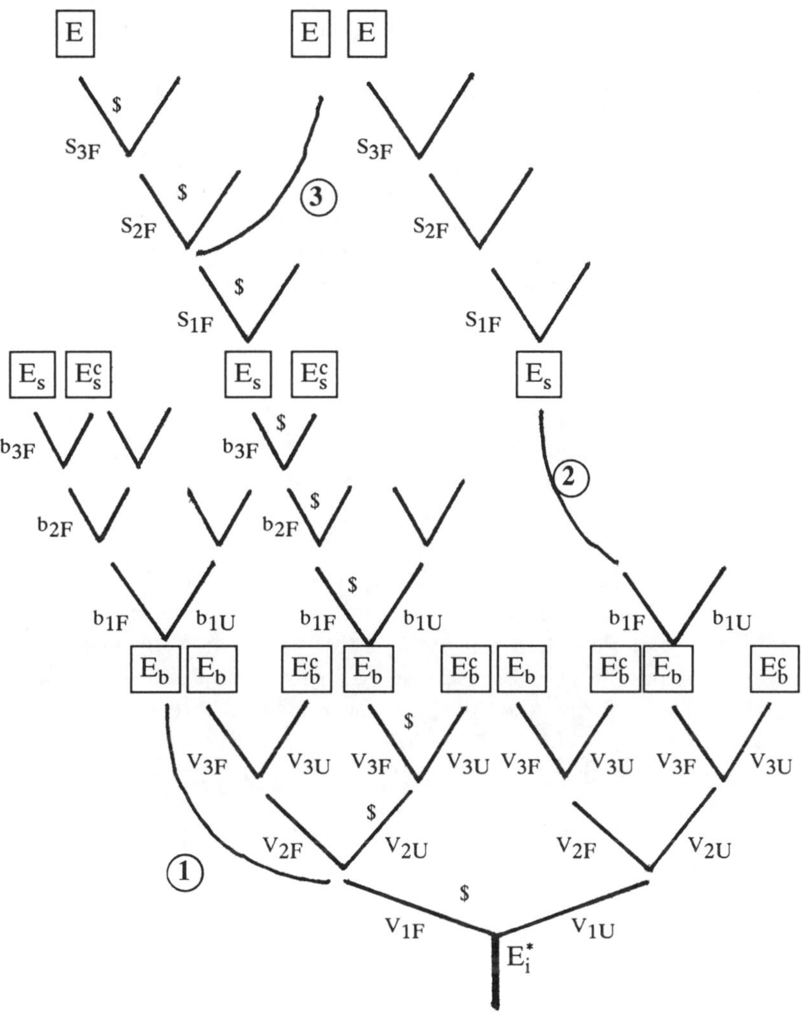

Figure 3: A Closer Look At Baconian Inference.

results on the other two veracity variables that have not been tested, we detach a belief that E_b is true. But now notice that our evidence at the next stage, involving objectivity in belief-formation, is solid, as indicated by the heavy evidentiary branches involving favorable evidential tests of W_i's belief formation processs. This is what seems to defy gravity, heavy branches being supported by a weak one. However physically impossible this sounds, it seems to be a fact of life in all human inference for reasons I mentioned in 3.3 concerning the distribution of incompleteness of evidence across stages of reasoning, whether credibility-related or not. Such incompleteness may occur at any stage of reasoning as the "*ceteris paribus* limbs" (2) and (3) in Figure 3 indicate. Given these three choices, it seems that a preference could be expressed for having the stituation involving (3); here we have no defiance of gravity since we have a strong evidential basis at the first two reasoning stages upon which to base our somewhat weaker inference at the third stage. I believe this corresponds to Cohen's remark that we should prefer probable inference based on known facts over certain inferences based on probable facts (1977, p. 73); the former is not gravity-defying but the latter is.

As I noted, "defiance of gravity" in cascaded inference is not a peculiarity of Baconian reasoning; all Jonathan Cohen asks is that we recognize it and then account for it in our inferences from one stage to another. Here is an example of gravity defiance involving statistical evidence, which often seems so compelling; this example is based upon the celebrated "cab problem" that has been a source of so much controversy in connection with behavioral studies of human reasoning capabilities. On Pascalian analyses of this problem, experimental data suggest that people ignore base rates or prior probabilities in inference (e.g. Tversky & Kahneman, 1982); Cohen has a different interpretation of these data (e.g. 1981, p. 328; 1983, p. 17). In the following analysis I am unconcerned about base rates; I wish to focus on the evidence.

A cab in a certain city, having just blue and green cabs, was involved in a hit-and-run accident and eye-witness W_i asserts E_j^* at trial that the cab involved in the accident was blue [E] rather than green [E^c]. The court allows a test of W_j's color discrimination under essentially the same conditions that existed at the time of the accident. The result of this carefully-supervised test is that W_i correctly identified each one of the two colors 99% of the time and failed 1% of the time. What we have here is *ancillary* statistical evidence on just one credibility-relevant variable, namely whether or not this witness can tell the difference between blue and green. Suppose we take these results as estimates of $P(E_j^*|E)$ and $P(E_j^*|E^c)$ as has been common in Pascalian analyses of this problem. Their ratio says that E_j^* favors blue over green 99:1; apparently, this is compelling evidence that a blue cab was involved in the accident. There are two problems: first, this construes the inference from E_i^* to [E, E^c] as being a direct one; second, these statistics are just backing evidence for an inference about what W_i observed *on the day in question*. Suppose Plaintiff's attorney rests upon this apparently compelling evidence.

A Baconian attorney, representing the "Blue Cab Co." and using the cascaded inference in Figure 1-C, has a different view about the weight of this evidence. As Figure 4 shows, absent any veracity evidence, we can only go out on the unsupported *ceteris paribus* limb (1) in inferring that W_j believed the cab was blue on the night of the accident. Since we have no belief-formation evidence, we must also go out on unsupported *ceteris paribus* limb (2) in our inference that W_j's senses recorded blue on the night in question. We finally come to the *ancillary* statistical evidence given at trial relevant in testing W_j's observational sensitivity. Since, as shown in 2.3.2. above, there are other tests of observational sensitivity, we need yet another *ceteris paribus* limb (3) in our final inference about E. So, what the Baconian can rightly conclude is that E_j^* and the ancillary test result, by themselves, have almost no weight at all. We have a weak argument constructed on a very shakey foundation. In fact, if this ancillary result was all we had, we would not be much worse off if we chose to go out on *ceteris paribus* limb (4) where we consider only the global generalization but no evidence at all.

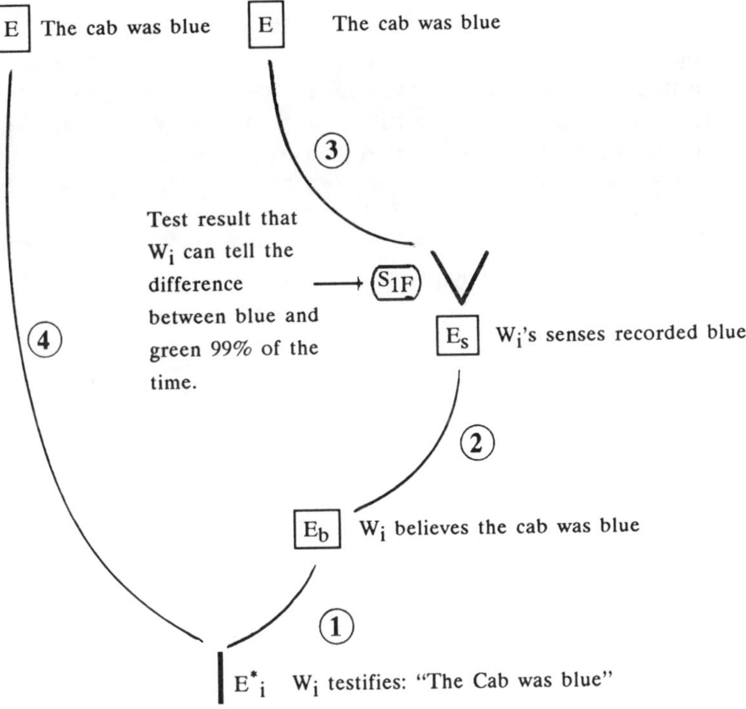

Figure 4: The Weight vs The Probability Of Evidence.

130

The fact that we cannot do algebraic combinations of Baconian probabilities seems no serious impediment to cascaded inference construed in Baconian terms. Here follows an analysis that I believe preserves the essential properties of Baconian probabilities. This analysis is based on the special case in which our reasoning route takes us through events E_b, E_s, and terminates at E; it would apply equally well for any other route we could have taken.

1). Let $I(E_b, E_i^* \& V \text{ evid}) \geqslant (j + 1)/(n_v + 1)$ indicate the inductive probability of E_b on testimony E_j^* and the veracity evidence we have considered. Suppose that j of the n_v veracity test possible have been made and that the results are, on balance, favorable to our believing that W_i believes that E happened. At this stage we are entitled to detach a belief in E_b at level j; let the belief so attached be $E_b(j)$.

2). Let $I(E_s, E_i^* \& E_b(j) \& B \text{ evid}) \geqslant (j + k + 2)/(n_v + n_b + 2)$ indicate the inductive probability of E_s on E_i^*, our detached belief $E_b(j)$, and on belief formation evidence B. The 2's in this expression result from a benefit of the doubt being given initially to both the veracity and objectivity generalizations. Suppose that k of the n_b possible tests have been performed and, on balance, favor the conclusion that W_i actually sensed E. Here we can detach a belief in E_s at level k; let this be symbolized as $E_s(k)$. This inductive probability reflects the fact that this second stage of our inference is now based upon $(j + k)$ results in the $(n_v + n_b)$ tests possible in both of these stages.

3). Let $I(E, E_i^* \& E_b(j) \& E_s(k) \& S \text{ evid}) \geqslant (j + k + m + 3)/(n_v + n_b + n_s + 3)$ be the inductive probability of E on E_i^*, our detached beliefs in E_s and E_v, on our ancillary evidence about W_i's observational sensitivity, and on a benefit of the doubt for the observational sensitivity generalization. At this third level, m of the n_s observational sensitivity variables have been tested and are, on balance, favorable to our concluding that E did happen. For anyone unconvinced about the completeness of credibility attribute coverage of my reasoning chain, we could add my "catch-all" and consider the collection of "all possible" credibility-related relevant variables $N = (n_v + n_b + n_s + n_c)$. The resulting Baconian probability of E would then be lower since N now contains those new unanswered questions of which rhere are n_c.

Notice in all of these formal expressions that the *difference* between numerator and denominator in a Baconian probabilistic expression always indicates the number of recognized relevant variables that have *not* been included in the test of credibility-related generalizations. Stated another way, the difference between numerator and denominator in a Baconian probability shows: (i) the number of relevant credibility-related queries left unanswered, (ii) the content of one or more *ceteris paribus* assumptions, or (iii) the "weight" of the ancillary evidence resulting from the number of tests actually performed when this number is compared with the total number of tests that might have been performed.

At each stage of reasoning we have to consider the strength of the foundation provided by earlier stage. Thus, for example, even though m might be very close to covering all of the n_s observational sensitivity variable, if j and k do not come close to n_v and n_b, then our final inductive probability $I(E_j^* \& E_b(j) \& E_s(k) \& S \text{ evid})$ cannot be very large. The reason is that the detachment of beliefs we have made at these earlier stages rests upon *ceteris paribus* assumptions having very rich content. Care should be taken not to suppose that the above Baconian probabilities behave as percentages. For example, it must be true that the inductive probability of E_b and E_s be no less than the smaller of $(j + 1)/(n_v + 1)$ or $(k + 1)/(n_b + 1)$ to preserve the conjunction rule property of Baconian probabilities, and the inductive probability of E_b or E_s must be at least equal to the larger of $(j + 1)/(n_v + 1)$ or $(k + 1)/(n_b + 1)$ to preserve the Baconian disjunctive rule.

My final word about Baconian cascaded inference involves the often-expected weakening of an inference as it passes through several stages. As an illustration, consider Figure 3 in which it was supposed that there were exactly three relevant variables at each of the three reasoning stages. Suppose evidence on two of the three tests at each of these stages is presented. At the first stage our inference leaves one question unanswered, the inference at stage two leaves two questions unanswered, and the one at stage three leaves three questions unanswered. Absent complete evidentary coverage, the higher we go up an inference tree, the more questions we leave unanswered. In short, the effect of evidential incompleteness is cumulative as one reasons from one stage to another. In discussing inference-upon-inference in legal contexts, Maguire et al (1973, p. 875) discuss a case in which an appelate judge remarked that piling one inference on another has no more substance than the soup Abraham Lincoln described which was made by boiling the shadow of a pigeon that had been starved to death. Inference-upon-inference need not be this weak. Jonathan Cohen argues that it need not be weak if our evidential coverage of matters relevant at each stage is as complete as we can make it. Bayesians, now to be heard from, also argue that cascaded inferences are not necessarily weak.

4.0. A Bayesian Account Of Credibility Assessment

Thomas Bayes and his present heirs have no shortage of things to say about the analysis of chains of reasoning. However, my remarks about a Bayesian approach to cascaded inference in general and credibility assessment in particular will be brief: the major reason is that there has been so much already written about applying Bayes' rule to chains of reasoning. The only Baconian account of chains of reasoning I have ever seen is the very brief one Cohen has provided (1977, pp. 68-73, 267-269). In a recent paper (1989) I have applied a Bayesian approach in the analysis of inferences depicted in Figures 1-C and 1-E above; following is a brief summary of the results of this analysis.

4.1. Bayes' Rule And The Combination Of Uncertainties About Credibility

I remind the reader again of my present focus on chains of reasoning in which the events of concern are singular, unique, or non-replicable. Thus, it should come as no surprise that the Pascalian probabilities I have in mind are subjective or personal in nature. One matter at issue is how the subjective probabilities in a Bayesian analysis are to be supported; I cannot suppose that a Bayesian is entitled to any advantage not accorded the Baconian in 3.0 above. I have to suppose that the subjective probabilities I will identify at various stages of resoning are somehow supported by the very same ancillary evidence that I have already identified in terms of Cohen's relevant variables. A Baconian and a Bayesian use exactly the same evidence in a chain of reasoning, but they use it differently.

I said earlier that decomposing the inference from E_j^* to events $[E, E^c]$ in terms of three credibility-related attributes exposes sources of uncertainty associated with these attributes: we have uncertainty about what W_i believed and "sensed" following his/her observation in additon to our uncertainty about whether or not event E happened, as W_i now claims. What a Bayesian analysis encourages is expression of these uncertainties about veracity, belief formation, and observational sensitivity according to Pascalian rules. As it turns out, the inference from E_i^* to events $[E, E^c]$ can be posed in more than one way consistent with Pascalian rules (Schum, 1979b); here is the construal I have favored. Using the odds-likelihood version of Bayes' rule, we can express:

$$P(E|E_i^*)/P(E^C|E_i^*) = [P(E)/P(E^c)] \ [P(E_i^*|E)/P(E_i^*|E^c)]. \qquad (1)$$

If we let $L_{E*i} = P(E_i^*|E)/P(E_i^*|E^c)$, then

$$P(E|E_i^*)/P(E^c|E_i^*) = [P(E)/P(E^c)]L_{Ei} \qquad (1\text{-}a)$$

Bayes' rule says that we cannot ignore the prior likeliness of event E; in fact, Equations 1 or 1-a capture the interesting relationship, noted by Keynes (1957, p. 180-185) and others, between the rareness of a reported event and the credibility of its source. As it will turn out, the likelihood ratio term L_{E*_i} can grade the credibility of witness W_i in terms of the three attributes I have identified. If $P(E)$ is very small, then it takes a very large value of L_{E*_i} to make $P(E|E_i^*)$ greater than $P(E^c|E_i^*)$. The reader may have noticed that I have already mentioned the two terms, $P(E_i^*|E)$ and $P(E_i^*|E^c)$ in connection with SDT; the connection is not coincidental since a theoretical SDT observer is "Bayesian".

In Bayesian terms, L_{E*_i} grades the inferential or probative value of testimony E_i^* in an inference about whether or not event E occurred. I have just said that L_{E*_i} concerns the credibility of W_i in the present inference about event E; let us see why this is so. On the conventional definition of a conditional probability, $P(E_i^*|E) = P(E_i^* \ \& \ E)/P(E)$ and $P(E_i^*|E^c) = P(E_i^* \ \& \ E^c)/P(E^c)$. Using the general product rule for conditional probabilities we can expand $P(E_i^* \ \& \ E^c)$ and $P(E_i^* E^c)$ in terms of the event classes $[E_b, E_b^c]$ and $[E_s, E_s^c]$ shown in Figure 1-C. Now, we can make this expansion in a large number of formally-equivalent ways. The one that has always made most sense to me has the following results, the derivation of which I have provided elsewhere (1989):

$$P(E_i^*|E) = h$$

$$= b[\{d(v_1 - m_1) + m_1\} - \{g(v_3 - m_3) + m_3\}] + \{g(v_3 - m_3) + m_3\}, \quad (2\text{-}a)$$

where:

$$b = P(E_s|E); \ d = P(E_b|E_sE); \ g = P(E_b|E_s^cE); \ v_1 = P(E_i^*|E_bE_sE);$$

$$m_1 = P(E_i^*|E_b^cE_sE); \ v_3 = P(E_i^*|E_bE_s^cE); \ m_3 = P(E_i^*|E_b^cE_s^cE).$$

Similarly,

$$P(E_i^*|E^c) = f$$

$$= c[\{e(v_2 - m_2) + m_2\} - \{r(v_4 - m_4) + m_4\}] + \{r(v_4 - m_4) + m_4\}, \quad (2\text{-}b)$$

where:

$$c = P(E_s|E^c); \ e = P(E_b|E_sE^c); \ r = P(E_b|E_s^cE^c); \ v_2 = P(E_i^*|E_bE_sE^c);$$

$$m_2 = P(E_i^*|E_b^cE_sE^c); \ v_4 = P(E_i^*|E_bE_s^cE^c); \ m_4 = P(E_i^*|E_b^cE_s^cE^c).$$

134

The first thing to observe in Equations 2-a and 2-b is that there are conditional probabilities relevant to each of the credibility-related attributes at the stages of reasoning in Figure 1-C. Terms b and c link E_s with E and E^c at the *observational sensitivity* stage; terms, d, g, e, and r link E_b with events E_s and E_s^c at the *belief formation stage*; and all the v and m terms link testimony E_i^* with events E_b and E_b^c at the *veracity* stage. But, these direct linkages are not the only ones possible as the conditioning patterns in these terms indicate; they require us to revise Figure 1-C to include other possibilities. A revised representation of our cascaded inference is shown in Figure 5-A. Three additional linkages labeled with question marks have been added in Figure 5-A; these linkages concern possible patterns of *conditional nonindependence* among events at various stages in our cascaded inference from E_i^* to event {E, E^c}. For example, the v and m terms ask us to consider whether or not W_i's testimony depends, not just upon what W_i believes, but also upon what W_i "sensed" and whether or not E did occur. There are some very interesting subtleties lurking behind these possible conditional noninde-pendencies. Perhaps, for example, the likeliness of W_i's testimony depends, not only upon his/her beliefs, but also upon whether or not E did happen. In short, our beliefs about the veracity of W_i might be different under the assumption that E happened than under the assumption that it did not. State-dependent credibility attributes are certainly not unheard of; a radiologist might be better at detecting a certain X-ray anomaly for patients having tuberculosis than for those having pneumonia. Equations 2-a and 2-b are general, given the inference depicted in Figure 5-A; a wide variety of special cases are possible depending upon which patterns of conditional idependence and nonindependence we assume. Following is the simplest possible special case.

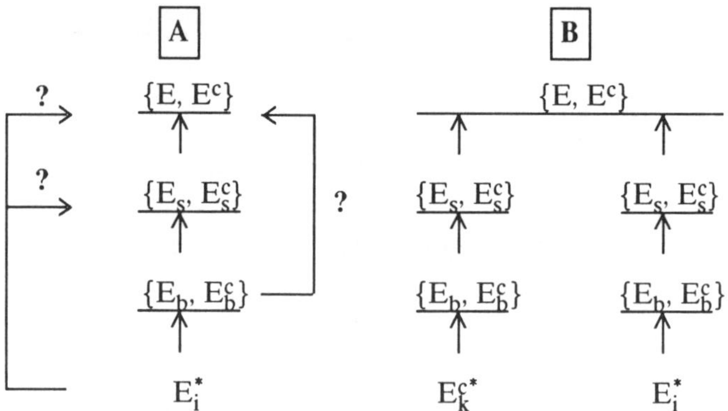

Figure 5: Some Considerations Exposed In A Bayesian Analysis.

I have believed for quite some time that the concept of conditional independence/nonindependence is the best example I know of a simple idea that has profound consequences. This concept figures in every one of the now large array of evidential subtleties I have studied. Given even the "simple" chain of reasoning I have presented in this work, there are many conditioning patterns possible in a Bayesian construal. It is sometimes said, nonsensically, that the use of Bayes' rule assumes conditional independence. Equations 2-a and 2-b follow from Bayes' rule and, in fact, identify various patterns of conditional independence/nonindependence. Here is just one possible pattern and it captures the following beliefs. Suppose, on the evidence we have, we believe that W_i's testimony depends just upon what W_i believes, what W_i believes depends just upon what he/she "sensed", and what W_i sensed depends just upon whether or not event E happened. This special case I have labeled the "Markov" case because the likeliness of an event at one stage of reasoning depends only upon events at the next stage higher in the inference tree. In Figure 5-A we answer all three question marks with the word "no", and so we are back to the inference depicted in Figure 1-C. In this special case, we can greatly simplify the two terms in L_{E*_i}:

$$P(E_i^*|E) = h \qquad\qquad (3 - a)$$
$$= [P(E_s|E)\{P(E_b|E_s) - P(E_b|E_s^c)\} + P(E_b|E_s^c)]\{P(E_i^*|E_b) -$$
$$- P(E_i^*|E_b^c)\} + P(E^*|E_b^c),$$

and,

$$P(E_i^*|E^c) = f \qquad\qquad (3 - b)$$
$$= [P(E_s|E^c)\{P(E_b|E_s) - P(E_b|E_s^c)\} + P(E_b|E_s^c)]\{P(E_i^*|E_b) -$$
$$- P(E_i^*|E_b^c)\} + P(E^*|E_b^c).$$

Notice that h and f in this special case differ by only one term: $P(E_s|E)$ in 3-a and $P(E_s|E^c)$ in 3-b.

In summary, in determining how much more likely E is than E^c, given testimony E_i^*, a Bayesian can combine prior uncertainties about E and E^c with uncertainties about attributes of the credibility of W_i. Taking the Markov case as an example, the credibility-related uncertainties are as follows:

(i) Is this testimony E_i^* more probable if W believed E occurred or if W believed E did not occur? This veracity-related question is answered in terms of the likelihoods $P(E_i^*|E_b)$ and $P(E_i^*|E_b^c)$.

136

(ii) Is W_i's belief that E occurred more likely if W_i sensed E or sensed E^c? This objectivity question is answered in terms of the likelihoods $P(E_b|E_s)$ and $P(E_b|E_s^c)$.

(iii) Is W_i's sensing E more likely if E occurred or if it did not? This observational sensitivity question is answered in terms of $P(E_s|E)$ and $P(E_s|E^c)$.

So, a Bayesian is not necessarily required to reason from one stage to another as I pictured the Baconian doing in 3.0 above. Using Equation 1 and some appropriate case of Equations 2-a and 2-b, a Bayesian can combine prior uncertainty about event E with whatever uncertainty exists about attributes of the credibility of the witness who tells us that event E happened. Other Bayesian formulations are possible in which we can form and combine Pascalian probabilities analogous to the Baconian ones I have considered. However, I find these other methods of posing cascaded inference in Bayesian terms to be considerably less informative.

4.2 Supporting The Burden Of Likelihood Assessment

There are many questions we have to answer about the ingredients in the equations considered in 4.1. The first question arises when we consider that these equations, as stated, represent an oversimplification of the evidential situation faced in credibility assessment. Equation 1 involves the likeliness of E relative to that of E^c, given testimonial evidence E_i^*; Equations 2 and 3 concern the likeliness of events at various credibility-related stages of reasoning. But, in none of these equations appears any representation for the three categories of ancillary and other evidence of relevance in credibility assessment. I shall first consider evidence from other sources that either conficts with or contradicts E_i^*. If such evidence exists, we have a much more complex situation than the ones depicted in either Figure 1-C or in 5-A. Some hint of the additional complexity we face is indicated in Figure 5-B in which I show a previous witness W_k, who asserts contradictory evidence E_k^{c*}, that event E did not occur; we have W_k's credibility attributes to consider as well. In such instances we can formulate likelihood ratio expressions for the probative value of E_i^* on E and E^c, given any pattern of contradictory or corroborative testimony form other witnesses, as I have shown elsewhere (Schum & Kelly, 1973). Such analyses are interesting and lead to the same conclusion now accepted by jurists, namely that counting heads on both "sides" (i.e., "majority rule") is not an appropriate device for resolving evidential conflict or contradiction;

what matters is the aggregate credibility on both "sides" of a matter in contention (e.g. VII Wigmore, 1940, p. 259).

Confining attention to *ancillary* credibility-relevant evidence about W_j, I must be concerned about where such evidence fits into the conditioning patterns the above equations reflect. Let V_j represent the j items of veracity-relevant evidence we have, B_k represent the k items of belief formation evidence we have, and S_m represent the m items of observational sensitivity evidence we have. When we build in these additional classes of evidence we have more complex terms in Equation 1. For example, we will now have $P(E|E_i^* \& V_j \& B_k \& S_m)$ and $P(E_i^*|E \& V_j \& B_k \& S_m)$. How these additional classes of ancillary evidence influence the required credibility-relevant likelihood assessments depends, predictably, upon some conditional independence issues we have to resolve. As an example, we have to decide whether or not our beliefs about the veracity of W, based on veracity-related evidence V_j, are in any way further contingent upon other ancillary evidence B_k and S_m we have relevant to the other two credibility attributes. I take this to be a very interesting matter for both the Baconian and the Bayesian to consider. In Baconian terms, it is interesting to ponder, for example, whether or not evidence regarding the process of belief formation could ever counteract unfavorable veracity-related evidence.

So, a more complete posing of the problem presents us with more complex likelihood terms which I shall illustrate using the Markov case under the assumption that V_j is the only veracity-relevant evidence, B_k the only evidence relevant to belief formation, and S_m the only evidence relevant to observational sensitivity. What a Bayesian has to provide in such instances are credibility-related assessments of the following sort. On ancillary evidence, is W_i's testimony E_i^*, on ancillary evidence V_j, more likely on the assumption that W_i believes E happended or on the assumption that W_i believes E did not happen. In symbols, we must have assessments of $P(E_i^*|E_b \& V_j)$ and $P(E_i^*|E_{\bar{b}} \& V_j)$. Similar pairs of assessment are required for the belief formation and observational sensitivity attributes. Bayes' rule tells us that simple ratios of these pairs of ingredients will not suffice; what we need are specific values; this fact has considerable behavioral consequences. The L_{E^*i} equations how that Bayes' rule responds to the differences as well as to the ratios of L_{E^*i} ingredients. There is a very recent discussion of some of the consequences of this fact (Edwards, Schum, & Winkler, 1990). Unfortunately, as psychophysicists tell us (e.g. Stevens, 1975, p. 18), people find relative judgments easier and more natural than the absolute judgments Bayes'

138

rule requires in cascaded inference. Our innate capacity for making relative judgments was noticed much earlier by Helmholtz (e.g., Warren & Warren, 1968, pp. 137-168).

I some situation suppose you hear me assert that $P(E_i^* | E_b \ \& \ V_j) = 0.89$ and that $P(E_i^* | E_b^c \ \& \ V_j) = 0.25$. You are entitled to ask how I managed such precision based upon the possibly imprecise or "fuzzy" V_j evidence I have. So, I might hedge a bit by giving a range of values for each of these ingredients. But now comes the question the Baconian is sure to ask: how are these numbers, however expressed, related to the amount of evidence in V_j, B_k, and S_m and to the adequacy of the coverage each collection represents on matters relevant to the testing of the three credibility attributes? I might say, in response, that my interval hedging took into account the incompleteness as well as the imprecision of the ancillary evidence I had. In so doing, however, I would find myself unable to discern how much of this interval was due to imprecision and how much was due to incompleteness.

4.3. The Heuristic Value Of Bayesian Analyses

These difficulties acknowledged, I will claim that Bayesian analyses of chains of reasoning and the evidential subtleties they expose can be extremely valuable, provided we do not take any specific collection of numerical ingredients too seriously as representing the "true" state of our uncertainties on any given occasion. The equations I have listed, as well as others representing other evidential problems in cascaded inference, can be exercised by trying out any possible combination of Pascalian probabilities. This process of "sensitivity analysis" sheds much light on some very complex situations and helps to suggest as well as to answer many questions about complex human inference. Here are just a few results that bear upon our present interest in credibility assessment.

A Bayesian analysis shows that there are several species of "bias" to be considered in assessing the credibility of a source; a witness can be as easily biased for [against] giving testimony E_i^* as he/she can be biased for [against] believing that E occurred, or biased for [against] sensing E in the first place. This is a consequence of Bayes' rule attending to differences as well as to ratios of likelihood ingredients. Bayesian analyses, in common with Baconian analyses, show the distinction that has to be made between a witness having "knowledge" and the testimony of this witness having probative value. Bayes rule and Cohen's Baconian probabilities give somewhat different answers to transitivity-related questions. A Bayesian analysis shows that the location of a "weak link" in a reasoning chain does not matter, provided that the reasoning chain does not involve the additional linkages depicted in Figure 5-A. When such linkages do occur, Bayesian analyses show that the behavior of a source of evidence can supply probative value in excess of that supplied by the event he/she reports. There are many species of interesting witness behavior that can be trapped for study and analysis. Finally, an analysis along the lines I have been following shows that it makes no sense to attempt to grade the "credibility" of a human source, or any surrogate of this term, by means of a single number. Indeed, it makes no sense to try to grade any single attribute of credibility by means of a single number. For this reason, previous probabilistic accounts of credibility-testimony issues have been more entertaining than useful. To answer questions about a person's credibility we have to acknowledge that it is a multiattribute characteristic; this comes from recognition that credibility assessment involves a chain of reasoning.

5.0 The Societies Of Baconian And Bayesian Trial Lawyers

I have never had any reputation as a clairvoyant and so, when I make an accurate prediction, I will usually try to advertise the making of it as widely as possible. Upon reviewing *The Probable And The Provable* I predicted that Jonathan Cohen's work would be influential in jurisprudence and in other disciplines (Schum, 1979a, pp. 450, 483). This prediction has certainly been accurate as far as jurisprudence is concerned; I have already mentioned that Cohen's work provided one major focus in a recent symposium on probability and inference in the law of evidence (Green & Tillers, 1986). It is fair to say that some of the participants in this symposium were (and still are) hardened and shameless Bayesians. One person who fits in this category is my colleaque and close friend Ward Edwards. In his summary of the debates during this symposium, Edwards (1986, pp. 937-941) concluded that Bayesian views occupied the status of orthodoxy in jurisprudence and that one purpose of the symposium was to provide opponents of this orthodoxy opportunity to take their best shots at it; he concluded: "no one laid a glove on us" and later suggested that the reason was that opponents didn't know where to aim their punches. Edwards went a bit farther in advocating the establishment of *"The Society Of Bayesian Trial Lawyers"*, whose members would have knowledge of various Bayesian decision and inference analytic strategies that could allegedly provide great benefit in the conduct of litigation.

Apparently, Edwards has a stouter jaw and fancier footwork than I do, for I have felt Cohen's intellectual gloves on more than one occasion and I did not regard his blows to be in any sense glancing. In addition, I never thought Cohen to be flailing about in the delivery of his punches. In fact, I will recommend that Jonathan Cohen establish *"The Society Of Baconian Trial Lawyers"* on the promise that, if I ever get in trouble with the law, I will employ a member from each of these two societies. I remember reading how the Romans used to incorporate into their pantheon of deities those worshipped by the peoples they conquered, on the off-chance that one of them might be genuine. Some scholars of evidence in jurisprudence have, to be sure, found Bayesian ideas useful. At the same time, however, I have to differ with Edwards about the extent of the orthodoxy of Bayesian views in jurisprudence and so I cannot agree that Bayesians should think of themselves as the conquerors of Baconians on the field of jurisprudence. I believe I have reasons that are better than the one adopted by the Romans for attempting to incorporate Baconian views in my own thinking about cascaded inference.

In the "ungodly" cascaded inferences I have included in my analysis, the Baconian judges the favorableness or unfavorableness of ancillary and other evidence on generalizations that license the reasoning from one stage to another. The differences among the covering generalizations and the ordinal nature of Baconian probabilities preclude combination of Baconian probabilities from stage to stage and so we have to suppose that we can "detach" beliefs about events at the various stage. I do not find this detachment unnatural in any way and I mentioned empirical evidence of its existence in studies conducted by persons who cannot be regarded as Baconian. The Bayesian uses ancillary evidence as a basis for assessing uncertainties at various stages of reasoning and can then combine them in assessing the probative or inferential strength of the entire chain. The basic idea is that what is common across all stages is uncertainty, something that varies in intensity and not in kind; therefore, it is combinable across stages. There is no basis here for Bayesians to sneer at Baconians for the ordinal nature of their probabilities. In the nonreplicable evidential situations I have assumed, we must suppose subjective or personal probabilities as gradations of uncertainty. The rules of combination I have presented assume that these numbers, for any person, have ratio-scale properties (i.e., they can meaningfully be added, subtracted, multiplied, and divided). I would not like to take on the task of demonstrating that the subjective probabilities anyone asserts indeed have these properties. However, I will not argue that subjective likelihood assessments have no value or meaning. In a series of empirical investigations Anne Martin and I (1982) observed that likelihood assessments, made by persons facing even more difficult cascaded inferences than those I have discussed in the present work, captured an array of evidential subtleties these same persons overlooked in other kinds of assessments.

I believe it fair to say that people often draw conclusions on the basis of what are actually small amounts of evidence. This does not necessarily mean that more evidence is not available; we certainly have a vast array of methods for collecting and transmitting evidence. Even in law courts, where it may wrongly be assumed that evidential coverage will be always be complete on matters in contention, there are to be found the distributions of incompleteness across reasoning stages I mentioned. Time and resources for evidence collection, presentation, and deliberation are limited in most contexts I know about. I believe that the Baconian wins the day in being able to grade the strength, and often the length, of the inferential limbs on which we so often find ourselves because of evidence we have not considered. Shy of interval hedging on likelihood assessments, I can find no way in which a Bayesian can grade the extent to which ancillary evidence covers matters relevant at some stage of reasoning. At the same time, I cannot ignore the rich array of evidential subtleties that can be captured for study and analysis by Bayesian means, and I have often been impressed by the extent to which conclusions drawn from such study match those drawn on the basis of centuries of experience in jurisprudence. So, if I ever get into legal trouble and employ a member from each of the societies of Baconian and Bayesian trial lawers, I will be assured of some reasonable coverage of inferential matters of great importance.

Baconians face a much tougher row to hoe in the field of psychology where, I will agree, the Pascalian view of probability is orthodoxy. I believe Cohen discovered this himself when he challenged views claiming essential human inferential incompetence that are also regarded, in may quarters, as orthodoxy (e.g. 1977, 1981). I believe it a mistake for psychologists to ignore the Baconian view of probability. Psychologists are typically aware of the distinction between "descriptive" models of human behavior that account only for data and those that identify plausible behavioral processes that might have produced the responses people make. I believe there to be considerable process virtue inherent in Baconian cascaded reasoning which, unlike the Bayesian view, does not suppose we mentally crunch large amounts of numerical information. What the Baconian view supposes is that, in reasoning from one stage to another, we can judge the favorableness or unfavorableness of evidence on generalizations we make, that we are capable of keeping track of question we leave unanswered, and that the number of unanswered questions we recognize influences the strength of the final conclusion we draw as well as do the questions for which we have obtained answers.

One of Cohen's papers, in which he expressed his views about behavioral research (1981, pp. 317-330), drew a large number of replies, 43 by my count. The reply that has remained most strongly in my mind is the one by Baruch Fischhoff (1981, pp. 336-337). With characteristic eloquence, Fischhoff discusses the relationship between philosophers and psychologists and how this relationship is so frequently marred by misunderstanding, lack of communication, and even indifference. In the process, Fischhoff inquires about what philosophers and psychologists could or should be learning from one another and he suggests that they ought, at least, to entertain the thought that they would be better off extending a warm smile and a deaf ear to one another. Speaking as a psychologist, I must say that I have received many a warm smile from Jonathan Cohen, but never a deaf ear on any matter, whether it involved a philosophical or a behavioral point. In return, I have taken his views very seriously; he has so much to say to anyone willing to listen. It seems to me that psychologists ought to be among the first to recognize that no single formal system can provide guidance for all elements of the rich behavioral task of probabilistic inference; I hope they will not be among the last to do so.

Department of Operation Research and Applied Statistics.
George Mason University
Fairfax, Virginia 220350
USA

FOOTNOTES

* This research was supported by The National Science Foundation under Grant SES – 8704377 to George Mason University.

[1] Discussions of Wigmore's evidence charting scheme are found in Twining (1985) and in Tillers & Schum (1988).

[2] I do note, however, that Glymore (1980) uses the term "cascades" in connection with the problem of whether or not evidence confirming some sentence also confirms others that are entailed by this sentence.

[3] As Peter Tillers and I noted on another occasion (1988), generalizations made in law are stated in "fuzzy" or imprecise terms. Thus, for example, we might assert the generalization: "people usually report what they believe". Particularly good examples of the generalizations made in legal contexts are to be found in Binder & Bergman (1984). These are exactly the sort of "commonplace" generalizations that Cohen discusses (1977, p. 275) and that the Anglo-American legal system credits jurors with being able to apply in their deliberations on the evidence in a case. So, we actually need not state these three credibility-related generalizations so strongly.

4 1988 Federal Rules Of Evidence With Advisory Committee Notes And Legislative History (Mueller, C., Kirkpatrick, L.), Boston. Little Brown & Co., 1988.

5 I am aware of the controversy surrounding the view of knowledge as justified true belief, as I have discussed in another work (1989).

6 A Midsummer Night's Dream, Act III, Scene 2.

7 I hedge here by saying "usually" because, on Bayesian analyses, it can happen that the second of two items of redundant testimony has more and not less value than the first.

8 I might have provided a more accurate latin rendition here by saying that the assumption of further favorable evidential tests is a "ceteris prosperus" assumption. I decided, however, to retain Cohen's labeling of this assumption.

REFERENCES

Binder, D., Bergman, P., (1984). *Fact Investigation: From Hypothesis To Proof.* St. Paul, Minn.; West Publishing Co.,

Charniak, E., McDermott, D., (1985). *Artificial Intelligence*, Reading Mass., Addisson-Wesley

Chisholm, R., (1957). *Perceiving: A Philosophical Study*, Ithaca, N. Y., Cornell University Press,

Cleary E., et al. (1972). *McCormick On Evidence*, St. Paul, Minn.: West Publishing Co.

Cohen, L. J. (1975). Probability–the one and the many. *Proceedings Of The British Academy*, **LXI.**

Cohen, L. J., (1977). *The Probable And The Provable*, Oxford: The Clarendon Press,

Cohen, L. J., (1980). Bayesianism versus Baconianism in the evaluation of medical diagnoses. *British Journal Of The Philosophy Of Science*, **31**, 45-62.

Cohen, L. J., Hesse, M., (1980). *Applications Of Inductive Logic*. Oxford: The Clarendon Press.

Cohen, L. J., (1981). Can human irrationality be experimentally demonstrated?, *The Behavioral And Brain Sciences*, **4**, 317-370.

Cohen, L. J. (1983). Freedom of proof., In: W. Twining, (ed.) *Fact in Law*, Wiesbaden, Franz Steiner Verlag GMBH.

Cohen, L. J., (1985). Twelve questions about Keynes's concept of weight. *British Journal Of The Philosophy Of Science*, **37**, 263-278.

Cohen, L. J., (1986). The role of evidential weight in criminal proof. *Boston University Law Review*, **66**, 635-649.

Edwards, W., (1986). Summing up: The society of Bayesian trial lawyers. *Boston University Law Review*, **66**, 937-941.

Edwards, W., Schum, D., Winkler R. (1990). Murder and (of?) the likelihood principle: A trialogue. *Journal of Behavioral Decision Making*, **3**, 75-89.

Egan, J. (1975). *Signal Detection Theory and ROC Analysis*. New York: Academic Press.

Fischhoff, B., (1981). Can any statement about human behavior be empiricically validated? *The Behavioral And Brain Sciences.* **4**, 336-337.

144

Freeling, A., Sahlin, N. (1983). Combining evidence. In: Gardenfors, Hansson, Sahlin (Eds.), *Evidentiary Value: Philosopical, Judical, And Psychological Aspects Of A Theory*, Lund, Sweden: C. W. K. Gleerups.

Glymore, C., (1980). *Theory And Evidence*. Princeton, N. J.: Princeton University Press,

Good, I. J. (1983). *Good Thinking: The Foundations Of Probability And Its Applications*, Minneapolis, Minn: University Of Minnesota Press,

Green, D., Swets, J., (1966). *Signal Detection Theory And Psychophysics*, New York: J. Wiley & Sons.

Green, E., Tillers, P., (1986). *Symposium on probability and inference in the law of evidence, Boston University Law Review*, **66**, Nos 3 & 4, [entire issue]. Selected papers from this special issue are published under the title: *Probability And Inference In The Law Of Evidence: The Uses And Limits Of Bayesianism*, Tillers & Green, eds., Dordrecht, Holland: Kluwer Publishing Co., 1988.

Keynes, J. M., (1957). *A Treatise On Probability*, London: Macmillan & Co,

Lempert, R., Saltzburg, S., (1977). *A Modern Approach To Evidence*. St. Paul, Minn.: West Publishing Co.,

Maguire, J., et. al. (1973). *Evidence: Cases And Materials*. Mineola, N. Y: Foundation Press,

Nozick, R., (1981). *Philosophical Explanations*, Cambridge, Mass: Belknap Press of Harvard University.

Peterson, C., ed. (1973). Special issue on cascaded inference, *Organizational Behavior And Human Performance*, **10**, No. 3,

Reiter, R., (1980). A logic for default reasoning, *Artificial Intelligence*, **13**, 81-132.

Schum, D., (1979a). A review of a case against Blaise Pascal and his heirs. *Michigan Law Review*, **77**, 446-483.

Schum, D. (1979b). A Bayesian account of transitivity and other order-related effects in chains of inferential reasoning. *Research Report 79-04. Rice University Psychology, Dept.*, Houston, Texas, December,

Schum, D., (1986). Probability and the processes of discovery, proof, and choice. *Boston University Law Review*, **66**, 825-876.

Schum, D., (1987). *Evidence And Inference For The Intelligence Analyst* [Two Volumes], Lanham Md.: University Press Of America,

Schum, D., (1989). Knowledge, probability, and credibility. *Journal Of Behavioral Decision Making*, **2**, 39-62.

Schum, D., Kelly, C., (1973). A problem in cascaded inference: determining the inferential impact of confirming and conflicting reports from several unreliable sources, *Organizational Behavior And Human Performance*, **10**, 404-423.

Schum, D. Martin, A., (1982). Formal and empirical research on cascaded inference in jurisprudence, *Law & Society Review*, **17**, 105-151.

Stevens, S. S., (1975). *Psychophysics*, New York: J. Wiley & Sons,

Swets, J., (ed.), (1964). *Signal Detection And Recognition By Human Observers*, New York: J. Wiley & Sons,

Tillers, P., Schum, D., (1988). Charting new territory in judical proof. *Cardozo Law Review*, **9**.

Tversky, A., Kahneman D., (1982). Evidential impact of base rates. In: D. Kahneman, P. Slovic, A. Tversky (Eds.), *Judgment Under Uncertainty: Heuristic And Biases*, Cambridge: Cambridge University Press.

Toulmin S. (1958). *The Uses Of Argument*, Cambridge: Cambridge University Press,

Twining, W., (1985). *Theories Of Evidence: Bentham & Wigmore*, Stanford: Stanford University Press,

Venn, J., (1973). *The Principles Of Inductive Logic*, New York: Chelsea, reprint of the 1907, edition.

von Winterfeldt, D., Edwards, W., (1986). *Decision Analysis And Behavioral Research*, Cambridge: Cambridge University Pres.

Warren, R., Warren, R. (1968). *Helmholtz On Perception: Its Physiology And Development* New York: J. Wiley & Sons,

Wells, C. (1914). Early opposition to the petty jury in criminal cases. *Law Quarterly Review* Vol. XXX, January

Wigmore, J. H., (1937). *The Science Of Judicial Proof.* 3rd Ed. Boston: Little Brown & Co.,

Wigmore, J. H., (1983). *Evidence*: Vol 1-A, Tillers Revision. Boston: Little Brown & Co,

Wigmore, J. H., (1940). *Evidence*: Vol, VII. Boston: Little Brown & Co.

*Poznań Studies in the Philosophy
of the Sciences and the Humanities
1991, Vol. 21, pp. 147-172*

James Logue

WEIGHT OF EVIDENCE, RESILIENCY
AND SECOND-ORDER PROBABILITIES

Suppose that I am uncertain about some proposition or the outcome of some event. Plausibly, my state can be represented as one of partial belief, where "partial" means that such beliefs, if not measurable on some scale with full belief and full disbelief (or full belief in the proposition's negation — which may not coincide) as endpoints, are at least rankable or gradable in some way. Of course, it may be that some such states are not comparable with some others; or, where belief-intensity is measurable, it may only be possible to assign it to an interval rather than a precise value. These problems surmounted, though, for some class of beliefs, we can take it that those beliefs come in numerical degrees of intensity; under suitable constraints, such degrees of intensity may then serve to interpret the uninterpreted measure "probable" of the standard, or a non-standard, probability calculus. An objectivist about probability would then characterise these beliefs as rational estimates of an unknown 'true' probability; a subjectivist would say that, if rational, such belief *are* probabilities.

But is such an enterprise, carried through with whatever degree of mathematical sophistication, sufficient to characterise uncertainty? Consider an analogy with ethical attitudes. You and I agree, let us suppose, that the issue of abortion on demand is a moral issue. You have brooded long and hard over the question, searching your conscience, comparing possible attitudes here with your attitudes on other moral issues; but, in the end, your attitude is just about as far from enthusiastic commendation as from outright condemnation. I agree that there is an important moral issue here, but I just have not bothered to form an attitude towards it. If our attitudes could be represented on some scale from commitment to abortion on demand's being a Very Good Thing to commitment to its being a Very Bad Thing, our positions might be assigned the same place near the middle. Yet no-one would want to say that there is no important difference between you and me. They, most of them, will insist on commending you and disapproving of me. Similarly, it would seem that there may be important differences between two belief-states, each of which can be represented as degree of belief 1/2 in some proposition, where one is the result of blank ignorance while the other is the result of possessing equal grounds for belief in the proposition as in its negation. We want to be able to claim that a preference for better-evidenced belief-states can be rationally supported, indeed is a factor in rational decision. Let us recall that Runyon's character The Sky lost most of his money as a young man through betting that St. Louis was the largest city on earth. Given the evidence of his experience such a bet was not unreasonable; but we feel that a rational gambler ought to have weighed up how extensive his experience had been before taking on such a bet.

So the problem from which I start is that considerations such as these suggest the need for a parameter other than probability or rational degree of belief as part of a proper characterisation of states of uncertainty. This is of course an old problem. §1 presents it in the form in which it appears in Popper's 'paradox of ideal evidence', in Keynes' 'weight of evidence', in Jonathan Cohen's Baconian development of Keynes' notion and in the 'epistemic reliability' idea advocated by Gärdenfors and Sahlin. In §2 I outline the consequences of admitting such a parameter — which are dangerous for any univocal interpretation of probability, but potentially deadly for a strongly subjectivist position. In particular, I outline the force of the concession such an admission is to one part of Cohen's elaborate network of pluralisms — that concerning the role of probability in inductive inference and in judicial decisionmaking. In §3 I change the subject: I present a summary of the positive programme I advocate for understanding probability — a quasi-realist version of univocal strong coherentist subjectivism (SCS). But here, too, a problem emerges. To carry this interpretation through requires reliance on the notion that resiliency (roughly, stability) of our judgments is a factor constraining the rationality of those judgments — one which, contrary to the theory, does not seem to emerge out of coherence. In §4 I examine ways in which subjectivism can incorporate the notion of second-order probability, subject to a generalisation of a coherence constraint. §5 establishes a connection between second-order probabilities and resiliency of first-order distributions. Then, it turns out in §6 that these connected ideas make sense of weight of evidence while defusing its threat to subjectivism. §7 represents a necessary digression; it may seem that introducing resiliency will lead to a concept of objective chance distinct from that of subjective credence: I examine and reject Lewis's (1980/1986) arguments for this claim. §8 then considers the import of this analysis for the relationship of subjective probability to induction. §9 claims, *contra Cohen,* that judicial decision does not rely upon probabilities which are not conformable to the standard mathematical calculus, but rather is to be understood in terms of second-order, as well as first-order, mathematical probabilities.

Much of this discussion will be programmatic, even sketchy. My intention is by no means to attempt either a complete or a compelling analysis of the concept of weight of evidence. To have pretensions of the former kind, an analysis would at least need to deal with the detailed questions raised in Cohen's (1985), which I do not tackle; and, while the suggestions I offer later for interpreting weight and higher-order probability are, I claim, intuitively plausible, I make no claim to establish them as compelling. Nor, especially, should this essay be seen as of a piece with typical expressions of Bayesian zealotry, where a siege-like defence of strong subjectivism so often issues in efforts at refutation of non-probabilistic accounts of confirmation such as Cohen's. For one thing, the internal cohesion and wealth of detail of Jonathan Cohen's system render ludicrous any notion of a quick direct refutation. But, more importantly, there is much in that system which both indicates the style, as well as sets the terms, of fruitful debate on these issues. Defenders of a univocal interpretation of probability should do so with a proper respect for the anti-dogmatism which Cohen has been at pains to urge, accepting his parallels with non-Euclidean geometry and his warnings about the folly of treating the formal calculus as some "preconceived norm", and with a recognition that the breadth of his concern with scientific, everyday and forensic reasoning draws up the problem-set (Wittgenstein's *Aufgabe*) for any comprehensive theory. Rather, this paper exemplifies a strategy which it accepts stands in need of wider application: in response to one (but an important) element in Cohen's system, it indicates how a univocal interpretation (strong subjectivism) can be, and must be, extended — not to reduce weight to probability, but to show that an analysis — even a Baconian one — of that concept does not necessitate a Baconianism about probability in general.

1. Recognition that there is a problem here goes back as least as far as Peirce: "to express the proper state of belief, not one number but two are requisite, the first depending on the inferred probability, the second on the amount of knowledge on which that probability is based" (1932, vol. 2. p. 421).

Keynes, too, in the course of developing his logical interpretation of probability, noted the existence of the difficulty, though, since he could not see any way to develop the concept of weight, he felt safe in declaring it to be of no practical consequence for action and so no threat to his account of the relation between evidence and the probability of an argument. He writes (1921, #9, p. 77).

> One argument has more *weight* than another if it is based upon a greater amount of relevant evidence. It has a greater *probability* than another if the balance in its favour, of what evidence there is, is greater than the balance in favour of the argument with which we compare it; ... As the relevant evidence at our disposal increases, the magnitude of the probability of the argument may either decrease or increase, according as the new knowledge strengthens the unfavourable or favourable evidence; but *something* seems to have increased in either case, — we have a more substantial basis upon which to rest our conclusion. I express this by saying that an accession of new evidence increases the *weight* of an argument. New evidence will sometimes decrease the probability of an argument, but it will always increase its "weight".

Popper, in an appendix to *The Logic of Scientific Discovery* written in 1958, puts the point by means of a forceful example (which derives from Peirce):

> Let z be a certain penny, and let a be the statement "the nth (as yet unobserved) toss of z will yield heads". Within the subjective theory [which Popper takes to include logical theories], it may be assumed that the absolute (or prior) probability of the statement a is equal to 1/2 ... Now let e be some *statistical evidence ... ideally favourable* to the hypothesis that z is strictly symmetrical ... We then have no other option concerning $P(a,e)$ than to assume that $P(a,e) = 1/2$. This means that the probability of tossing heads remains unchanged in the light of the evidence e; ... Now this is a little startling; for it means, more explicitly, that our so-called *"degree of rational belief"* in the hypothesis a, ought to be completely unaffected by the accumulated evidential knowledge, e; that the absence of any statistical evidence concerning z justifies precisely the same "degree of rational belief" as the weighty evidence of millions of observations which, *prima facie*, support or confirm or strengthen our belief. (pp. 407-8).

This, he claims, is paradoxical — the "paradox of ideal evidence".

In recent years, L. Jonathan Cohen (1977, 1985, 1986, 1989) has taken up Keynes' notion in order to draw supporting parallels with his own theory of inductive support and inductive probability. As in Keynes, the weight of an argument and its probability are independent of one another. For Cohen, probability — of whatever kind — is a generalisation of deductive inferability, the familiar mathematical "Pascalian" probability being the result of one set of ways in which to generalise. The weight of an argument is "the degree of derivability of an implicitly singular monadic probability judgment, such as the derivability of $p(Hs) = n$ from premisses $p(Hx|Ex) = n$ and Es", p here being Pascalian probability. Then $p(Hx|Ex) = n$ is necessarily equivalent to $p(not\text{-}Hx|Ex) = 1-n$ and $p(Hs) = n$ is necessarily equivalent to $p(not\text{-}Hs) = 1-n$. From this it follows that the derivability of $p(not\text{-}Hs) = 1-n$ from premiss $p(not\text{-}Hs|Ex) = 1-n$ and Es must be just the same as the derivability of $p(Hs) = n$ from premisses $p(Hx|Ex) = n$ and Es, even when both derivabilities are very low, or very high, and therefore not mutually complementary. Correspondingly the weights of the arguments from Es to Hs and from Es to $not\text{-}Hs$ must be reckoned the same. Hence the additivity of Pascalian probability in effect ensures the non-additivity of weight. Weight is therefore a non-Pascalian parameter of evidential strength (1989, p. 104).

Recently, too, what I might call the Lund school (Gärdenfors, Sahlin and others) have advanced models of decision situations as containing two components: an agent's set of subjective probability distributions and the epistemic reliabilities for that agent of those distributions. The simplest cases will be those where only two outcomes of each event are under consideration, so that the probability distributions can be described simply by the probability of one of the outcomes for each event. As an illustration (1982), they ask us to consider an agent whom they call "Miss Julie"[1], who is asked to bet on the outcomes of three different tennis matches. For each match, her partial beliefs are to be represented, not by a single subjective probability but by the set of all the measures over possible outcomes which are epistemically possible for her (i.e. are internally coherent and do not contradict items which she knows). In match A she is extremely well-informed about both players and predicts it will be a very even match with some minor factor determining the winner. In match B she knows nothing at all about either contestant. In match C the only thing she knows is that one of the contestants is an excellent player, the other a novice, but she has no idea which is which. If her beliefs were to be represented by a single probability measure — so that she had to choose one set of odds at which to bet on each match — then she would have to choose odds of 50:50 in each case. If they are represented instead by a set of epistemically possible measures, in each case it is clear that the mean of those measures will be 0.5. But, Gärdenfors and Sahlin argue, there is an element in each situation not captured by any such set of probability measures — the epistemic reliability of each measure. So, they say, "Miss Julie ascribes a much greater epistemic reliability to the probability distribution where each player has an equal chance of winning in match A where she knows a lot about the players than in match B where she knows nothing relevant about the players. In match C, where she knows that one player is superior to the other, but not which, the epistemically most reliable distributions are those where one player is certain to win." (p. 366). Epistemic reliability here is clearly the same parameter as we have been calling weight, or very closely related to it; at any rate, it differs from the probability measure.

2. One could multiply such examples of arguments for the need to employ a decision parameter other than probability and utility (perhaps the Levi-Shackle measure of potential surprise, or Carnap's requirement of total evidence, for example). But let us enquire instead into the consequences of admitting any such concept into an account of partial belief.

Keynes, as we have seen, thought weight did not matter. Popper, on the other hand, was convinced that, and how, it did:

> The fundamental postulate of the subjective theory is the postulate that degrees of the rationality of beliefs in the light of evidence exhibit linear order ... all attempts to solve the problem of the weight of evidence within the framework of the subjective theory proceed by introducing, in addition to probability, another measure of the rationality of belief in the light of evidence. Whether this measure is called "another dimension of probability", or "degree of reliability in the light of the evidence", or "weight of evidence" is quite irrelevant. What is relevant is the (implicit) admission that it is not possible to attribute linear order to degrees of the rationality of beliefs in the light of the evidence; that there may be more than one way in which evidence may affect the rationality of a belief. This admission is enough to overthrow the fundamental postulate on which the subjective theory is based. (1958, p. 407).

The force of Popper's point may be seen by considering Jeffrey's attemped resolution of the paradox (1965/1983, pp. 195-202). Jeffrey claims that, in focussing attention on the probability to be assigned to one toss, the puzzle misdirects attention away from the locus of difference between our prior and posterior belief-states: that the latter, given ideal evidence, must assign degree $(1/2)^n$ to any proposition that all of n tosses will yield heads ($n > 1$), but the former may assign a different value to such conjunctions — indeed, it must do so if we are to hope to learn from experience. "Any decision about the rate at which you hope to learn from experience", he goes on, "corresponds to same intitial set of estimates about where the objective probability is likely to lie". A committed subjectivist can achieve the same results, he argues, if for "objective probability" is substituted "the subjective probability measure that in fact we should all come to in the end". But this reply will not do. The reliance on our hoping to learn from experience is question-begging:

the point of the paradox is that a single-parameter theory cannot account for such hopes, since it takes the effect of evidence upon us to be exhausted by its effects on the measure of that parameter. Calling an additional parameter, instead of "weight", "objective probability" or "the subjective probability we should all in fact come to in the end" is no help either (if that *in fact* implies objective convergence): again, the point is that respect for evidence is being accounted for in some way independent of the ground-level degrees attaching to our beliefs and consequently a constraint on rationality of probability-judgments must be admitted which does not emerge from constraints on the internal consistency of those beliefs.

It may be possible to evade the problem in a manner suggested by the Lund school, treating desired level of epistemic reliability as a measure of the risk preference of an agent. This measure will not then function as an additional probability measure; but it will become necessary to define utility partially with reference to epistemic reliability. In effect, the problem is cleared from the realm of judgment by being shunted into that of decision. But this is a very unhappy shift. Firstly, however weight or reliability is to be explained, it is undoubtedly an epistemic concept, and attempts to account for it or measure it in terms of non-epistemic criteria must be suspect: risk preferences may measure an agent's motives, but surely the import of evidence is primarily a matter of reasons for belief, not motives for action. Secondly, there are strategic grounds for wanting to avoid such a move: one of the main merits of the de Finettian subjectivism which I support is that, by contrast with preference theories, the complete explication of probability which it promises allows us safely to leave utility as an undefined primitive term until probability has been defined. If we admit that considerations about weight impose some structure on utility much of that advantage is dissipated.

The most damaging consequences of all, though, emerge from Jonathan Cohen's position. In bald outline (which does no justice at all to the complexity and persuasiveness of Cohen's theory) the argument runs as follows. Cohen begins by finding what he calls "monocriterial" accounts of probability (e.g. a pure frequency theory, or strong subjectivism) defective rather than wrong — applicable to some types of judgment only; equally, though, casual pluralist tolerance which fails to explain why such diversity emerges from the applications of one concept rather than many is an unsatisfactory dodging of the main task of theorising, in philosophy as it would be in science. He finds the unity of the concept in its resemblance to provability: probability is a measure of partial inferability, an indicator of the degree to which inferences of some sorts are licensed. Then, to quote Cohen's most recent work (1979):

154

> Probability-judgments may be either necessary or contingent. Probability may be viewed as a function either of terms or of propositions. Statements of probability may be either extensional or non-extensional. They may be either general or implicitly singular, and either counterfactualisable or non-counterfactualisable. Differing analyses of probability are made possible by the variety of possible combinations of such options that they can exploit. So all this pluralism may be viewed as operating within the framework that is constituted by the unifying conception of probability as a gradation of inferability (p. 112).

We have here a pluralism about probability interpretation which is, so to speak, internal to the standard "Pascalian" calculus. But Cohen's position involves a more radical external pluralism too. The partial provability framework generates non-Pascalian probabilities too, if the generalisation is from incomplete rather than complete deductive systems[2]: for such "Baconian" or "inductive" probabilities will not conform to a complementational principle for negation or a multiplication rule for conjunction. Weight of evidence, Cohen argues, has just such a Baconian structure. Its importance lies in the contribution it makes to demonstrating that Baconian systems are not just a theoretical possibility but rather the most natural way to understand important patterns of non-deductive reasoning. Cohen concentrates on two such areas. Firstly, he develops a theory of inductive support which takes it to be a gradation of the legisimilitude (closeness to being a natural law) of universally quantified conditionals, parallel to the gradation of their substitution-instances by inductive probability — which is, or at least closely resembles, weight of evidence. So weight here is crucial to the claim that neither inductive support nor inductive probability can be reduced to functions of mathematical, Pascalian, probabilities. Secondly, in one important and reasonably rigorous framework, that of judicial decision, the principle of conformity to the standard calculus will not fit the mode of reasoning which juries are enjoined to pursue, which instead conforms to a Baconian structure. While in some inductive contexts it may be the case that "... weight-orientated modes of reasoning are not intrinsically in any kind of conflict with probabilistic ones but can serve to complement them" (1985, p. 277), the verdicts of triers of fact are, and ought to be, supported by Baconian assessments of evidential weight which render Pascalian probabilities of ultimate issues otiose.

3. Let us think instead for a while about the successes of one univocal approach to probability, deriving from the work of Ramsay and de Finetti — strong coherentist subjectivism (SCS).

Probabilities are the (sets of) degrees of belief — measurable, at least in principle, by idealized betting quotients — which an agent can hold coherently with his or her other beliefs. There is no other constraint than internal coherence; in general, the probability distributions of coherent agents need not coincide. A set of bets is coherent if no subset of it can be selected which will give rise to a Dutch book, that is, a betting arrangement where the agent is bound to lose overall no matter what the actual outcomes of the events bet on. The requirement of coherence extends across not only the agent's simple, or monadic, probability judgments, but also dyadic judgments of the probability of one event given that another has occurred (or occurs — this relation may be synchronic or diachronic). Such dyadic judgments are what are usually called "conditional probabilities," though I prefer to avoid that term because of the temptation to identify, dubiously $p(Q|P)$ with $p(P \to Q)$. Dyadic probabilities are defined in terms of the decision procedures for setting bets upon them, procedures representable by payoff tables which need not correspond to the truth-tables for any particular logical connective. If one is betting on a complete class of incompatible events, it is a necessary and sufficient condition for coherence that the sum of one's betting-quotients be 1. A set of dyadic evaluations, $p(E_i|E_j)$ is coherent iff for all i and j, $p(E_i|E_j) = p(E_i$ and $E_j)/p(E_j)$ — (de Finetti, 1974). Note that this is a theorem of the interpreted calculus, not a definition of $p(E_i|E_J)$ — that is accomplished via payoff tables. Coherence is not, though de Finetti thought it was, a normative demand. It is a regulative ideal of the same nature as deductive consistency; both derive their force because, when we codify certain areas of actual untutored reasoning, we find that they arise as regularities of competence in our practice as we strive for certain ends.

How can rational beliefs rationally change?

Suppose that at time zero you hold a set of beliefs so distributed as to be both monadically coherent and dyadically coherent for all subsets of those beliefs. And suppose that at time 1 you come to know for certain that event E_j, to which occurrence you at time zero attached probability $p(E_j)$, has in fact occurred. Your distribution will remain coherent if you now alter all your judgments $p(E_i)$ to the value at time zero you attached to $p(E_i|E_j)$, and, of course, make the corresponding alterations to your dyadic judgments. More generally, if at time 1 you acquire evidence which alters $p(E_j)$, without making you certain of it, you will remain coherent if you alter all your judgments $p(E_i)$ in accordance with a generalised version of Jeffrey's rule (1965), conditionalising, in effect, by means of a weighted average of conditionalising on E_j and on not E_j, with the weights derived from the ratio of $p_1(E_j)$ to $p_1(not\text{-}E_j)$.

Coherence alone, then, requires for rational belief change that it operate in accord with these rules of Bayesian conditionalisation. Crucially, this constraint can be shown to ensure that, as evidence accrues, the posterior distributions obtained from widely divergent prior distribution will converge. This furnishes a wholly subjective explanation of widespread intersubjective consensus about a large range of probabilities, defusing the worry that the coherence constraint is implausibly tolerant. Finally, suppose we introduce one more subjective concept, exchangeability: informally, we define a set of events as exchangeable, for some agent, if any coherent probability that agent can assign that h of them occur will depend only on h and not on which events are chosen. Two results follow. Firstly, any coherent assignment of probabilities to a sequence of exchangeable events can be modelled by a sequence of Bernoulli trials with fixed probability — what an objectivist would call independent events. Secondly, wherever a posterior distribution across exchangeable events is arrived at by Bayesian conditionalisation on the outcomes observed of trials of those events, coherence requires that the posterior distribution for successes of all trials will have as its mode the observed "success frequency": the greater the number of observations, the closer to the mode our probability-values must be.

But why change one's beliefs at all? If coherence is the only constraint on the rationality of belief why not simply ignore the evidence and retain one's prior distribution? The analysis so far has offered no reason to prefer opinions conditionalised upon a great deal of evidence to those, no matter how extreme, which have not yet been much conditionalised. It offers no reason to prefer to judgments which are, we might say, on a rational track those which have gone further along that track.

A partial answer can be given in terms of resiliency. The onus is on the radical subjectivist to explain how our preference for more heavily conditionalised judgments will precipitate out of coherence, rather than representing a hard-core realist preference for "truer" approximations to a unique unknown correct probability-value.

We can identify the preference more precisely as a preference for judgments with higher resiliency over larger domains and with smaller scope. Suppose we have conditionalised $p(H)$ over a domain of sentences $(P_1, P_2, P_3, \dots)$. Resiliency is the complement of the maximum "wiggle" of $p(H|P_i)$ about some value α: resiliency of $p(H)$'s being α over $(P_1, P_2, \dots)$ = $= 1-\max|\alpha - p(H|P_1)|$.[3] If we consider the monadic probability $p_t(H)$ at time t as having derived via conditionalisation over the P-domain from an initial monadic $p_o(H)$, high resiliency over the P-domain will represent a great degree of stability of $p(H)$ during that conditionalisation.

But that is not quite what is represented by our practical preference. What interests us is whether we can expect our judgments to remain stable over possible further conditionalisations over some further set of sentences $(Q_1, Q_2, ...)$ — call that the scope of $p(H)$. And that depends on the extent, to our judgment, to which the P-domain has incorporated the features causally relevant to H, leaving few such factors "untested" in the scope.

This analysis offers a characterisation wholly in terms of subjective probabilities of our stability preferences. They do not threaten subjectivism; but do they threaten the coherentist element of SCS? Are we not admitting here a rationality constraint additional to that of coherence? What is left dangling here is the justification for that psychological preference: and what that comes down to is the question of the justification of our inductive habits, of our expectations of stability within our projections. And this, alas, is just the problem of weight of evidence all over again.

4. Direct consideration of the concept of weight suggests that subjectivism is mistaken and incomplete; and our best efforts at developing subjectivism do not seem to yield a full explanation within the theory of the effects of evidence on our judgments. What can be done about this?

Let me raise what may seem to be a quite unconnected question: can SCS be expanded to incorporate second-order personal probabilities? Given that I can have beliefs about my beliefs, and those beliefs can come by degrees, can I have partial beliefs about my partial beliefs which, under some constraints, can be treated as probabilities? And could that constraint be no more than a generalisation of coherence?

Subjectivists have traditionally fought shy of the notion of higher-order personal probability, usually on philosophical grounds rather than mathematical: there is no mathematical difficulty in treating probability as itself a random variable. De Finetti seems[4] to take this line. In his (1972), he writes: "Any assertion concerning probabilities of events is merely the expression of somebody's opinion and not itself an event. There is no meaning, therefore, in asking whether such an assertion is ... more or less probable".

And in his (1974) he claims that talk of probabilities of probabilities resembles, when conducting a survey, recording not only the "don't knows" but those who don't know if they are "don't knows", which is patently absurd. Such rapid dismissal of higher-order probabilities as meaningless is plainly untenable. Suppose that one day in three, at random, I become so bored with assessing probabilities that I take the outcomes of all binary guesses as equally probable. You ask me to take up an attitude to some future event — say, that tomorrow's noon temperature will be over 120°F — which, on my unbored days, I would regard as vanishingly unlikely; and you do not allow me time to decide, whether by introspection or otherwise, which type of day today is. Cannot I now meaningfully, even justifiably, assert that the probability that my personal probability that tomorrow's noon temperature will be over 120°F is 1/2, is 1/3?

Perhaps, though, there are coherent higher-order probabilities but they are always trivial: they take on the values zero or one only (so the example I've just given is a meaningful judgment, but could never form part of a coherent set of judgments). A bad argument for such triviality is that partial beliefs are introspectively transparent — you always can be certain whether you have them or not. If partial beliefs are dispositions to act in particular way — say, to bet at some odds rather than others — it is surely entirely possible that you do not know your own mind with certainty.[5] There is, however, a much better argument which emerges out of the nature of coherence. A coherent set of bets is a distribution of risks which cannot encounter a Dutch book. If you are challenged to bet that your bets are wrong, you must put the odds on this at zero. If you did not, you could always avoid a Dutch book by laying off your bets against yourself: so, if coherence is to constrain judgments at all, bets on the rightness of bets must be avoided — that is, probabilities of probabilities must vanish into the limiting values of 0 and 1. A slightly different version of this argument appears in van Fraassen's (1984) paper *Belief and the Will*. If you assign any value other than 0 to the probability that your future probability of A will differ from your current dyadic probability of A given that future value, you will be incoherent — a Dutch strategy can be set up against you. Considerations such as these give rise to the natural worry that second-order personal probabilities must simply collapse onto first-order ones.

There must, in fact, be a constraint upon second-order probabilities if paradoxical results are to be avoided: a generalisation of coherence so that second-order dyadic probabilities will be coherent with the first-order probabilities forming the second term of the dyad. (Just because they are probabilities, of course, an agent's set of monadic and dyadic second-order partial beliefs must be internally coherent, as must his or her first-order beliefs.) Skyrms (1980a) demonstrates how this generalisation can be accomplished. Suppose we have a language L_1 comprising non-probabilistic propositions p, q, r ... on which is defined a probability measure p_1; extend L_1 to L_2 by adding propositions of the form $p_1(r) = n$ or $p_1(r) \in I$, I being a subinterval of $[0, 1]$, and defining a measure p_2 over the propositions[6] of L_2. To avoid immediate paradox we must require that:

$p_2(r|p_1(r) = n) = n$ or, for interval probabilities,
$p_2(r|p_1(r) \in I) \in I$.

The dyadic second-order probabilities do indeed collapse onto first-order monadic ones. But this does not entail that second-order monadic probabilities will similarly collapse.

Applying Bayes' theorem[7] to $p_2(r|p_1(r) = n)$ yields:

$p_2(r|p_1(r) = n) = p_2(p_1(r) = n|r) \times p_2(r)/p_2(p_1(r) = n) = n.$

Treating $p_2(r)$ as identical to $p_2(p_1(r) = n)$ — see fn 6 — this gives us a definition of the intuitively obscure $p_2(p_1(r) = n|r)$ as identical to $p_2(r|p_1(r) = n)$. The monadic second-order probability, which we are expressing indifferently as $p_2(r)$ or $p_2(p_1(r) = n)$, is not however determined by the first-order values, although any dyadic probabilities formed from it must satisfy the generalised coherence constraint.

So second-order subjective probabilities are conceptually legitimate and non-trivial. But how can they be determined? Once an agent has accepted a book of bets which measures his or her first-order judgments, what further possibilities exist which would enable us to measure the second-order jugments?

One way to respond to this question would be to query or reject its operationalist presuppositions. Skyrms (1980b, p. 115) complains that one "need not be so rigidly operationist as to assume that the *only* way that one can gain evidence for a degree of belief is by making a wager". Second-order probabilities are a refinement of the intuitive notion of attitudes towards first-order odds, and function "as theoretical parts of an imperfect but useful psychological model, rather then as concepts given a strict operational definition". Another, somewhat less Pilate-like response, is that of Mellor (1980), who treats first-order beliefs as unconscious, inarticulate dispositions measurable in principle by bets while second-order beliefs are inner states of conscious assent determinable by introspection. But we can surely use the notion of coherence to do rather better than either of these responses. The collapsing coherence constraint on dyadic second-order probabilities of the type considered above means that they can readily be determined by combining payoff tables for appropriate first-order bets. The problem is how to get at the detached monadic judgments $p_2(r)$, i.e. $p_2(p_1(r) = n)$ for some n, and at dyadic judgments such as $p_2(p_1(r) = n|s)$ where s contains no probability-assignments but may be evidence bearing on $p_1(r)$. I shall return to this latter case at the end of §6. The former can be resolved at once by noting that an agent's monadic p_2 distribution must be internally coherent too, but that what corresponds here to the gain/loss outcomes of a simple bet is finding oneself in a bet at particular odds. Suppose an agent can choose to enter bets on a binary-outcome (r/not-r) single event at a finite[8] range of odds $p_1(r) = \alpha_1, p_1(r) = \alpha_2, ..., p_1(r) = \alpha_k$. Then $p_2(p_1(r) = \alpha_i)$ represents the odds that agent can choose, coherently with the other bets entered — and not necessarily uniquely determined — in a bet where the favourable outcome is that the agent can now bet on r at odds α_i, the unfavourable outcome being that no such bet is open. This is equivalent to the fraction of a unit stake which the agent can, coherently with other wagers, put up to buy entry to that p_1 first-order bet.

As an illustration, consider a restricted version of Miss Julie's decision situation. Making some obvious approximations, and assuming that in each case the only p_1-values open are 0, 0.5 and 1, one natural — but not the only rational — way for her to distribute her p_2-values would be: in match A, allocate her whole stake of 1 to $p_1 = 0.5$, nothing to 0 or 1; in match B, she could allocate one third of her stake each to the p_1-values 0, 0.5 and 1; or she could allocate her stake in some other coherent way; in match C, she could allocate half of her stake to $p_1 = 0$, half to $p_1 = 1$ and nothing to $p_1 = 0.5$; but we would expect her to be unhappy about betting here at any odds, if it can be avoided.
Clearly, further analysis is needed here; §6 will reconsider this case.

5. How might second-order probabilities connect with the resiliency of first-order conditionalised probabilities? We should expect there to be a link: intuitively, as evidence is conditionalised upon, it not only affects the odds but also concentrates our attitudes to particular odds. What we would expect is that for a given domain, the higher the resiliency of a first-order distribution the more "concentrated" is any second-order distribution coherent with it. That is to say, the second-order distributions coherent with poorly resilient first-order distributions will be "flat" (i.e. uniform) while those coherent with highly resilient ones will have high second-order probabilities attaching to a few points and almost zero probabilities at other values. Our practical preferences then represent a preference for more concentrated second-order distributions which will stably remain so over further conditionalisation on any interesting scope of $p(H)$. That is psychologically very plausible. If we take something on which we have no decided views at all — say, the content of next year's brithday present, or on which we should have no views — the guilt of the defendant at the start of a case where we are jurors — our state can be described as one of no particular commitment to any ground-level probability, no strong attitude to any particular set of odds.

Clearly, we do in general find a state of suspended judgment an uncomfortable one; we value the greater concentration deriving from greater stability because it gives us a commitment to stand by in reasoning or acting.

This is, of course, a very rough and ready account of the matter. It is important, I believe, not to become too besotted with the desire for numerical precision here: it would be tempting to try to present these notions in such a way that p_2 values represented Neyman-type levels of confidence in interval estimation of p_1-values, for example — but then our account would be applicable only to situations where sampling methods were appropriate. What can be done is to clarify our general intuitions and, in the process, delineate the limits of the cases where precise calculation is possible.

Suppose that we are evaluating a single L_1, proposition r and that the domain of resiliency of $p_1(r)$'s being α is a sequential feedback set $(P_1, P_1, \& P_2, P_1 \& P_2 \& P_3...)$ fixed and identical for each α in $[0,1]$. Suppose that the uniform coherent p_2-distribution gives, for all α, a value of β for $p_2(p_1(r) = \alpha)$. Then I suggest that one plausible representation of the intuitions of the previous paragraphs is to take $p_2(p_1(r) = \alpha) = k \times \beta \times \text{res}$ $p_1(r)$'s being α, over the fixed domain, for all α. The constant of proportionality k here is determined by the coherence constraint upon p_2 which requires that the sum of all the permissible p_2-values be 1.

In most decision situations this suggestion will not produce precise p_2-values. The resiliencies may not be well-defined or may be indeterminate for some α's. It may be more appropriate to conditionalise over different domains for different α's. Conditionalisation may be identifiable only as some vague constraint on the direction of change from prior to posterior distribution. Caveats such as these will apply to most everyday and forensic decisions. But, as I have indicated, such imprecision is to be welcomed here: in most situations first-order probabilities are similarly difficult to assess, and one ought to be suspicious of any theory which assumed a greater ease of assessment for second-order judgments.

The cases where greater precision is possible are those where the domain of resiliency comprises linear combinations of exchangeable first-order L_1 propositions. This is, of course, typically the situation which an objectivist would describe as one of repeated independent trials of an event. In such cases resiliencies about the possible α values approximate smoothly increasing or decreasing functions of the domain size. The convergence result presented in §3 generalises into a version of Savage's famous (1954) theorem, that as conditionalisation is based on more and more evidence, for exchangeable events, the probability that the probability of the hypothesis' truth will exceed α approaches 1, for all $\alpha < 1$. Generalised, this becomes: increasing resiliency, given first-order exchangeability, tends to concentrate p_2 onto fewer p_1-values and the p_2 of these values approaches as close as coherence permits to 1. (Normally p_2 will concentrate onto a value near 1 attaching to one p_1-value). Where the domain of resiliency includes feedback of information about the relative frequency of forecast success, this result will yield a solution to the problems of calibration of subjective probabilities: but I cannot pursue that issue here.

6. There is one blatant omission from the previous section's analysis of second-order probabilities: I have so far (deliberately, for the sake of presentation) ignored the point made earlier that our preferences for better evidenced judgments can only be explained by reference to the scope, as well as the domain, of resiliencies. Stability is not just a matter of how much conditionalisation "wiggle" has already taken place but of whether we expect any relatively sizable further wiggle over the scope propositions. Once again, a precise evaluation of this factor will be possible only in the exchangeable cases; there, since order of conditionalisation does not matter and wiggle approximates to a monotonically decreasing function of domain size, we can evaluate (via a law of large numbers) the precise effect of domain size on p_2-values. In general, though, the best we can do will be to estimate, based on our judgments about relevant causal or projectible factors, the weighting to attach to the domain-scope relationship — a matter of pragmatic judgment as to how easy or difficult it would be for our judgments to be much altered by new evidence.

Avoiding spurious attempts at precision, then, my full proposal for assigning second-order probabilities, with the previous constraints and an added factor w $(0 < w < 1)$ to represent what I shall call the *weighting* an agent attaches to domain versus scope, is: $p_2(p_1(r) = \alpha) = = k \times \beta \times$ res $p_1(r)$'s being $\alpha \times w$.

Let us look at how this might work out in the pared down Miss Julie example from §4. Here, r is the proposition, "the first player wins"; $p_1(r)$ can take only the values 0, 0.5 or 1. The uniform $p_2(p_1(r) = \alpha)$ distribution likewise contains only three points, so $\beta = 0.\dot{3}$. In match A, Miss Julie's well-informedness about the even matching of the players can be represented as very high resiliency ($\simeq 1$) of $p_1(r)$ about 0.5 when conditionalised over the sequential feedback set of the evidence available to her: likewise, res $p_1(r)$ about 0 and about 1 will each be close to 0. Fullness of information implies a value of w near to 1. Then

$$p_2(p_1(r) = 0) \simeq k \times 0.\dot{3} \times 0 \times 1 = 0;$$
$$p_2(p_1(r) = 0.5) \simeq k \times 0.\dot{3} \times 1 = 1;$$
$$p_2(p_1(r) = 1) \simeq k \times 0.\dot{3} \times 0 \times 1 = 0.$$

(k must here be taken as 3 to ensure $p_2(r)$ coherence).

In match B, her total ignorance can be represented by taking both res $p_1(r)$ and w as indeterminate for all α: consequently, she may rationally adopt any coherent p_2-values — the uniform distribution is most natural, but it is not rationally compelled; which is just what should be expected from a coherence constraint in the absence of conditionalisation.

In match C, the domain of resiliency contains only the two propositions "the first player is an expert" and "the first player is a novice"; w will be rather low. Res $p_1(r)$ about 0.5 and about 0 and 1 will be indeterminate. (Because the sequential feedback set immediately contains a contradiction, the conditional probabilities are undefined: the only odds which can be chosen to generate a non-negative payoff are 0/0. So the resiliency is indeterminate.) So, just as in B, any coherent distribution of p_2-values is open to her. Here, I am in conflict with the Lund analysis. On my view a piece of evidence which is worthless for conditionalisation should not alter one's p_2-values from what they were in a state of complete ignorance. It can only seem that they should if p_1 is taken to be an objective probability. Taking it instead as rational degree of belief, the evidence available to Miss Julie here gives her no reason to change her attitudes to any bets offered; she should continue to have no more confidence in any one bet than any other, if that was her previous position. This offers both a clarification and a refinement of our earlier intuitions.

So the analysis offered here accounts for weight of evidence not as a parameter independent of probability but not, either, as determined wholly by first-order probabilities: rather, it is a function of second-order probabilities or, which is the same, of the resiliency of first-order probabilities and domain/scope weighting. Thus, the weight of evidence for $p_1(r)$ being $\alpha = f[p_2(p_1(r) = \alpha), \beta] = g[\text{res } p_1(r)\text{'s being } \alpha, w]$; one plausible candidate for f would be the variance of $p_2(p_1(r) = \alpha)$ about β. The choice of function is relatively unimportant: the issue is whether this analysis can deal with the arguments of Popper, the Lund schoool and Cohen suggesting that weight could not be a function of probabilities.

Popper's "paradox" disappears once we appreciate that the posterior second-order distribution for his coin differs from its prior, even though the first-order distributions are identical. It is perfectly correct to claim, as Popper does, that the introduction of p_2-values means that degrees of rational beliefs in the light of evidence can no longer all be linearly ordered. But that would be a problem for subjectivism only if coherence entailed linear order and the generalisation of it to p_2 conditional on p_1 shows that it does not.

The Lund school offer some support, in fact, to my account, since they concede (1982) that epistemic reliability can be interpeted as second-order probability. As I have argued, such a step is preferable to taking weight into risk preferences. The main difference between their analysis and this one is that they seem to wish, implausibly and unnecessarily, to treat epistemic reliability as objective rather than part of an agent's judgment.

Cohen's internal pluralism loses one of its chief motivations if weight can be accounted for within a univocally subjectivist theory: an important plank is removed from the platform supporting identifying probability with partial inferability. Insofar as that platform supports his external pluralism that, too, is weakened — but, as a later section will suggest, only in its claim that induction should be understood in terms of non-Pascalian inductive probabilities, not in its claim that induction involves factors other than Pascalian probability. The account given here of weight, if successful, undermines one argument for the central partial provability notion of Cohen's system; that notion is intimately bound up with the argument that weights must be non-additive — indeed, were it not for Cohen's independent arguments for the central tenet, his argument would move in a circle from probability as partial inferability to weight as non-additive to weight as inferability in incomplete systems to probability as partial inferability. I cannot consider those other arguments here. My point is, however, that the subjectivist analysis of weight as a function of second-order probabilities given here prevents the concept of weight from offering separate support to the Cohen schema. Weight is not shown to be "Baconian" probability; but this still leaves open the prospect that the structure of the concept, in some contexts at least, can be analysed in something like a Baconian way — for it is not simply a function of Pascalian first-order probabilities only.

Finally, to tidy up an issue left dangling from §4: occasionally we are interested in establishing the weight of evidence for $p_1(r)$ being α given s, where s is a proposition neither in the domain nor scope of res $p_1(r)$ about α. There is no difficulty in carrying through a similar analysis for conditional resiliency (since no paradoxes loom here), thus: the weight of evidence for $p_1(r)$ being α, given $s = f[p_2(p_1(r|s) = \alpha), \beta] = g[\text{res } p_1(r|s)$'s being $\alpha, w]$.

7. Although I have utilised here some of the ideas of 'probability kinematics' theorists (particularly Jeffrey, Skyrms and Lewis) I diverge

from them in one crucial respect: I do not accept that the notion of resiliency under conditionalisation generates a notion of objective chance. That, though, is a natural as well as influential worry, one worth a digression to explore and assuage. I take Lewis's (1980/1986) article *A Subjectivist's Guide to Objective Chance* as a comprehensive statement of the view I oppose[9].

Lewis attempts to capture our fundamental intuitions about chance within a " Principal Principle" which identifies chance at a world with the unique degree of belief deriving from any reasonable initial degree of belief conditionalised on the complete history and theory of chance at that world.

Starting from the notion of chance as objectified credence, the strategy is to characterise admissibility in such a way that H_{tw} (the complete history of world w up to time t) and T_w (the complete "theory of chance" for w) are admissible evidence upon which to conditionalise; the conjunction $H_{tw}T_w$ will be partitionable, and chance P_{tw} will be the uniquely determined objectified credences obtained by conditionalising on the cell of the history-theory partition which is true at w and t. This analysis will carry us as far as possible towards an account of present chances as supervenient upon history up till now and the totality of history-to-chance conditionals.

Consider first the interplay within it of the notions of resiliency and admissibility. Resiliency, as we have seen, is a measure of invariance of credence functions under further conditionalisation. It behaves as an idiolect of consensus: the invariance resembles the way in which divergent prior distributions are brought towards agreement when each is conditionalised upon agreed items of evidence. But, just as it is always possible for a new item of evidence to emerge which, conditionalised upon, will radically alter the distributions, so the invariance of highly resilient credence functions is contingent upon the set forming the basis of further conditionalisation — the domain of resiliency — not containing some radically transforming item.

If objectified credences are to be uniquely determined chances, their domain of resiliency must be the maximum possible — it must be the case that no further item of evidence could emerge which would alter these credences and that no alternative partition of the domain could yield variation of credence. The former could be achieved by conditionalising on the whole of history, or at least, on the complete causal description of the event which is the object of the credence: the chance of the event would then be its credence conditional on the complete description of the whole backwards light cone of that event. However, even that is not enough to satisfy the second requirement. Credences maximally conditionalised in this way could still yield different objectifications if H_{tw} remained open to different partitionings. Lewis hopes to prevent this by adding to the domain of resiliency T_w, the complete set of history-to-chance conditionals at world w (pp. 95-6). Then, since H_{tw} and T_w together link chances uniquely to history, however the giant conjunction $H_t T_w$ is partitioned, any reasonable credence conditional on the cell of it true at t and w will have the same value — the chance P_{tw}.

But the problem is that this latter manoeuvre will only have this effect if it imports into the analysis a realist construal of T_w, treating it not as a theory which might be "an incredible miscellany of unrelated propositions about chance" (Lewis, p. 96) but as the conjunction of all history-to-chance conditionals that *hold* at w, as the specification of the truth about chance — as Lewis immediately does (p. 97). This slide from all possible history-to-chance conditionals to those true at w is a slide form contingency into necessity of the relation of chance to ontologically complete evidence, and thus into surreptitious realism about T_w. Lewis's analysis presupposes the objectivism it hopes to establish.

8. Enthusiastic proponents of a subjectivist interpretation of probability have often claimed that its account of consensus arising out of Bayesian conditionalisation will dissolve the practical and sceptical problems of induction. Thus, de Finetti writes (1974 vol. II, p. 202): "In our formulation, the problem of induction is, in fact, no longer a problem ... Everything reduces to the notion of conditional probability". And Dorling has argued (1981) that the theory provides us with a rational recipe for modifying our pre-existing expectations about the future in the light of incoming evidence — in other words, a direct answer, granted only that we do not expect inductive inference to yield certain knowledge, to the practical problem of induction. With the Bayesian method in our armoury, the traditional sceptical arguments are bound to strike us as misguided. Such Bayesian enthusiasm overlooks much which an adequate account of inductive practice ought to include and which the conditionalisation model certainly does not: the value of variety of evidence, questions of relevance, the relation of theory to observable evidence and so on. More importantly, it mistakes the fact that de Finetti's representation theorems create a central place for frequencies within a subjectivist model for a justification of our inductive habits as arising out of no other constraint than coherence.

The analysis which I have developed here should, I hope, demonstrate that such enthusiasm is misplaced. Coherence requires conditionalisation; conditionalisation yields highly resilient distributions in exchangeable contexts and there concentrates second-order probability onto observed frequency. But, as we have seen, even to get as far as accounting for our respect for weight of evidence without threatening to bring a wholly independent parameter to bear on decision has required a number of assumptions which are plausible in a description of our inductive habits but in no way displace induction from its central place in our inferences. The generalisation of coherence I offered allowed me to advocate as plausible taking p_2-values as functions of resiliencies of p_1's — but not to establish it as compelling; and, crucially, we cannot avoid in characterising our preferences for better evidenced judgments introducing a w-factor representing our judgment of how little resiliencies might alter if the untested scope were to be tested. The inductive habit is just the preference for as high a value of w as possible, achieved by increasing the domain size and/or, as far as we can pragmatically determine, ensuring the absence from the scope of factors causally relevant to the conditionalisations. My account here offers standards of proper projection, emerging from coherence alone, which account for rational consensus in probability assesment. But it does not, and does not attempt to, *justify* the inductive habit, the preference for evidence increase. What it does do is to show that weight of evidence can be accounted for as a function of Pascalian probabilities and of w, so that understanding and justifying our respect for evidence does not require the introduction of non-Pascalian measures of belief but does demand a justification of our w-preferences: which is precisely the problem of induction.

Nothing I have said here sets limits to the form such a justification might take, whether anti-sceptical, naturalist or whatever. Nor does it constrain the analysis of our w-preference criteria: it may be perfectly possible to transpose into it, more or less wholesale, Cohen's Baconianism, method of relevant variables, gradation of legisimilitude and all. None of that need threaten the claim that SCS provides a complete univocal interpretation of probability and that all probabilities are Pascalian. Cohen's arguments demonstrate that the whole-hog Bayesian approach to induction is mistaken; a mistake, I believe, arising out of an over-zealous effort at protecting the univocal character of subjectivism. But it would be equally over-zealous in the opposite direction to extend a non-probabilistic account of induction into an attack on the claim that Pascalian probabilities fill the whole space of judgments which can properly be termed "probable".

9. It follows from this that subjectivist capacity to tolerate Baconianism in inductive logic will extend to the judicial context only if its pretensions there are localised: Cohen's claim that Baconian probability there displaces Pascalian must be resisted. It would be impossible to do justice

168

here to the array of arguments Cohen has assembled to support that claim; all that can be attempted is to outline how this analysis of p_2, resiliency and weight might apply to some of the main points. (Of course some counter-arguments to Cohen need not rely on these notions — they might be based on consequentialist considerations, for example).

Cohen's general claim is that, while mathematical probabilities enter into judicial processes in vavious ways (e.g. in the presentation of statistical evidence), they do not bear upon juridical proof where it relates directly or immediately to ultimate issues, neither do they, nor should they, constitute the criminal or civil standard of proof — all of which are a matter for Baconian judgment. The notion of weight is central to this claim: he argues that trials, in an adversarial system, should be seen as contests of case-weight rather than attempts to persuade the trier of fact to allocate, on a basis of Pascalian probability, some measure of case-merit. It follows that neither of the two main standards of proof in Anglo-American law — proof beyond reasonable doubt and proof on the balance of probabilities — can be interpreted as given by mathematical probability (henceforth, p_m) without serious anomalies emerging. In his (1977) he put forward (apart from problems about inference upon inference which it seems he no longer wishes to rely upon) five such problems, which I shall consider in turn:
(a) that none of the generally accepted criteria of p_m could be applicable to evaluating juridical proof;
(b) that Pascalian explanations of the effect of testimonial corroboration involve, inappropriately, an assignment of positive prior probabilities;
(c) that proof beyond reasonable doubt cannot just be proof to a high level of p_m;
(d) that there is a difficulty with the conjunction rule for p_m's in civil cases;
(e) that there is a difficulty with the negation rule for p_m's in civil cases.

(a) For the strong subjectivist, the only criterion of p_m is, ultimately, rational betting. Cohen complained in (1977) that that is inappropriate for two reasons: forensic matters deal with past events where all relevant evidence is assumed to be in, so bets would be unsettlable, and, since stake size generally affects betting behaviour, we cannot make much of stakeless betting. He now [post (1986)] seems to accept that these difficulties may be overcome, though at the cost of demanding extremely sophisticated introspection from jurors as to how they would bet in imagined circumstances. The real problem, he now argues, is that "on a subjectivist interpretation each person's judgment of the probability of a particular proposition on given evidence does no more than describe that person's state of mind" — when, of course, it should describe the case-strength. But only a very naive subjectivism will fall foul of this criticism; as we have seen, SCS does establish projective standards of rationality for partial beliefs such that to claim that some proposition is probable is not to describe a belief but to express it, with the implicit claim that it meets such standards and with a commitment to its corrigibility if it does not. Furthermore, with the extension of coherence proposed here SCS now includes under such standards p_2 as well as p_1, and judgments of weight of evidence and resiliency: the leading idea is that we now have available p_2 criteria as well as p_1 ones.

(b) Cohen is right in claiming that testimonial corroboration will not raise the probability of a particular verdict unless non-zero prior probability of that verdict is admitted, and he would be right to protest that such a non-zero prior is legally inadmissible — if p_1-values exhausted probability judgments. On my view, jurors should and do start a case with any (set of) coherent priors (so long, in direct opposition to Cohen's view, as none is zero) but with a uniform p_2-distribution — that is, no commitment to any of these odds: there is no particular preferred prior p_1-value of guilt, rather than that prior p_1 having to be zero. This is judicially quite admissible: the p_2-dimension allows another way of representing having an open mind. And then there is no difficulty in seeing how the evidence affects p_1 while its weight affects p_2, nor of seeing how corroboration of testimo ny and convergence of circumstantial evidence can raise p_1-values.

(c) One cannot, Cohen claims, interpret "beyond reasonable doubt" simply as meaning some high level of p_m. How could we say precisely — judges never try to — how high that level was to be? Instead what is needed, he argues, is some measure of how the evidence bears on the alleged guilt of the accused considered as individual, not as a member of a statistical reference class — so that p_m is an inappropriate measure. But it is far from obvious that is what a court requires: consider the use of genetic "fingerprinting", for example, which is increasingly eroding former exclusions of expert evidence on ultimate issues. Nor is it obvious that the absence of numerical guidance in judges' directions is to be taken as urging the jury that no particular high p_m is sufficient for conviction rather than as just avoiding specifying a precise line of demarcation. A refinement does need to be made to incorporate p_2 but it can be done just as in the civil case — see (e) below.

(d) If the civil standard were simply $p_m > 0.5$ then in situations where one party's case rested on a number of independent elements each of these would have to be established at a much higher level for their conjunction to remain above that standard. This, Cohen argues, is contrary to actual normal judicial practice. This argument is, I believe, fully met by Blackburn's (1980) point that such conjunction effects are not part of normal practice simply because truly independent elements of a case are so rare; but that when they exist, then the conjunction effect does yield a perfectly reasonable criterion of decision — accept the conjunction if its probability is better than 1/2.

(e) The difficulty about negation comes out most vividly in Cohen's example (1977, p. 75) of the rodeo gatecrasher. Here (although Cohen is rather too ready to ignore the plight of the organiser, who, after all, is entitled to his day in court too) it can scarcely be adequate to take proof on the balance of probabilities as just $p_m > 0.5$, even if it is substantially greater. A clue to the correct analysis here is seen in the fact that English courts do sometimes take account of what Denning, in *Bater vs. Bater* (1951), called "degrees of proof within that [civil] standard": a recent immigration case, *Khawaja vs. Secretary of State* (1984), had the Law Lords speaking of "the flexibility of the civil standard of proof" in satisfying a court that "the facts which are required for the justification of the restraint put on liberty do exist" (this was, unusually, a case where the civil standard was being applied while incarceration was at stake).[10] "Degrees within a standard" could mean nothing if only thresholds of first-level p_m were involved — but nor could it if only thresholds of Baconian probabilities were involved. Evidently, what is needed here, and what such comments are groping towards, is the notion of a two-dimensional measure: we want $p_1 > 0.5$ and resiliency of p_1 to be high compared to p_1-0.5 (so that the p_2-distribution is concentrated above 0.5). In general it is enough in such cases that the α of greatest resiliency be such that res of $p_1(r)$'s being α is large compared to α-0.5 provided that w is acceptably large: variation in how large a resiliency and w-value are demanded is what is meant by degrees within the same standard — and that is equivalent to variations in requisite degrees of p_2 while p_1 is above 0.5. (In the criminal case just the same applies with $p_1 > n$ for arbitrarily large n.) This is the judicial practice here. If a justification of that practice is sought, we can find it by noting that the purposes for which trials are held are not satisfied merely by ensuring that more cases are decided correctly than incorrectly (as relying only on p_1 would ensure); stability of decisions, avoidance of revocation of penalties — which are what the high p_2 guards against — are necessary to our confidence in the catharitic as well as fact finding effectiveness of judicial process, and justifiable on such consequentialist grounds. Both the reliability and the dramatic power of the process demand that evidence should be judged sufficient to convict only if it meets appropriate standards of first-order probability and of weight, taken as a function of second-order probability.

Somerville College
Oxford
OX2 6HD
U.K.

FOOTNOTES

[1] I take the Strindberg reference to be flippant – though I am not sure I see the point: as though we were to imagine Macbeth playing marbles, perhaps?

[2] A complete system, in this sense, is one where any sentence of the system if inferable from the system's axioms just in case its negation is not inferable.

[3] The notion of resiliency may usefully be refined in at least two ways. Firstly, we could stipulate that the domain include no self-contradictory sentences, inconsistent sets of sentences or   sentences which are logical consequences of H or not-H: this would simplify the §6 discussion of the "Miss Julie" match C situation. Secondly, we could adjust the definition so that whatever the value of α the resiliency ranged over the whole of $[0,1]$: this might make identification of an appropriate weight-function (§6) more straightforward. I have avoided cluttering up the main issue with these refinements.

[4] In his earliest work de Finetti was less hostile to the notion; see §28 of his *Probabilismo* (1931), which hints at a treatment of "subjective values" of probability statements in terms of the relative stability of probability-judgments. It is this hint which the next two sections follow up; de Finetti himself seems later to have ignored or rejected these early musings.

[5] Skyrms [1980b] provides a thoroughgoing defence of the concept of higher-order partial belief against such arguments. In   a very recent work (which I have seen too late to discuss here) Vickers (1988) argues the importance of iterated, "embedded" (including second-order) probabilities to any account of probabilistic judgment which treats it as more than a measure of ignorance: contrary to the approach here, he takes subjectivism to be too psychologistic to deal properly with such iteration, which instead, he claims, demands a sophisticated logicism about probability.

[6] I gloss over here the issue of whether p_2 should be defined over L_2 or only over the complement of L_1 in L_2. There is an oddity about attaching second-order probabilities to p, q, r ... but it is unimportant for present purposes; one can as well think of every L_1-proposition as having attached to it two numbers, a p_1-value and a p_2-value, as of its having a p_1-value attached to it to which a p_2-value is attached.

[7] This derivation makes it easy to see why the objectivist takes second-order probability as trivial. If $n = 1$ or 0, the Bayes expansion will go through. But if $0 < n < 1$, so that the $L. H. S.$ is positive, there will be a contradiction: the $R. H. S.$ will be zero, since, for the objectivist, $p_2(p_1(r) = n \mid r)$ must be zero if $n = 1$. But the subjectivist has no such

172

problem: the "given" of dyadic probability does not mean "given knowledge of the true value", so there is no difficulty in constructing payoff tables which allow $p_2(p_1(r) = n \,|\, r)$ to be n, whatever the value of n.

[8] The infinite case may raise difficulties if all the p_2-assignments are to be real-valued.

[9] Much of this section has emerged out of discussion with Simon Blackburn of an unpublished response of his to Lewis's article.

[10] These cases are discussed in Cross & Tapper (1985, pp. 142-3)

REFERENCES

Blackburn, S. (1980). Review of *The Probable and the Provable*. *Synthese*, **44**, 149-159.

Cohen, L. J. (1977). *The Probable and the Provable*. Oxford: Clarendon Press.

Cohen, L. J. (1985). Twelve questions about Keynes's concept of weight. *British Journal for the Philosophy of Science*, **37**, 263-278.

Cohen, L. J. (1986). The role of evidential weight in criminal proof. *Boston University Law Review*, **66**, 635-649.

Cohen L. J. (1989). *An Introduction to the Philosophy of Induction and Probability*. Oxford: Clarendon Press.

Cross, R. & Tapper, C. (1985). *Cross on Evidence* (sixth edn.). London: Butterworths.

De Finetti, B. (1931). Probabilismo. *Logos*, 1931, 163-219.

De Finetti, B. (1972). *Probability, Induction and Statistics*. New York: Wiley.

De Finetti, B. (1974). *Theory of Probability*. New York: Wiley.

Dorling, J. (1981). Bayesian personalism, falsificationism and the problem of induction. *Proceedings of the Aristotelian Society*, Supp., 109-141.

Gardenfors, P. & Sahlin. N. E. (1982). Unreliable probabilities, risk taking and decision making. *Synthese*, **53**, 361-386.

Jeffrey, R. C. (1965, 2nd. edn. 1983). *The Logic of Decision*. Chicago: University of Chicago Press.

Keynes, J. M. (1921). *A Treatise on Probability*. London: Macmillan.

Lewis, D. K. (1980, postscripts 1986). A subjectivist's guide to objective chance, In his *Philosophical Papers*, vol. **II**, 83-132. Oxford: Oxford University Press.

Mellor, D. H. (1980). Consciousness and degrees of belief. In: D. H. Mellor (Eds.), *Prospects for Pragmatism*, 139-1973. Cambridge: Cambridge University Press.

Peirce, C. S. (1932). *Collected Papers*. (ed. Hartzhorne & Weiss).

Popper, K. R. (1958). Appendix *ix to *The Logic of Scientific Discovery*, 406-419, London: Hutchinson.

Savage, L. J. (1954). *The Foundations of Statistics*. New York: Wiley.

Skyrms, B. (1980a). *Causal Necessity*. Yale: Yale University Press.

Skryms, B. (1980b). Higher order degres of belief. In D. H. Mellor (Ed.), *Prospects for Pragmatism*, 109-137.

Van Fraassen, B. C. (1984). Belief and the will. *Journal of Philosophy*, **81**, 235-256.

Vickers, J. M. (1988). *Chance and Structure*, Oxford: Clarendon Press.

*Poznań Studies in the Philosophy
of the Sciences and the Humanities*
1991, Vol. 21, pp. 173-180

Czesław S. Nosal

NEUROBIOLOGY OF SUBJECTIVE PROBABILITY

Motto:

"Man stands upon giddily high scaffold made of improbabilities: the scaffold that rises by one storey along with each new achievement..."

P. Teilhard de Chardin

1. How many sources of subjective probability exist? Why does neurobiology seem to be important?

The interpretation of empirical data about estimating probabilities which is carried out under the Norm Extraction Method (NEM) (Cohen, 1982) has some interesting heuristic value, because this interpretation leads to the posing of a few interesting questions. Obviously the first of these questions is concerned with that concept's source of heuristic power. Why do the interpretative frames of the norm extraction approach seem to be more psychologically adequate and generate more questions? One answer to the question lies in the penetrating analysis of the Preconceived Norm Method (PNM) carried out by L. J. Cohen (1981, 1982). However it seems useful to develop the analysis initiated by Cohen in a few directions. The heuristic power of this analysis is due to the fact that probability estimation under the norm extraction is actually a processual description of how judgments are formed in the situation of uncertainty. For that important reason the proposed description seems to be psychologically sounder. The processual description has more than one determinant of norm contents and more than one grade of certainty of the subjects' probabilistic estimations.

The Norm Extraction Method proposes a structural description of cognitive competence rather than a diversity of performance. Man is assumed to produce a single model (the norm for estimation) for each definite probabilistic situation. This description however does not exclude the fact that the norm as a cognitive standard is formed under the influence of various sources of information. Since they depend upon the subject's intentions, direct information and the structure of individual experience stored in LTM (long term memory), subjective probability estimations can be biased within various dimensions. So it is possible that what we know as a final — verbal or numerical — judgment or an act of behavioural choice is based upon a few sources of information and different representations. Then, during the formation of the probability estimates cognitive representations are compared and integrated on the basis of the assumed norm. From the processual point of view the final estimation seems to be a multicomponential result of the performance process which is regulated by the higher norm of subjective probability. The performance course does not depend only on cognitive representation of the norm but also on many other factors such as, for example, personal capabilities, knowledge, the type of task, the cost of consequences.

The main goal of this paper is to pay attention to neurobiological support for a processual interpretation of probabilistic norm formation. Neurobiological research seems to speak in advocacy of the existence of sorts of "probability centres" in the brain of mammals. On the other hand these results extend the cognitive basis for the interpretation of the essence of probability, thus making it interesting for neurobiology and neuroscience as well as for logic and methodology. Neurobiological findings seems to be important for understanding individual differences or ways of probabilistic thinking resulting in a variety of cognitive inclinations.

The author of the present considerations is aware of the risk due to the search for neurobiological correlates of probability. Most people find the notion to be a pure mental product. However the risk assumed can be accepted in the context of some other heuristic question: from what information basis is the probability extracted? Does the neurobiological basis underlie the process?

Looking for an answer to the first question one can content oneself with a functional description of the process of forming the subjective probability of cognitive representations and external cues. The latter of the above questions focuses our attention on biological and behavioural aspects of the genesis and differentiation of "probability centres" in the brain.

2. New and old "probability centres": neocortex versus hippocampus

Interesting neuropsychological results show that in the brain there are separate neuroanatomical structures correlated with the learning of a conditioned reaction to signals with various objectively scheduled reinforcement probabilities (Hirch, 1974; Pigarieva, Simonov, 1983; Stenens, 1973). In general these results point to some relationship between the frontal cortex and the hippocampus and a susceptibility to high and low probabilities. Hippocampectomized animals reacted only to very probable signals.

Particularly interesting in regard to the role of the hippocampus are those investigations in which experimental procedures consisted of teaching animals the "switching" of conditioned reactions (cf. Pigarieva, Simonov, 1983). The same signal (bell ringing) was reinforced twice: with food in the morning and with an electric shock in the evening. The animal acquired in this way some probabilistic experience. Then after this training an intervention of hippocampectomy was performed and it was checked how the animal reacted to signals with various probabilities. The experiment employed four reinforcement frequency levels (10%, 33%, 50%, 75%) for various groups of animals. The analysis of learning curves done separately for each level showed that hippocampus destruction narrowed the range of signals down to very probable ones.

The animals in these experiments resembled "automata" reacting only to very probable signals. There appeared a strong effect of behaviour fixation (rigidity). The reason for this fixation is assumed to lie in disturbances in the events frequency coding mechanism and a lack of confronting current information with traces stored in LTM (cf. Hirch, 1974).

The above results of investigations of the hippocampus role throw — as it seems — some interesting light on the dual nature of behaviour flexibility which must be subject alternatively to high and low probabilities. Coping with and mixing the two probability intervals seem to be the important aspect of intelligent, creative and adaptable behaviour.

Looking for food or having already perceived "a victim" in the visual field the animal incessantly behaves on two levels because it deploys its attention between the central and peripheral parts of the visual field. Each of these parts is concerned with different sources of probability. There are fluctuations of attention woven into the animal's behaviour programme which are subject to high and low probabilities. The programme must be flexibly modified depending on what happens in the centre of the visual field (high probability) and on its peripheries (low probability). The role of fluctuating vigilance programmes in performing various adaptive functions is underlined by the biological analysis of cognition (cf. Lorenz, 1973). The efficient adaptation of the variable conditions needs two anatomically separated but functionally interrelated "probability centres". In the behaviour called rummaging by Lorenz each of the centres exerts its specific influence. Among animals environment scanning is directed by the currently dominant need which widens the high probability spectrum and raises its value. Using a classical language we deal here with the stimulus genrealization mechanism described by I. Pavlov. In a contemporary language this process can be described as an extension of a set of events with similar probabilistic values. An example of that sort of extension are anxiety generalizations.

Scanning is also of dual nature. The need-directed animal reacts as if "something could still happen". So in general one behaviour programme contains operations (theoretical acts) heterogeneous in respect of probabilities. Particular attention is paid here to the subjective probability rise mechanism. Among animals this mechanism is hardly of primary character. However among people the mechanism takes a developed form owing to symbolic processes, making use of numbers, measuring scales and communicability. Regarding that it is characteristic for the human being to underestimate as well as overestimate probability in a number of areas of activity.

Human behaviour does not confine itself only to real and reachable goals. Ideals and visions of future states play a very important role (cf. Ackoff, Emmery, 1972). For the human being transgressive behaviour which expresses itself by going beyond results attained, in characteristic, i.e. that which is entirely certain or almost certain (cf. Kozielecki, 1987). Aristoteles paid attention to the fact that for a human being this was leading (directing) which seemed to be possible and also improbable. So changing the probability limit and raising its overestimation of subjective value seems to be one of the leading motives of human behaviour. I have named the mechanism a sailing inclination (cf. Nosal, 1986).

The subjectively overestimated probability and the positivity bias accompanying it make those behaviour directions which were intitially weak and unstable stronger and stabler — analogously to handling a sail in a weak wind. Owing to that reinforcement "weak" ideas, associations and mental chains gain a status of stable structures in a man's mind. That sailing inclination aspect seems to be particularly important for the description of creative thinking.

It is worthwhile noticing that the mechanism of subjective probability modification seems to be related closer to changes in estimations assumed within the Norm Extraction Method than to quite one-sided conditions assumed within the Preconceived Norm Method. In the course of the norm constructing (current estimation) the probability may be determined to be high or low. From the behavioural point of view the estimations correspond to risky or cautious judgments. The relatively stable inclinations of definite sorts must relate to the neurobiological behaviour basis. From the psychological point of view these relationships may be examined and operationalized as relationships between a sort of temperament and sensation seeking need on one hand and probabilistic behaviour aspects on the other.

3. *Left and right half of the subjective probability in the brain*

So far we discussed in this paper the neurobiological arguments speaking for the existence of "centres" to control "new" and "old" as well as "upper" and "lower" sources of subjective probability. They were associated with functional differences concerning the role of neocortex and hippocampus structures. In these considerations the argument related to results after hippocampectomy. However it has been generally known for a long time that hemispheres of the brain are functionally asymmetric. That is why, regarding the main problem discussed in this paper, it is important to concentrate on differences between hemispheres for probability estimation. Other aspects and theories of hemisphere differences will not be considered here since there is a rich bibliography in this area (cf. Cohen, 1983; Springer, Deutsch, 1981).

Investigations of the role of hemispheres for probability estimation have not been widely carried out yet. However some of the results testifying in favour of the differences may comprise a basis for the discussion of neurobiological roots of subjective probability. It is clear that these results must be coherent with the present knowledge about hemisphere differences.

Information processing by the left hemisphere is supposed to be sequentially organized, linear and analytical. Whereas the processing by the right hemisphere is configurative, synthetic and based on intuitive gestalts. Under what form are the hemisphere differences manifested in estimations of the subjective probability and of the response time for signals appearing with various frequencies? Are there two sorts of probability (linear versus gestalt) and what corresponds to them in the subject's convictions and when do they change?

A rough answer to the question posed above may be given by taking into account interesting results of an experiment on frequency estimation and anticipation of signals with various (0,25; 0.75) objectively scheduled frequencies (cf. Shiriayev, 1986). This experiment resulted in a functional hemisphere dissociation making use of the well known foveal visual input method.

Persons under examination were acquiring probabilistic information by observing sequences of signals (150 items) and anticipating which of the two signals (frequent vs rare) would appear. By exposing the signals

178

in the left and right field of vision it was possible to check up on the hemisphere differences.

Results of the experiment show that estimation and anticipation by the left hemisphere are analytical and related to the short range (threshold) of the anticipation which attains the certainty threshold earlier. In other words the hemisphere is governed by the "law of small numbers". However the lack of significant differences in reaction time between frequent and rare signals testifies to a generally slower probablistic information processing rate for the left hemisphere.

The right hemisphere works as if it was an "organ" to increase uncertainty and incredulity. Results from the quoted experiments show that frequent signals were evidently underestimated by the right hemisphere while rare signals were overestimated. In short, to use another term, this is a probability equalization process. However that equalization is not accompanied by entropy manifested by a reaction time structure for frequent and rare signals. Regarding the right hemisphere the reaction times are proportional to the frequency of signals, i.e. shorter for a more frequent signals and longer for rare ones.

Such paradoxical results seem to testify that the process of increasing uncertainty in the right hemisphere is rather a form of probabilistic experience "opening" represented in global form. The previously mentioned probability equalization mechanism may distinguish an inclination for using the law of big numbers and of long series for right hemisphere processing.

4. Integrating subjective probability (final remarks)

Considerations about neurobiological correlates of the subjective probability norm presented in this paper seem to be interesting for the description of possible inclinations which occur during the process of the probability norm extraction. Obviously one should remember that empirical data on the correlates are fragmentary. The main thesis of this paper is that the subjective probability norm extraction process is based on a few actually complementary cognitive strategies resulting from different sources of internal and external information. Some of them may have neurobiological correlates.

The subjective probability extraction process may be described by two dimensions. The first dimension refers to the probability estimations on a riskiness-carefulness scale. As is shown by a number of experiments, that value determines the significant regulating standard which affects changes in subjective estimations and susceptibility to current information. The main cognitive strategies related to the first dimension concern the direction of probabilistic inference, i.e. concentration on high vs low probabilities. What is crucial is which of these alternatives becomes the information norm and which is a sort of "background". Under uncertainty both strategies may be used one-sidedly or also integrated into a programme that leads to the unbiased estimation of chances. Impulsive vs reflexive behaviours are an example of that one-sidedness. Another more paradoxical one-sidedness representation is the sailing inclination which consists in overestimation of low probability.

The second dimension describes changes in the subjective probability in terms of the differentiation threshold for the changes within an assumed set of events and along a scale of time. Strategies related to this dimension refer to hemisphere differences. The strategy characteristic of left-brain processing is relatively slow and based upon low differentiation threshold values. It is the analytical strategy of small steps and frequent changes. In terms of the class of formal models this strategy goes along with linear probabilistic information integration (Bayesian model). The right-brain probability estimation strategy refers more to the determination of a set of possible events than to the analytical estimation of chances. Left hemisphere processing is based on elementary cognitive units (micro-structures). Right hemisphere processing grounds seem to be dominated by programmes for uncertainty generation and expectation finding out some global tendency — a general formula. Probability estimation under uncertainty needs a contribution of various cognitive strategies to determine the initial probability quantity as well as threshold of its differentiation. The subjective probability extraction mechanism is undoubtedly quite complicated and depends on a number of factors such as the subject's experience, behavioural intentions and procedural skills. As it seems some neurobiological correlates — discussed in this paper — throw interesting light on the nature of its complexity.

FOOTNOTE

*The work presented in this article was supported by Ministry of National Education, Grant RPBP.III.29.

REFERENCES

Ackoff, R., Emery, F., (1972). *On purposeful systems*. New York: Aldine.
Cohen, G. (1983). *The psychology of cognition*. London: Academic Press.

Cohen, L. J. (1981). Can human irrationality be experimentally demonstrated? *Behavioral and Brain Sciences*, **4**, 317-370.

Cohen, L. J., (1982). Are people programmed to commit fallacies? Further thoughts about the interpretation of experimental data on probability judgment. *Journal for the Theory of Social Behaviour*, **12**, 251-274.

Hirch, R. (1974). The hippocampus and contextual retrieval of information from memory: a theory. *Behavior and Biology*, **12**, 421-432.

Kozielecki, J. (1987). *Koncepcja transgresyjna człowieka*. [A transgressive model of man] Warszawa: Państwowe Wydawnictwo Naukowe.

Lorenz, K. (1973). *Die Ruckseite des Spiegels: Versuch einer Naturgeschichte menschlichen Erkennens*. München: R. Piper Verlag.

Nosal, C. S. (1986). Mózg, prawdopodobieństwo i transgresja [Brain, probability and transgression]. *Studia Filozoficzne*, **174**, 17-27.

Pigarieva, M. L. and Simonov, P. V., (1983). Need – information organization of behavior and brain anatomy. In R. Sinz and R. Rosenzweig (Eds.), *Psychophysiology* (pp. 207-212). Amsterdam: Elsevier.

Stenens, R. (1973). Probability discrimination learning in hippocampectomized rats. *Psychophysiology of Behavior*, **10**, 1023-1025.

Springer, S. P. and Deutsch, G. (1981). *Left brain, right brain*. San Francisco: W. H. Freeman and Company.

Shiriajev, D. A. (1986). *Psychological mechanisms of probabilistic anticipation*. Ryga: Zinatne (in Russian).

*Poznań Studies in the Philosophy
of the Sciences and the Humanities
1991, Vol. 21, pp. 181-195*

Ryszard Stachowski

ON STANDARD AND NON-STANDARD MODELS IN THEORIES OF PSYCHOLOGICAL MEASUREMENT

Introduction

In his paper on the paradox of anomaly Cohen (1973) conceives the anomalous fact as such a fact which, although in conflict with a certain theory, none the less is incapable of causing the theory's rejection because it is well substantiated by other evidence. The situation is characterised by three features: 1) A statement A (describing the anomaly) is thought to be observably true, 2) A apparently conflicts with a theory T that is thought to be otherwise satisfactory, and 3) the counter-evidence to T that is described by A is thought to be quite unimportant in relation to the merits of T (cf. Cohen, 1973, p. 79).

In the present paper I shall show that every theory of a psychological quantity (e.g., utility, perceived risk, and subjective probability), being bound to construct such a quantity in terms of standard (Archimedean) measurement theory, has to deal with the same situation, for because of the non-standard (non-Archimedean) nature of presumably all the psychological quantities, the aforementioned process of the construction of a psychological quantity inevitably results in the paradox of anomaly. By anomalous fact I conceive here a fact which is in conflict with experience because it is the result of an essential for a standard-measurement-theory-approach process of abstracting from the real possibilities of the construction of a standard psychological quantity no matter how large or small. And although nobody rejects the theory the anomalous fact can be explained in terms of non-standard measurement theory and made reconcilable with it. A logical solution to this paradox of anomaly will be based on the process of abstraction from actual constructing of such quantities and hence allowing for the existence of non-standard quantities. The solution does not mean resigning from the principle of abstraction, the main approach to a scientific study of the relation between a phenomenon and its essential nature, since the new theory of a psychological quantity is based now on another principle of abstraction which lends itself better to the description of a fragment of psychological reality under study.

Some general remarks on the use of mathematical structures in psychology

Many years ago Hermann Ebbinghaus remarked that psychology has a long past but only a short history. I am of opinion that these words hold good for mathematical psychology as well, because — although it is generally accepted that a history of the use of mathematics by psychologists in psychological theory and measurement goes back at least to David Bernoulli's famous paper in 1738 on the measurement of risk — I hold that the very idea of the mathematical treatment of psychological phenomena goes back into the fifth century B. C., to the Pythagoreans, for whom the numbers were the basis not only of lines, surfaces, solids or physical bodies, but also of justice, opinion, as well as soul and mind.

In my paper I do not intend to search the Pythagorean roots of the mathematisation of psychological thought as the problem requires a separate treatment, but perhaps we should realise at the onset in what sense one speaks of the use of mathematics in science. Broadly speaking, it rests on an analogy: something about the way the symbols are related must resemble something about the relations among the observed phenomena (cf. Miller, 1964). Indeed, the symbols and the objects themselves become insignificant, and what is important is the pattern of interrelations among them, or their structure. There are two kinds of mathematical structures as a source of models for diverse psychological phenomena.

The simplest structures are spatial in character and Miller (ibidem, p. 224f.) summarises some efforts that have been made to represent a great variety of psychological phenomena in Euclidean spatial structure, such as colour mixing, feelings, emotions, attitudes, and the most general and ambitious attempt — factor analysis. As an example of a psychological theory that makes use of nonspatial mathematical structure — the second kind of structures that psychologists have also considered — Miller points to the work of Piaget who in his searching for some way to represent the operations that a person can perform has turned to the more abstract structures of modern mathematics — to the structure of logic, set theory, lattice theory, and so on. Yet another attempt to exploit nonspatial mathematical structures is a quantitative explication of some qualitative concepts such as utility, perceived risk, and subjective probability in terms of standard measurement thory. As a result of this procedure these concepts acquire status of psychological magnitude. This approach goes back to the classical paper of Helmholtz *Zählen und Messen* (1887) in which he developed the representational idea of extensive measurement, pointing out extraordinarily useful applications of a "symbolic system of the numbers": by means of this numerical relational system we can give descriptions of the relationships between real objects. Helmholtz's idea was given axiomatic form in Hölder (1901) and was further developed by Suppes (1951), Ajdukiewicz (1965), Krantz, Luce, Suppes and Tversky (1971), and Narens and Luce (1975), but the underlying philosophy of measurement has remained basically unchanged: measurement is conceived as the construction of iso-morphisms (or at least homomorphisms) from empirical relational systems (or empirical structures) into numerical relational systems (or numerical structures) providing the former with adequate numerical representations.

Construction of psychological magnitude in terms of
standard measurement theory

Under this approach, in the modern standard theory of measurement, an ordered triple

$$\mathcal{A} = (A, \succsim, o),$$

where A is a non-empty set of empirical objects, $\succsim$ is an ordered binary relation on A, and o is a binary operation of concatenation on A, is called *an extensive empirical measurement system* if and only if (1) all $a, b, \in A$ fulfill a certain set of algebraic assumptions (relational and Archimedean axioms), and (2) there is a real-valued function f embedding A into the positive numbers Re^+ such that for all $a, b, \in A$ the following conditions are satisfied:

(i) order preserving: $a \succsim b$ iff $f(a) \geqslant f(b)$,
(ii) additivity: $f(a \; o \; b) = f(a) + f(b)$.

What this representation theorem asserts is that if a given empirical rela-tional system $\mathcal{A}$ satisfies certain axioms, then there exists a homo-morphism f of this relational system into a numerical relational system $\mathcal{N} = (Re^+, \geqslant, +)$ in the sense that a function f takes A into Re^+, $\succsim$ into $\geqslant$, and o into $+$ in such a way that $\geqslant$ preserves the properties of $\succsim$ (condition (i) above) and $+$ preserves the properties of o (condition (ii) above).

In psychology such a semiformal[1] axiomatisation of empirical relational systems has proved most adequate for a quantitative explication of some qualitative concepts such as perceived risk (cf. Pollatsek and Tversky, 1970). Von Neumann and Morgenstern (1944) are known to have for the first time in the social sciences made effective use of this approach to explicate the concept of utility, with the only

184

difference being that their empirical measurement system is an intensive rather than extensive one, i.e., the function f satisfies only the first of the two conditions mentioned above.

The Archimedean axiom as an eliminating assumption

We have said that the axiomatisation of empirical measurement systems requires that all $a, b \in A$ fulfill a certain set of algebraic assumptions which in turn fall into two groups: (1) relational (algebraic, first-order) axioms and (2) Archimedean axioms (cf. Narens and Luce, 1975). In this paper we will concentrate on the second group of assumptions only as they are relevant for our considerations.

Being variously formulated from situation to situation, Archimedean axioms are used in any single case to guarantee, in part, the existence of numerical representations f. In one of these formulations *the Archimedean axiom* asserts as follows:

(I)   If for all $a, b \in A$ the sequence $a, 2a, ..., na$ ... is an infinite standard sequence[2], then there exists some positive integer n such that $na \gtrsim b$, where na is defined recurrently as $na = = [(n-1)a]\ o\ a$ for $n = 2, 3, ...$

The Archimedean axiom implies the following *eliminating theorem:*

(II)   It is not the case that $na < b$ *for all* $b \in A$ and for each postive integer n.

The Archimedean axiom (I) is a strong eliminating assumption since it means the exclusion of quantities that are infinitely small and infinitely large in comparison with others. In the positive sense it postulates the construction of arbitrarily small and arbitrarily large objects regardless of the practical possibility of such constructions. This type of abstraction, called *abstraction of potential realisation*, is related to the concept of potential infinity which applies to the process of constructing such arbitrarily large and arbitrarily small objects and to the objects themselves. These objects at each moment of their increasing and decreasing are always finite. Consequently such an infinite magnitude is always potential and not actual. This is the only correct understanding of infinity of any Archimedean magnitude. The essence of Archimedeaness consisting in continuous, infinite repetition of the same operation of concatenation of quantity is reflected in the expression na defined as $[(n-1)a]\ o\ a$. It is this idea of potential infinity that Anaxagoras had in mind when he said 2,500 years ago: "Neither is there a smallest part of what is small, but there is always a smaller (for it is impossible that what is should cease to be). Likewise there is always something larger than what is large". In other words, there is the infinite divisibility of matter just as there are no limits to its expansion. While ascribing infinity to qualities, Anaxagoras "constructed a bewildering image of matter, similar perhaps only to the system of Leibniz, younger by 2,500 years, where each particle of the world reflects the world in its entity" (Tatarkiewicz, 1968, p. 50). Hegel will later call this quantitative infinite progress "bad infinity", i.e., constant transgression of limits which is the inability to annihilate them and the eternal return to them (Hegel, 1967, p. 343).

The prototype for all infinitistic processes is repetition. To comprehend the sense of infinity is to understand that one can always repeat what one has once said or done. This understanding is tantamount to awareness of one's own life: our consciousness of life results from the fact that we hope to repeat our actions at will (cf. Ooka, 1951). Poincaré (1902) believed that the human mind is capable of imagining infinite repetition of the same act when the act is actually possible only once.

The Archimedean condition used to axiomatise empirical relational systems is an immediate analogue of the Archimedean property of real numbers: given positive real numbers a and b, no matter how small a is and no matter how large b is, there is an integer n such that $na \geqslant b$, where na means $a + a + ... + a$, with n *terms* a. And since the reals are Archimedean[3], then it follows that to be homomorphically embedded into an intended numerical representation, the empirical relational system must of necessity (in the sense of mathematical, not practical, necessity) be also Archimedean; otherwise proof of the numerical representation theorem could not be possible. Then the Archimedean axiom in standard measurement theory is *necessary*: it guarantees the existence of a numerical representation and at the same time it results from the homomorphism accepted a priori. And although the construction of the empirical measurement system is left to the ingenuity of the researcher, his freedom to select the axioms determining the quantitative contents of these systems is restricted by the necessity to prove an existence theorem of the prior numerical representation and at the same time limits the range of quantification of qualitative concepts in the social sciences based on such measurement theories.

The representation theorem implies that the (mathematical) infinity of the Archimedean relational system $\mathcal{N}$ is embedded into the empirical relational system $\mathcal{A}$ (by means of homomorphism f) along with the structural-relational properties of the numerical system $\mathcal{N}$. Thus if the empirical relational system must be Archimedean — and it must be because of the necessary nature of the Archimedean axiom — then it is evident that infinity of the Archimedean psychological magnitude must be interpreted in terms of the infinity of the Archimedean mathematical quantity. So, irrespective of whether it is in accordance with empirical

186

facts or not, we will have to admit that, e.g., there is no risk, regardless of how large, which could not be "overrisked"; it suffices to concatenate any other risk, no matter how small, a sufficient number of times.

The Archimedean eliminating theorem and empirical evidence

The previously presented method of quantification of qualitative concepts in the social sciences based on standard, i.e., Archimedean measurement theory, absolutises the transgressional nature of human beings. In such a world there is no room, metaphorically speaking, either for Heaven or for Hell since one can always imagine something respectively even more delightful or even more terrifying. But irrespective of the fact that von Neumann and Morgenstern themselves were strongly convinced that non-Archimedean orders are completely incomparable with our normal understanding of the essence of utility and preference, one can point to examples showing that the elimination theorem (II above) implying such an image of a state of affairs and of human nature, and moreover — the only image — is not always fulfilled. Such examples are supplied by:

a) *mathematical model theory*: no formal interpretation of the Archimedean axiom is verifiable empirically since it can not be written as universal sentences in an appropriate formal language (in this case in a first-order language), i.e., free of predicate variables. Indeed, there is no possibility of constructing an experiment which falsifies this axiom because it is practically impossible to exhibit an infinite standard sequence. As a matter of fact, the Archimedean axiom can not be falsified merely because $b > a, b > 2a, ..., b > ma$, as there may be some $n > m$, for which $na \gtrsim b$ (cf. Krantz et al., 1971).

b) *geometry:* there exist such infinitely small entities as so-called horn angles, i.e., angles between two curves with a common tangent (see Fig. 1). The characteristic feature of such an angle α determined by the arc

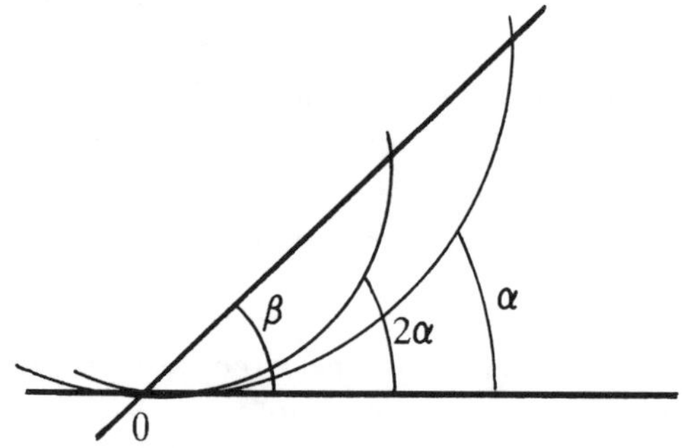

Figure 1: Non-Archimedean horn angles.

of the circle and its tangent at one of its ends is that, even if we increase it any number of times, we will never exceed an angle β between the tangent and any line cutting it at the point of osculation 0. Horn angles are non-Archimedean quantities because $n\alpha < \beta$.

c) *dialectic law of measure*: any quantitative change is neutral in relation to an entity only up to a certain limit, after which a new quality appears; there is no potential infinity in nature, only infinity of measure ratios for which Hegel's nodal line of measures on the scale of such referents as "more" and "less" is the "prototype" (Hegel, 1967, p. 343). Such a nodal line of measure ratios is characteristic for:

α) *natural phenomena*, e.g., transition of water from one state to another due to changes in temperature, phenomena of superfluidity and superconductivity, etc.,

β) *human behaviour:* "The something 'more' or 'less' — writes Hegel (ibidem, p. 547) — can be seen in that which causes the limits of rashness to be transgressed giving way to something completely different, i.e., crime; which causes lawfulness to give way to lawlessness, and virtue to vice". For a starving coolie there is probably no amount of trinkets which could make him part with one bowl of rice (with no resale value) (cf. Chipman, 1960). Piaget observed that children at the preoperational stage of intellectual development (up to 7–8 years of age), who are only intuitively able to generate the concept of space and have not yet acquired the concept of preservation of quantity, are unable to generate the iterative concept of Archimedeaness according to which the length of a concatenated segment is always constant in the process of iteration. For example, let the child be presented with a line segment AB and a point C on the right of AB. According to the Archimedean axiom, if the length AB is added to itself a number of times, it will always exceed C no matter how far AB and C are apart. But for the preoperational child, who is unable to conserve this concept at all, the displaced segment does not remain constant in length. Therefore he will draw a number of segments AB, in principle, as follows: $A'B' > AB$, $A''B'' > A'B'$, and so on, since he assumes that the new intervals are longer and longer because they have been added to the former ones. For him "longer" is the same as "farther away" (cf. Piaget, 1968; 1973; Stachowski, 1978). Borch (1968, p. 21-22) tentatively writes: "Questions of life and death and ethical principles like an absolute aversion to gambling would ... be considered as belonging to the more general social sciences — outside the narrow subject of economic proper", and — let us add — according to which, given a particular quantity of some commodity of which person is desirous, it is always possible to find some quantity of other commodity sufficiently great to compensate him for loss of part of his consumption of the given commodity (cf. Chipman, 1960). In other words, considering that some kind of "trade-off" will always be possible, one takes no account of the context of human values, motivation, etc.

The paradox of abstraction of potential realisation

If we assume that, on the one hand, the main approach to a scientific study of the ralation between a phenomenon and its essential nature is abstraction consisting in the disregard or elimination of factors qualified by the given theory as nonessential, and that, on the other hand, this kind of abstraction, called abstraction of potential realisation is a necessary condition for constructing model objects in psychology based on the standard measurement theory, then we must conclude — in the light of these few examples presented above and showing that *there are* situations where false empirical censequences are implied by the eliminating Archimedean theorem (II) — that, in at least such cases the abstraction of potential realisation underlying the construction of empirical quantity must lead to conclusions at odds with experience. Let us call such a construction *the paradox of potential realisation*. When applied to the measurement scale of any Archimedean magnitude, e.g., perceived risk as conceptualised by Pollatsek and Tversky (1970), such a paradox will consist in the fact that if we abstract from its "qualitative nodes" we cannot consider it a scale for measuring risk (for magnitudes outside the nodes) since a new qualitative definiteness has appeared: risk has turned into rashness which is now within the realm of catastrophe. This new quality requires a new measurement unit different from the one used for the standard risk scale[4].

The logical solution of the paradox of abstraction of potential realisation

If we express the paradox of abstraction of potential realisation in logical terms we obtain:

$$(p \rightarrow q) \land \text{non-}q,$$

where p stands for the Archimedean axiom (I) and q for the eliminating theorem (II).

The logical solution to the paradox of abstraction of potential realisation, non-p, results from the law of modus tollens. In other words, the solution of the paradox of abstraction of potential realisation is reducible to negation of the truth of the Archimedean premise (I) and construction of a new empirical measurement system more adequate to the reality, i.e., without the Archimedean axiom. However, this solution of the paradox is itself paradoxical, since, as we know, the Archimedean axiom is needed[5] in standard theories of measurement. But whether it is needed or not depends on the context within which the term "needed" is used, i.e., what kind of a model of the reals you have in mind.

Naturally, removing Archimedean axioms does not necessarily imply that we are to resign from the method of abstraction in science. On the contrary, elimination of the Archimedean premise does not end abstracting; we go on abstracting but this time not only the real possibilities of the construction of arbitrarily large (and arbitrarily small) mathematical objects, but also the very process of its construction. Let us call this type of abstraction *an abstraction of absolute potential realisation*. This introduces into consideration the concept of *the actual infinity*, i. e., some persistently existing formation, an infinite set in the sense of actual infinity, i.e., given all at once in finite form, if we know the property by which its elements differ from elements that do not belong to it (or we know how to construct such a set). In consequence, one accepts the existence of infinitesimal quantities over the continuum, where infinity is now understood, of course, as actual infinity. The infinitely large is a reflection of the internal contradiction within the concept of mathematical infinity approached syncretically; on the one side, it is an infinite quantity, i.e., a certain quantitative definiteness, while on the other side going beyond the boundaries of any quantitative definiteness is to be something large, i.e., a definite quantity, and also something infinite, since it is not a definite quantity (cf. Hegel, 1967, p. 342). Similarly, what is infinitely small, as small, is a definite quantity and hence remains something absolute, i.e., qualitatively too large for quantitative infinity to lead to its abrogation. This internal contradiction within mathematical infinity abolishes its split into "potential" and "actual" infinity (ibidem). Thus in our theory-construction we are going up to a hierarchically higher (a more general) level of idealisation which more adequately reflects the reality of interest.

The mathematical principles of a non-standard measurement theory

In removing the Archimedean axiom we do not state that the eliminating Archimedean theorem (II) always implies false empirical consequences, i.e., its elimination is by no means universal. Quite the reverse; we require that a far-reaching new theory corresponds with the "old" one. To put it another way, if we consider any theory of measurement as a system, then the "old" system (in this case a standard theory of measurement) is to be regarded as a subsystem of some extended (enlarged)

190

system, or supersystem. Specifically, our standard empirical measurement system should be a proper subset of this superset which we denote as $\mathcal{A}^*$, i.e., $\mathcal{A} \subset \mathcal{A}^*$, where $\mathcal{A} \subseteq \mathcal{A}^*$ but $\mathcal{A} \neq \mathcal{A}^*$. Let us call this generalised system $\mathcal{A}^*$ *a non-standard empirical measurement system* and the relative complement of $\mathcal{A}$, or $\mathcal{A}^* - \mathcal{A}$ a non-Archimedean context. The structure of the non-standard empirical measurement system can be shown graphically as in Fig. 2:

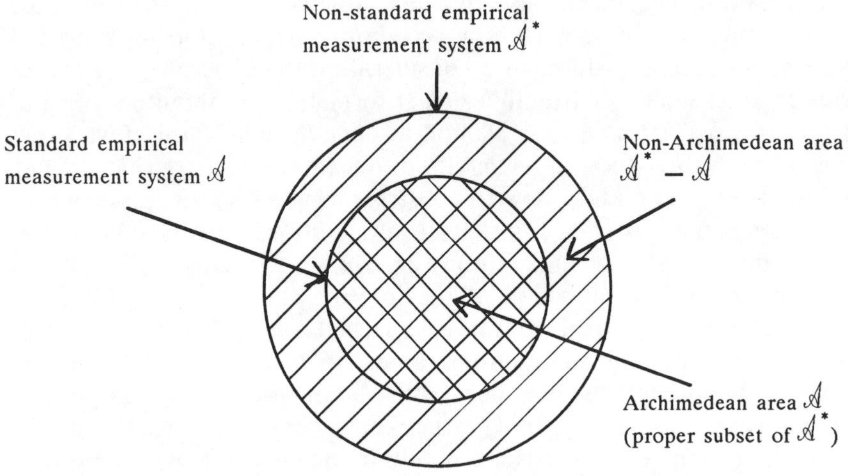

Figure 2: The structure of the non-standard empirical measurement system.

Thus our philosophical considerations on the nature of the characterisation of empirical measurement systems lead us to the conclusion that in order to solve the paradox of abstraction of potential realisation arising from the necessary nature of the Archimedean axiom in the process of construction of psychological magnitude, a solution which leads to the construction of a theory of psychological magnitude that would account for the possibility of non-Archimedean magnitudes, we should be in a position to embed such non-standard measurement system in a numerical system that would be the generalised or extended system of the reals. Fortunately, an idea of a non-standard model of the reals given by Abraham Robinson (1966) renders possible the solution to the problem. Since good expositions of Robinson's theory are available elsewhere, the coverage will be quite brief.

In short, the point is that typical (standard) numerical relational systems are built up of subsets of a continuously ordered Archimedean field of the reals which — from the standpoint of mathematical logic — is a model of the axiomatic theory of ordered Archimedean fields. To put it another way, in the standard theory of measurement such standard models of the reals are numerical systems. However, as has been shown (cf. Robinson, 1966; Chang and Keisler, 1973), there exist ordered fields elementarily equivalent[6] to the ordered field of the reals that are non-Archimedean. That set of the non-standard reals Re^* is in the simplest sense an enlarged numerical universe which contains not only the standard reals Re but also numbers that are greater than all numbers of Re. Thus, for some number $a \in Re^*$, $1 < a$, $1 + 1 < a$, ..., $1 + 1 + ... + 1 < a$, and so on, where we add 1 any (finite) number of times to itself. A formal proof of the existence of a function f embedding non-Archimedean extensive structures into non-standard model of the reals can be found, for example, in Narens (1974). (Cf. also Skala, 1975). Moreover, in Narens (1974) one can find an idea of constructing a measurement scale for non-Archimedean quantities.

Final remarks

The point of our discussion should be evident by now: so long as we are considering a standard model of the reals, the only "candidate" for a numerical representation of empirical measurement systems, the Archimedean axiom is necessary. However, this is not the case when a non-standard model of the reals is taken into consideration, since a "non-Archimedeaness" is the only characteristic that differentiates this enriched set of the reals from a set of standard reals. To put it another way, such a non-standard model of the reals has all the relational and algebraic properties of the reals except for the Archimedean property. So we arrive at the conclusion that non-standard models of the reals can be used as numerical relational systems.

The problem of the ontological significance of all sorts of infinitistic concepts has been intriguing scholars for a long time. As concerns the reality of infinitely small and infinitely large quantities, Aristotle rejected the actual infinity while accepting the concept of a potential or growing infinite: nothing that is actually infinite can exist, and nothing that exists can be actually infinite; the infinite is rather a power, a potential operation of certain processes (cf. Aristotle, *Physics,* III, 206b). For Aristotle, as for the Greek mind in general, the infinite as infinite is unknowable: it is matter, not form, hence it is not graspable by *nous*. And as such, the infinite is at the opposite pole from the complete, the knowable, the graspable, the divine (Randall, 1960, p. 194). On the other hand, Archimedes and his predecessors, e.g., Eudoxus of Cnidus, rejected the reality of the infinitely small and infinitely large in spite of the fact that already in Greek antiquity it was known that there existed such infinitely small entities as the mentioned above horn angles. Eudoxus's theory of proportion, being a solution to the problem of irrationals, or incomensurable magnitudes, was incorporated in the fifth book of Euclid's *Elements*. From Anaxagoras's conception of infinitely small and a denial of materialism of Democritus, Eudoxus developed his theory of infinitesimals, or method of exhaustion, which is often claimed nowadays as a forerunner of modern infinitesimal calculus (cf. Guthrie, 1978). On the opposite pole stood, for instance, Cusanus and later Bruno, who having taken seriously the notion that the universe is actually infinite, had caused the immediate destruction of the whole Aristotelian physical system and cosmology (cf. Randall, 1960).

In accepting the non-standard model of the reals as the basis for a new theory of psychological quantity, such as, for example, perceived risk or utility, we — according to Robinson (1966, p. 282) — must be aware that:

1) from the syntactic point of view the non-standard model of the reals introduces new deductive procedures, rather than new mathematical entities, and

2) if all measurements are performed in terms of integers or rational numbers, and if our theoretical framework goes beyond these and encompasses the reals, then there is no compelling reason why we should stay within an Archimedean number system.

But in the first place, if a discussion of the ontological significance of infinitary notions of any kind has to make any sense, we have to resign from a positivistic way of thinking.

By and large, it seems to me that taking into consideration the non-Archimedean quantities in the theories of measurement in social sciences, and especially in psychology, may turn out to be very promising because of the possibility of investigation of the specificity of human behaviour in situations toward which the traditional approaches based on the standard Archimedean measurement do not have access.

Our analysis has revealed that a structure of a psychological magnitude is the complex relational-operational system. This theoretical outcome reflects the widely accepted thesis that thinking, both in the evolutionary and in the ontogenetic sense, is derived from actually performed actions. The essence of the method of analysis employed here is the same reflective abstraction that underlies getting "general coordinations of activity", or interrelations common with all the activity patterns (cf. Piaget, 1968) out of the activity patterns that are characteristic of outward activity, and transforming these interrelations reflected on an internal plane into complex, internalised conceptual structures. What is characteristic of the reflective abstraction is that some property of thing or phenomenon is obtained from the activities which may be performed with them, mainly from such most general coordinations of those activities as jointing, ordering, etc., rather than from activities themselves. Hence the behavioural evidence of the construction of psychological magnitude by the subject is the extrinsic structure of activity, i.e., observable coordinations of activities into some overall systems of operations. However, the problem of both the genesis and internalisation of non-Archimedean conceptual structures goes beyond the scope of this paper.

Institute of Psychology UAM
Szamarzewskiego 89
60—568 Poznań
Poland

FOOTNOTES

[1] A semi-formal axiomatisation of some mathematical theory is understood here in the sense that the theory's basic concepts are not directly defined, but they are explicated indirectly by a set of related axioms. Moreover it is maintained that this system of axioms can have many interpretations or models (cf. Krantz et al., 1971).

[2] The standard sequence a, $2a = a \, o \, a$, $3a = 2a \, o \, a$, ..., in which a', a'', ... are perfect copies of the expressions a, is an ordered sequence of entities, e.g., measuring rods in which consecutive elements are equally spaced (cf. Krantz et al., 1971).

[3] Strictly speaking, an Archimedean property of the additive subset of the reals is implied by the axiom of the least upper bound or supremum which asserts that every nonempty set of real numbers that is bounded above has a supremum.

[4] Some of the best evidence for the existence of such "qualitative nodes" and a two-process scale of measurement comes from psychological investigation of the century-old question of the span of attention or, from the operational point of view, of the perception of number. One of the response variables that is of use in a typical experiment on the span of attention is the exact number of items in a collection that the subject is required to estimate. But, because of the brief exposure of sample collections of dots, the subject has no time to count them and consequently he makes his decisions on the basis of impression. There were in the forties several studies on the topic which are summarised by Woodworth and Schlosberg (1954, especially p. 98-99 and 245-246). In one experiment, by Taves, collections of dots sampling the range from 2 to 180 were exposed for only 200 ms and the subjects were asked to pick a collection that contained half as many rather than to ascertain the exact number of items in a collection. Taves found that up to 6 dots the estimates

194

were nearly all exactly correct but above 6–8 dots the subjects began to differ in their estimates. When the results are plotted, with the dots in standard on the abscissa and the
apparent half on the ordinate, the halving is seen to be arithmetically exact up to a whole
number of 6–8 dots, but beyond that level the estimated halves are about 20 percent too
large, half of 10 being (apparently) about 6, half of 20 about 12, half of 50 about 30, half of
100 about 58. There is a break or discontinuity in every function (or "curve") when the
collection to be halved reaches the neighborhood of 8 dots. With these short exposures
which do not allow time for counting, the subject's method shifts fairly abrouptly from
exact perception of the smaller numbers to approximate estimation of numbers that are
too large to be exactly perceived. "It is obvious – write Woodworth and Schlosberg
(ibidem, p. 99) – that we are dealing with two functions, one holding up to 6 dots and then
giving way rather abruptly to a different function" (italics mine R. S.).

Taves's data show that he obtained a regular scale for numerousness (the name coined
by Stevens for that property of a collection of items which one discriminates, without
counting, when one estimates the number of items) of dots with a break or discontinuity at
about 6 dots. "These experiments show that the basic formula, $R = f(N)$, is too simple.
When N is less than 6, one function is found. When N is geater than 6, one of two other
functions appears according as the conditions favor estimating or counting. These
distinctions hold good when the response measure is accuracy, confidence or reaction time.
Here we have about as clear evidence as one could desire that there is a distinct process
with an upper limit at about 6 units. It matters little whether we speak of the limit of
subitising [this coined word is derived from the Latin word for "sudden" and is used to
designate the rapid perception of small numbers exposed very briefly] or of the span of
attention; the important fact is that there is a process whose upper limit can be
determined" (ibidem, p. 100; italics mine R. S.).

[5] In the measurement literature it is generally agreed that "every system of measurement
that leads to an interval or ratio scale must satisfy the Archimedean principle in some form
in order for a numerical representation theorem to be proved" (Suppes and Zinnes, 1963,
p. 36).

[6] Two relational systems are said to be elementarily equivalent if and only if each sentence
of the first-order language that is true in one relational system is also true in the other one.
And since Archimedean axioms can not be formulated in the first-order language (the
elementary language) then the non-standard reals Re^* as a proper extension of the
standard reals must be non-Archimedean.

REFERENCES

Ajdukiewicz, K. (1965). *Logika pragmatyczna* [Pragmatic Logic]. Warszawa: PWN
Bernoulli, D. (1738). *Specimen Theoriae Novae de Mensura Sortis, Commentarii Academiae
    Scientiarum Imperialis Petropolitane*, Tomus V.
Borch, K. (1968). *The Economics of Uncertainty*. New York: Wiley.
Chang, C. C. and H. J. Keisler (1973). *Model Theory*. Amsterdam: North-Holland.
Chipman, J. S. (1960). The foundations of utility. *Econometrica*, **28**, 193-224.

Cohen, L. J. (1973). The paradox of anomaly. In: R. J. Bogdan and I. Niiniluoto (Eds.), *Logic, Language and Probability*. Reidel: Dordrecht, pp. 78–82.

Guthrie, W. K. C. (1978). *A History of Greek Philosophy*. Vol. 5: *The Later Plato and the Academy*. Cambridge: Cambridge University Press.

Hegel, G. W. F. (1967). *Nauka logiki* [Logic]. Warszawa: PWN.

Helmholtz, H. (1887). Zählen und Messen erkenntniss-theoretisch betrachtet. *Philosophische Aufsätze Eduard Zeller gewidmet*. Leipzig: Fues' Verlag, pp. 17-52.

Hölder, O. (1901). Die Axiom der Qantität und die Lehre vom Mass. *Berichte über die Verhandlungen der Königlich Sachsischen Gesellschaft der Wissenschaften zu Leipzig*. Mathematisch-Physische Classe. Leipzig: Teubner, pp. 1-64.

Krantz, D. H. Luce, R. D., Suppes, P., and A. Tversky (1971). *Foundations of Measurement*. Vol. 1. New York: Academic Press.

Miller, G. A. (1964). *Mathematics and Psychology*. New York: Wiley

Narens, L. (1974). Measurement without Archimedean axioms. *Philosophy of Science*, **41**, 374-393.

Narens, L. and Luce, R. D. (1975). The algebra of measurement. *Social Sciences Working Paper*. Irvine: University of California Press.

Õoka, S. (1951). *Nobi*. Tokyo: Orion Service.

Piaget, J. (1968). *Le structuralisme*. Paris: PUF

Piaget, J. (1973). *Introduction à l'epistémologie génétique*. La pensée mathématique. 2nd ed. Paris: PUF.

Poincaré, H. (1902). *Le science et l'hypothése*. Paris.

Pollatsek, A. and Tversky, A. (1970). A theory of risk. *Journal of Mathematical Psychology*, **7**, 540-553.

Randall, J. J., jr. (1960). *Aristotle*. New York: Columbia University Press.

Robinson, A. (1966). *Non-standard Analysis*. Amsterdam: North-Holland.

Scala, H. J. (1975). *Non-Archimedean Utility Theory*. Dordrecht: Reidel.

Stachowski, R. (1978). *Struktura i psychogeneza pojęcia wielkości psychologicznej* [The Structure and Psychogenesis of the Concept of Psychological Magnitude]. Poznań: Adam Mickiewicz University Press.

Suppes, P. (1951). A set of independent axioms for extensive quantities. *Portugaliae Mathematica*, **10**, 163-172.

Suppes, P. and Zinnes, J. L. (1963). Basic measurement theory. In: R. D. Luce, R. R. Bush, and E. Galanter (Eds.). *Handbook of Mathematical Psychology*. Vol. 1, New York: Wiley, pp. 1-76.

Tatarkiewicz, W. (1968). *Historia filozofii* [A History of Philosophy]. Vol. 1. Warszawa: PWN.

von Neumann, J. and Morgenstern, O. (1944). *Theory of Games and Economic Behavior*. Princeton: Princeton University Press.

Woodworth, R. S. and Schlosberg, H. (1954). *Experimental Psychology*. New York: Holt.

PART IV

RATIONALITY AND METHODOLOGICAL
PLURALISM

*Poznań Studies in the Philosophy
of the Sciences and the Humanities*
1991, Vol. 21, pp. 199-223

Lola L. Lopes and Gregg C. Oden

THE RATIONALITY OF INTELLIGENCE*

There are a great many words that refer to mental capacity. Of these, "intelligence" is at once the most widely used and the most hotly debated in academic circles. Controversy notwithstanding, intelligence is what we attempt to measure with intelligence tests and the complaints that surround such tests highlight basic features of the construct. Of central importance is that intelligence is supposed to be independent of conventional book learning and acculturation.

Debate also surrounds whether intelligence is one thing or many, an unorganized collection of separate capacities or an overarching superskill that draws on and coordinates the contributions of constituent talents and abilities. In either case, however, the attribution of intelligence to an individual or to a species suggests the existence of a generative capacity for solving problems and discovering patterns outside the scope of past experience and formal training. In other words, intelligence denotes a generic potential for adapting to whatever the environment brings.

Intelligence is also what computer scientists hope to build into their programs, yielding up a form of artifical intelligence (AI) that bears a family resemblance to the natural product. Central to the achievement of this goal is the spontaneous emergence of a resourcefulness or functionality that transcends the specific rules and algorithms that are built into the program in anticipation of particular demands. Though a program may execute prodigies of calculations and check quickly and compulsively through thousands of rules and instances, most observers will consider the program to be devoid of real intelligence until it surprises its creators by displaying a feature not foreseen in the construction of the program or (by an irony of the way people construe their own intelligence) until it makes a mistake that is human-like in its scope and circumstances.

"Rationality" is another term common in academic circles, though it resides mostly in the normative disciplines of logic, statistics, economics, and decision science. The term always presupposes mental capacity of considerable degree, and sometimes presupposes capacity levels that are unknown in the universe (Cherniak, 1986). We will not concern ourselves with notions of rationality that require infinite memory and processing speed or other capacities that are outside those that are commonly found in human beings. Instead, we focus on the theoretically achievable levels of rationality that one sees embodied by the Spock-like characters who inhabit science fiction and cognate publications in the normative sciences.

Although rationality has been subdivided into several different kinds — procedural, substantive, instrumental, and so forth — it refers generally to an internal calculus that is deductive in nature, proceeding from the general to the specific. One says that a person is rational if his or her thought processes follow an orderly sequence in which true conclusions are achieved via valid arguments from premises that are either taken to be true empirically or, better yet, are judged to be true by virtue of their self-evident reasonableness. Implicit in this view is the belief that rationality is at least necessary for intelligence or, as Baron (1985) has put it, "intelligent thinking is, among other things, rationally conducted" (p. 1). That is, although rationality is not a substitute for domain specific knowledge, it provides fundamental equipment for putting knowledge to work.

It seems that a semantic connection should exist between intelligence and rationality in the same way that terms like "cleverness" and "cunning", while signaling important differences in style and intent, denote comparable mental capacity. Though the label "intelligent" may conjure up a more human image than the label "rational" — Einstein, perhaps, instead of Spock — one does not expect to find the labels juxtaposed in contraries such as "human beings are intelligent but irrational".

As it happens, however, a contrary of exactly this sort is what one is likely to glean from scholarly research in cognitive psychology and cognitive science. On the one hand, there is a large body of research that documents striking failures of naive humans when confronting relatively simple tasks in probability theory, decision making, and elementary logic. On the other hand, there is the continuing belief of psychologists and computer scientists that by understanding human problem solving performance we will be better able to build machines that are truly intelligent. One can only wonder at the puzzlement of lay readers who come across articles like one published recently in *Newsweek* that spent several columns describing people's failures at decision making in scathing terms ("sap", "sucker", "woefully muddled") only to conclude that researchers hope to "model human decision makers' rules of thumb so that they can be reduced to computer software" and so speed the progress of artificial intelligence, the aim of which is "not to replace humans but to give them an important tool".

In what follows, we explore the quite different bases on which researchers ascribe rationality (or irrationality, as it happens) and intelligence to human action. We hope by so doing to demonstrate the large and ineradicable differences between these two ideals and to find some reconciliation between the conflicting claims that the two impose on human performance. In the endeavor, we will tread much of the same ground as Cohen (1981) in his well-known critique of the psychological literature on irrationality and will argue as he did that the normative criteria held in such high esteem by psychologists are often irrelevant for human action. Our goal, in fact, is to suggest that the apparent failures people manifest in many laboratory tasks actually signal the operation of a quite different kind of intelligence than is implied by conventional notions of rationality, an intelligence reflecting the properties of the massively parallel computational system that has evolved to meet the requirements of existence in a noisy, uncertain, and unstable world.

Rationality and Normative Models of Decision Making

In most cases, science begins with observation and moves toward theory. We see that rocks fall and seek an explanation. If early attempts to explain are inadequate, we persevere, revising theory at the same time that we add to the observational base. This being so, there is considerable truth in the scientific adage that "only a fool would propose a theory that fails to account for the known facts". It is curious, therefore, that psychological research in the areas of intuitive statistics and decision making has focused almost uniformly on why human beings can't or won't behave in accord with the theories that have been proposed to account for their behavior.

Have these theories been proposed by fools? We think not. But they have been proposed in an observational vacuum, motivated largely by interest in normative models drawn from probability theory, economics, and logic. Small wonder that these theories did not turn out to fit the facts, once some facts became known. A more substantial wonder is why these particular theories continue to guide virtually all research in behavioral decision making.

Our goal in this section is to sketch the major turns that have been taken by researchers in judgment and decision making in the last half century and attempt to extract some generalizations about the underlying commitments that the research makes to particular conceptions of the conditions under which human beings live their lives.

Behavioral decision theory first appeared on the academic scene in the late 1940s and early 1950s, stimulated initially by an axiomatization of the expected utility (EU) model by von Neumann and Morgenstern (1947) in their classic book on game theory. Although experimental psychologists had previously shown no systematic interest in how people evaluate and integrate information and bring it to bear on decisions or judgments, one immediate impact of these normative developments was to encourage mathematically minded researchers to use EU theory as a model of human behavior and test it in the laboratory. At the same time, Savage's (1954) reformulation of EU theory in terms of subjective (or personal) probabilities (SEU) focused interest on probability theory and Bayesian inference, leading researchers to test these models as well.

From the beginning, results were mixed. On the one hand, people were sufficiently adept at estimating simple statistical indices that a famous review paper was titled "Man as an Intuitive Statistician" (Peterson & Beach, 1967). On the other hand, tests of the normative models for statistical inference and decision making under risk revealed striking departures from predictions. In terms of the EU model, Irwin, Smith and Mayfield (1956) showed that probability and value do not influence decisions independently. At about the same time, Allais (1979/1953) and Ellsberg (1961) showed that human judgments under risk and ambiguity violate both EU and SEU axioms. In terms of inference, Edwards (1968) and others (e.g., Shanteau, 1970, 1972) showed that human probability judgments in simple urn tasks were qualitatively and systematically different from what Bayesian logic would dictate.

Much of the work on decision making done during the early period was inaccessible (or at least uninteresting) to a mainstream psychology that was becoming increasingly focused on the cognitive processes underlying behavior. Since virtually none of the normatively inspired work had concerned itself with process, the entire endeavor might have sunk from view had something not occurred to turn this research in more cognitive directions.

The nudge came in the early 1970s when Amos Tversky and Daniel Kahneman began to publish a series of papers on the heuristics (or rules of thumb) underlying people's judgments of probability. Their research strategy was a classic example of what Platt (1964) has termed "strong inference". In each of a series of experiments, they posed a probability problem couched in a familiar setting with parameters (givens) tuned exactly so that reasoning via the laws of probability theory would result in the correct answer while intuitive reasoning via heuristics would result in a qualitatively different and, therefore, necessarily incorrect answer.

Three major heuristics for judging probability were identified and will be discussed here in turn.

Representativeness

The *representativeness* heuristic refers to judgments of probability that are based on similarity. Thus, a sample is judged to be probable to the degree that it is similar to the parent population or reflects salient features of the process that generated it (Kahneman & Tversky, 1972).

Evidence demonstrating people's use of the representativeness heuristic is based on the responses of naive subjects to simple probability problems such as the following.

All families of six children in a city were surveyed. In 72 families the exact order of births of boys and girls was GBGBBG. What is your estimate of the number of families surveyed in which the exact order of births was BGBBBB? (Taken from Kahneman & Tversky, 1972, p. 432.)

The correct answer is 72 since all sequences of equal length taken from a Bernoulli process with equally probable outcomes have the same probability. In terms of representativeness, however, the string GBGBBG will appear more probable since it matches the parent population in terms of the 50/50 mix of boys and girls. Similarly, GBGBBG will also appear more probable then BBBGGG since it reflects the presumed randomness of the generating process.

A more subtle problem is given below:

There are two programs in a high school. Boys are a majority (65%) in program A, and a minority (45%) in program B. There is an equal number of classes in each of the two programs. You enter a class at random, and observe that 55% of the students are boys. What is your best guess — does the class belong to program A or to program B? (Taken from Kahneman & Tversky, 1972, p. 433.)

The correct answer to this problem is program B, the program with 45% males. This is due to the fact that the standard error of the sampling distribution of the binomial is larger for program B than for program A. Although the observed sample is equally distant from the means of the two target populations when measured in percentage terms, it is closer to program B in terms of standard error. For subjects reasoning in terms of representativeness, however, program A will seem more likely since the majority relationship in the sample

204

(55% males) corresponds to the majority relationship in program A (65% males).

Availability

The second heuristic, *availability*, refers to judgments of probability or frequency that are based on either the ease with which remembered instances of the events to be judged can be brought to mind or else constructed in the imagination (Tversky & Kahneman, 1973). Data supporting people's use of the availability heuristic comes from problems such as the following.

The frequency of appearance of letters in the English language was studied. A typical text was selected, and the relative frequency with which various letters of the alphabet appeared in the first and third positions of words was recorded. Words of less than three letters were excluded from the count ... Consider the letter R. Is R more likely to appear in the first position or the third position? (Adapted from Tversky & Kahneman, 1973, p. 212).

In this case, the correct answer (taken from available word count data) is that R is more likely in the third position of words than in the first, but subjects using the availability heuristic will choose the first position because it is generally easier to generate words beginning with R than having R in the third position.

Anchoring and adjustment

The final heuristic, *anchoring and adjustment*, is more general than the others in that it applies to the estimation of any quantity, not just probability or frequency. The idea here is that people estimate quantity by choosing some initial quantity that is available in the statement of the problem or easily calculable from available information and then adjust it upwards or downwards in order to take into account other relevant aspects of the situation. Typically, adjustments will be insufficient so that final estimates will be closer to the anchor point than they should be. An example of an anchoring and adjustment problem follows.

Consider two urns. Urn A contains 90 red marbles and 10 white marbles. Urn B contains 10 red marbles and 90 white marbles. You will draw 7 times from either urn, each time replacing the marble you have just drawn before drawing the next. Which event would you rather bet on? Event 1: you draw 7 red marbles from Urn A. Or Event 2: you draw at least 1 red marble in 7 tries from Urn B? (Adapted from Tversky & Kahneman, 1974, p. 1129; orginally from Bar-Hillel, 1973).

According to probability theory, the probability of Event 1, a conjunctive event, is just $.9^7$ or .48. The probability of Event 2, a disjunctive event, is just 1 minus the probability of the conjunctive event. This equals $1 - .9^7$ or .52, making the disjunctive event more likely than the conjunctive event. People using the anchoring and adjustment heuristic, however, will judge the conjunctive event to be more probable because they will use the urn proportions as anchors and then adjust insufficiently.

In all of the problems reported here as well as many others (the volume by Kahneman, Slovic & Tversky, 1982, presents a broad selection), predictions based on the hypothesized heuristic processes are borne out by data. Although there have been several serious criticisms of the validity of these experimental demonstrations (see, e.g., Cohen, 1981; Gigerenzer, Hell & Blank, 1988), many of which we consider to be cogent, we prefer for the moment to accept the experimental evidence as presented. In other words, we wish to stipulate (for the time being) that people's probability judgments reflect the three heuristics to at least a first approximation. However, this stipulation does not require us to accept the frequently cited conclusion (see Christensen-Szalanski & Beach, 1984, and Berkeley & Humphreys, 1982, for reviews) that people's judgments in these matters are irrational or even seriously deficient. Although we will not pursue the matter here, a case can be made (see Lopes, 1991) that the contentions of human irrationality based on this literature represent at best a misunderstanding of the experimental strong inference logic that Kahneman and Tversky have used and, at worst, rhetorical pandering in the interests of increasing the visibility and the fundability of individual research programs.

The point we do wish to pursue concerns the underlying assumptions that this type of research on heuristics and biases makes about the computational environment in which people live and the kinds of performances that are necessary for success in that environment.

Assumptions Underlying Heuristics and Biases Research

The first assumption is that elementary probability theory provides a good first-order approximation to understanding the uncertainty in natural situations. Almost certainly this is false, as can be verified by perusing some of the serious attempts by statistical experts to apply probability models to real-world problems (see, e.g., Schum's chapter in this volume). Research on heuristics tends to hide this assumption by carefully selecting word-problem domains in which some simple statistical idea does hold and then subtly switching the emphasis when the question is posed to subjects.

For example, the birth order problem conjures up a world in which, to a very good approximation, a Bernoulli process with independent and equally probable events does hold. So far so good. We might even suppose, as Kahneman and Tversky suggest we should, that daily acquaintance with this particular process should be sufficient to eliminate erroneous intuitions about the laws of chance, at least with respect to births of children. And so such acquaintance would behave if subjects were asked about family compositions. Almost everyone has reasonably sound qualitative intuitions about likely ratios of boys to girls in families of various sizes. But the experimental problem concerns *sequences* of births. Here people appear either not to know that all birth sequences with a given N are equally likely or not even to recognize that the problem is asking about sequences (it is impossible to tell which) suggesting that birth sequence in human families is unimportant enough that intuitions about sequences are undeveloped in this particular area.

When we try, however, to find some other domain in which sequence information is salient to people we run into difficulties. In gambling games such as roulette, sequence does appear to be salient, but erroneous beliefs about sequences cannot affect outcomes since roulette wheels are, indeed, random. This leaves people with neither incentive nor opportunity to improve their intuitions. In less artificial domains such as predicting the weather tomorrow based on the weather today, events are not independent and cannot be modeled by simple probability processes. Indeed, we cannot think of a single real phenomenon that is reasonably described as a Bernoulli process in which sequence actually matters to people.

A second assumption is that small differences in probability judgments produce big differences in real-world outcomes. In the urn problem, for example, the parameters of the problem have been tuned so that the two bets are as similar as they can be while still having the disjunctive bet (52%) be better than the conjunctive bet (48%). Had the red/white marble ratios been set at 91/9 versus 9/91 instead of 90/10 versus 10/90, conjunction would have been the better bet. Or, had subjects been asked about 8 draws (or perhaps 9) they surely would have preferred to bet on disjunction.

Setting up the problem this way is part of the strong inference logic. By balancing the problem at the margin between error and accuracy, the subject's response can be used to diagnose whether the judgment reflects probability theory or heuristic processing. The scientific difference is a large one but the actual difference, looked at from the point of view of the subject-as-bettor, is a small one. Even in this diabolically constructed problem, the 'cost' of choosing incorrectly is just 4%.

A similar case holds for the classroom problem. Here it is impossible to tell what the cost of being wrong is from the subject's point of view. However, we can consider the likelihood that the "majority rule" strategy will lead to error in general. To do this we must specify the size of the class, so let's choose 25 as a relatively common figure for high school groups. Were one actually to use the majority rule in classifying samples of classes of size 25 from a population of classes in which half were in program A (binomial with p = .65) and half were in program B (binomial with p = .45), the correct classification would be made 75% of the time for samples with male majorities (the more difficult case) and 92% of the time for samples with female majorities (the easier case), for an overall rate of 82% correct. In contrast, a statistically optimal rule would lead to 80% correct for samples with male majorities and 92% correct for samples with female majorities or 85% correct overall. It is not clear to us that the final difference of 3% in overall accuracy is large enough to warrant people preparing themselves educationally for answering questions of this kind.

The last assumption is that absolute knowledge of frequencies and probabilities is important for good decision making. For example, in the letter R experiment, subjects were actually presented with 5 different consonants (K,L,N,R,V), each of which is more frequent in the third position. Of the remaining 15 consonants, 3 others (D,X,Z) are also more likely in the third position and 12 (B,C,F,G,H,J,M,P,Q,S,T,W) are more likely in the first position. The data revealed that roughly 2/3 of the subjects judged the first position to be more likely for the majority of letters and 1/3 judged the third position to be more likely for the majority of letters. Moreover, a majority of subjects judged each of the individual letters to be more likely in the first position.

Now consider the problem from the subjects' point of view. Given 5 letters, subjects are likely to assume that some are more frequent in the first position and others are more frequent in the third position. If they chose between these hypotheses at random or if their different strategies are effectively at cross purposes, the data would look as they do given only a mild tendency to guess that first position is more likely. Considering that 12 of 20 consonants actually work this way, that is not a bad guess.

A more sensitive way to find out how much knowledge people have about relative letter positions would be to present subjects with equal numbers of the two sorts of consonants and find out whether they know which are which. Clearly this is something we cannot learn from Tversky and Kahneman's actual experiment. In the area of risk assessment, however, there are comparable data concerning people's knowledge of mortality rates for various kinds of accidents and diseases (Lichtenstein, Slovic, Fischhoff, Layman & Combs, 1978). These data can be interpreted two ways. The one provided by the authors is that people's estimates of mortality rates are seriously biased quantitatively, tending to overestimate the frequencies of newsworthy deaths such as tornados, flood, botulism, and homicide, and tending to underestimate the frequencies of deaths due to more invisible causes such as diabetes, tuberculosis, asthma, and emphysema. This account not only supports the availability notion but reinforces the idea that people's judgments are poor. The other interpretation is that people's estimates of fatality rates are surprisingly well related to actual numbers qualitatively (linear correlations being .89 and .91 for two groups of subjects), despite the fact that information about deaths in sources such as newspapers is more strongly related to interest value than to frequency.

The assumption that people need to have accurate absolute judgments of probabilities and frequencies grows out of a deeper commitment to the idea that rational choice between risks involves maximizing expected (or subjectively expected) utilities. In such computations, probabilities are traded off multiplicatively with utilities, so it is important for EU or SEU maximizers to have accurate input concerning both. But forty years' research on risky choice has demonstrated again and again that the EU and SEU models are not descriptive of human decision making under risk (for a review, see Schoemaker, 1982). Instead, people and even businesses usually have more tangible goals, such as putting caps on potential losses, achieving specific aspiration levels, avoiding future regret, finding acceptable balances of risk and return, and so forth.

Goals like these often require only comparisons of outcomes and probabilities, not ratio-based arithmetic. For example, even though someone may know that coal-based energy sources cost more human lives per year on average due to accident and disease than do nuclear-based energy sources, coal-based sources may still be preferred because they have less potential for catastrophe. This kind of judgment focuses on simple comparisons of worst case scenarios and treats as less important the very large probability that coal-based technologies will produce more deaths in a given year than nuclear-based technologies.

Of course, judgment rules based on factors such as aspiration level, regret, and loss limiting are considered to be irrational by most researchers in behavioral decision making, even those who have models that account for the discrepancies between human judgments and the EU model (see, e. g., Kahneman & Tversky, 1979, p. 277). This is because discrepancies between data and model are typically attributed to people's limited capacity to process information (see, e.g., the treatment in Schoemaker, 1982). There is, however, no support for the view that people would choose in accord with normative prescriptions if they were provided with increased capacity or even with 'right answers'. Published attempts to teach subjects to prefer normative choices have consistently failed because subjects prefer to pursue their own values and often offer cogent arguments to indicate why they do so (Lichtenstein, Slovic, & Zink, 1969; Montgomery & Adelbratt, 1982; Slovic & Tversky, 1974).

Intelligence and Computational Models of Thought

Thus, the dominant view within the decision area seems to characterize heuristics as an error-prone crutch of the limited minded, something to resort to if exigencies preclude a theoretically valid analysis, but then only at the price of being less than fully rational. And implicit in this view is that this is a steep price to pay because *it's dumb not to be rational*, or at least not to be all that you can be with respect to rationality.

It is ironic, then, that when we turn to the people whose business it is to make dumb machines behave intelligently, we find that they not only tolerate the use of heuristics by their systems, but actually seek them out and embrace them. That is, in the field of artificial intelligence, whose practitioners have the almost god-like power to construct systems to perform any way they choose, it turns out that the systems often are not made to be rational *by choice*. "Heuristic reasoning" is seldom said with a sneer in AI.

This is not to say that the rational approach plays no role in AI. Within certain subfields and with respect to certain problems, logical validity is sought and deductive methods are exploited. For example, some parts of the expert systems area makes use of theorem proving methods with a good deal of success in problem domains like complex computer system configuration. Considerable advances have been made in the efficient application of deductive techniques (e.g., resolution, Robinson, 1965) in such cases. Thus, it is not for want of effective computational implementations of deductive tools that large segments of AI choose to resort to heuristics. It is in no way a last resort.

Rather, the use of heuristics in AI is motivated by concerns that arise from the fact that AI systems must cope with the real world; that is, AI is inherently a practical field. There are at least three such concerns: robustness, generality, and practicability. Let us briefly consider each of these in turn.

Robustness. Rational methods are designed to satisfy the criterion of validity: to be guaranteed to be exactly right under ideal conditions.

210

Under less than ideal conditions, such as with noisy input and/or unreliable calculations, all guarantees are off and it's as possible to be exactly wrong as to be just slightly wrong (under ideal conditions, wrong is wrong, but in the real world, it is important to try to be reasonably close to right). In other words, rational approaches are brittle: they tend to break down when faced with realistic conditions. This can be the case even with probabilistic methods that have explicit provision for stochastic variability in the input data. This is because such methods rely on a single way of computing an answer so that even though they can be counted upon to be right on the average in the long run, they may be considerably wrong in a particular case without any means of detecting the error.

Generality. Guarantees of exact correctness are hard to deliver and deriving valid conclusions usually requires that numerous simplifying assumptions be satisfied. Thus, under the conditions in which these conditions are satisfied, rational methods can be applied, but such conditions are limited at least and more than likely very rare. For example, to use Bayesian techniques requires that the set of possible alternative hypotheses be (1) finite and (2) specifiable a priori. This in turn means that the hypotheses must be crisp, that is, discretely distinct one from another. These properties characterize little of the reality that intelligent systems must deal with. Rules of thumb do not promise exact results and, thus, do not need to be specified in exact terms. This greatly increases the generality of their application. Furthermore, it means that when they only sort of apply, they can still be expected to provide an answer that is at least sort of right. (This is in stark contrast to rational methods which can often be worse than worthless when they do not exactly apply.) Thus, in general, heuristics are general; that is, they can be applied more nearly without concern for satisfying stringent presumptions.

Practicability. Even when (or if) noise is not an issue, the reliance of rational methods on exact data means that their application may be unnecessarily laborious. Furthermore, rational approaches characteristically involve long inferential chains from a small set of input data (premises) to their conclusion. The processing of such chains is not especially amenable to making use of the massively parallel computational capabilities of the human brain and similar modern information processing systems. Heuristic systems usually start from a much broader base of input data and proceed to the conclusion in a fairly small number of steps. This is the smart thing to do if, contrary to the widely held belief in the decision area, man is not significantly limited in information processing capacity but is very much limited in time. (Life is short and stable states of the world are even shorter; it doesn't pay to be lost in thought when facing the tiger.)

This, of course, is not to say that rational methods have no place within intelligent functioning. There is no question that logic and probability theory are refined tools whose development was a significant advancement of human capability much as were the plow or Morse code. Mastery of these techniques provides an extension of intellectual power in those situations that enable their use. However, as already observed, the very specialization that makes tools well suited to the purpose for which they were designed tends also to make them only suited for a limited range of conditions. When there is, in addition, a significant cost associated with determining which method to apply, and/or when the performance advantage of the specialized technique over the general is slight, then even those who possess the tool may be better off sticking with ordinary methods. (As one of us knows from experience, there was a time when young boys felt an obligation to learn Morse code in case of an emergency, but although the use of the skill filled boys' books and the occasional human interest story, most waited in vain for a suitable emergency and had to content themselves with dit-dahing to friends over the telephone. If there had ever been a real emergency, valuable time might well have been wasted trying to determine if some distant observer knew code when a few sharply yelled "helps!" would probably have been more effective.)

In recent years, there has been a major development within AI and also psychology that has been largely motivated by the striving for robust, general, practicable intelligent systems: the new connectionist movement. In models subscribing to this approach (see, e.g., McClelland & Rumelhart, 1986), incoming information is encoded in terms of the activation of nodes corresponding to primitive sensory features, whether diagnostic of particular patterns or not. These nodes, in turn, are interconnected with successive layers of nodes through weighted links. The later, higher level nodes come to be activated as a function of the activation of the lower level nodes to which they are connected and of the weight (positive or negative, greater or lesser) associated with each such connection. The pattern of activation of the nodes in the final layer then corresponds to the system's identification of the stimulus encoded in the activations of the initial layer.

In order for a connectionist system to be able to perform interesting computation, the node activations must be non-linear functions of the weighted sum of activations input to each node. Often this function takes the form of an ogive, which effects contrast enhancement such that activation levels above a certain value are amplified whereas lesser degrees of activation are reduced.

There are many variations on the connectionist theme and many dimensions along which any particular connectionist model may differ from some other. One such source of difference concerns the learning process by which the weights in the system are determined. The major contrast here is between self-organization (e.g., Linsker, 1988; Kohonon, 1984; Rumelhart & Zipser, 1985) and feedback-guided learning (e.g., Rumelhart, Hinton, & Williams, 1986). In the former instance, through mere exposure to stimuli, the system extracts systematicities and co-occurrences to learn the features and patterns appropriate for encoding its environment. In the latter instance, through some process of reinforcement, the system comes to know appropriate labels or actions to attach to encoded categories or patterns. In any relatively complete system, both types of learning are almost certainly going to be required to produce a system capable of recognizing significant environmental objects and events and reliably making an effective response in the situation.

In such a model, every node performs the same sort of computation as every other. Whatever knowledge the system may have about the structure of the problem domain is distributed diffusely throughout the collection of connection weights. Any small change in the value of an input activation will have at least some effect on all later nodes to which it is directly or indirectly connected. Such models are often described as satisfying multiple "soft" constraints — the operative metaphor is of a bunch of elastic bands tugging in various respective directions, sometimes but not always against each other, cooperatively influencing the eventual state of rest of the object of contention.

Connectionist models are considered by their proponents to be more realistic than traditional rule execution models because they are more suitable given the kind of computers that we humans seem to be and given the type of stimuli that we have to deal with. First, the noisiness of the environment we inhabit means that local decisions (e.g., presence/absence of an individual stimulus feature) are unreliable so that any conclusions reached deductively from such shaky premises would be highly fallible. Connectionist models exploit the law of large numbers to defeat this noise by relying on the collective effect of overdetermined input so that the conclusion does not depend on the accuracy of analyzing any single piece of information. Likewise, the individual neural components of our brains are also unreliable, but the distributed nature of processing in connectionist models makes them substantially tolerant of piece-wise failures. Such distributed, cooperative processing is well-supported by the massively parallel structure of the brain which more than compensates for the weaknesses inherent in its flesh-and-blood machinery.

Furthermore, by virtue of their continuous nature, connectionist and related (see, e.g., Oden, 1988) models have a number of additional properties that people seem to have and that we would want people to have. One is the property of responding to similar situations in similar ways. This property, often referred to as automatic generalization, means that in a world having a reasonable degree of continuity, the system can know about things it has never directly experienced. This in turn allows such systems to exhibit graceful degradation under duress: rather than staying at peak levels of performance near to situations the system is familiar with and then falling apart completely once beyond that point, connectionist systems exhibit a gradual decline, which enables them often to maintain effective performance over a considerable range of situations.

Another important property of connectionism is the capacity for incremental learning. Because their knowledge is distributed across a large collection of continuous weights rather than encapsulated in individual crisp rules, it is easy to make small changes to this knowledge on an ongoing basis as a result of experience. Not only does this provide a powerful means of acquiring knowledge in the first place, it also allows for the kind of fine tuning needed to allow the system to effectively track a changing environment.

Pattern-based Reasoning

From this discussion of connectionist models, the reader might be led to think that such models are really only suited to perceptual problems, not the sort of problems with which rational methods are typically illustrated. However, the power of connectionist models to recognize pattern in the environment can be applied as well to other kinds of problems that must be faced under real-world conditions. In the context of inference and decision making, this approach may be referred to as pattern-based reasoning (see also Margolis, 1987; Massaro, 1987, pp. 272-277).

As in the perceptual domain, pattern-based reasoning takes advantage of overdetermined input, the continuousness of input information, and a large base of knowledge acquired incrementally over the course of experience with a wide array of situations. Rather than involving contentless rules that advance the reasoning process in small steps, a pattern-based system makes use of rich encodings of situation specifics gleaned from direct prior experience and tends to go relatively directly from data to conclusion and from there to the action that is called for under the circumstances. Reasoning, then, like other varieties of coping with the world, will involve a sequence of pattern-action episodes. Once again, we find that the attractions of this kind of reasoning are robustness, general applicability, practicability with brain-like hardware and so on.

With specific regard to practicability, this type of reasoning mostly involves simply being able to categorize the current situation into which of the many known situations it belongs to. Hence, pattern-based reasoning substantially reduces the inference problem to the kind of identification process that was discussed above. Again, the massive parallelism of the brain makes this an efficient, reliable process. And the automatic learning procedures of connectionist models make feasible the acquisition of the masses of specific information necessary to make this approach work.

Of course, the most interesting reasoning problems are those that are not just like any situation previously encountered. In a pattern-based system, the situation will match a number of familiar situations to various degrees and each will contribute cooperatively (in respective degree) to the arrived at solution. In this cooperative computation, the solution features that are supported within most or all of the relevant known cases will tend to be reinforced whereas those that would apply only under one or a few of the known cases will tend to be relatively suppressed. In effect, this amounts to making use of multiple rules of thumb to arrive at the conclusion (although, of course, the system does not involve explicit rule execution in any way), thereby having built-in checks and corrections. Such an approach is conservative and may be less than absolutely correct in cases where a single rule applies uniquely and exactly, but it is much less susceptible to corruption of the data.

With this all in mind then, let us turn back to the three heuristics and try to get a fix on how the use of heuristic rules might be related to the sort of reasoning just described. To anticipate our point a bit, and to set the stage for retracting our previous stipulation that Tversky and Kahneman's three heuristics operate essentially as they claim, we will suggest that what ordinarily is seen as reasoning following heuristic rules is more like pattern-based reasoning.

To motivate our discussion, we will introduce some additional problems that Tversky and Kahneman have used to contrast extensional types of reasoning (such as reasoning based on probability theory and set theory) with the use of heuristic rules. These problems all share a structure in which subjects are asked to judge the probabilities of expressions corresponding to p(A) and p(A&B). The basic result is that people tend to judge p(A&B) > p(A) even though the extension of A includes A&B and must, therefore, be larger.

Let us begin with the Linda problem:

Linda is 31 years old, single, outspoken, and very bright. She majored in philosophy. As a student, she was deeply concerned with issues of discrimination and social justice, and also participated in anti-nuclear demonstrations. Please rank the following statements by their probability: (a) Linda is a bank teller. (b) Linda is a bank teller and is active in the feminist movement. (Adapted from Tversky & Kahneman, 1983b, p. 92).

In the Linda problem, a majority of subjects judge that Linda is more likely to be a bank teller *and* a feminist than to be a bank teller. This happens, according to Tversky and Kahneman, because such subjects rely on representativeness in assessing the probabilities of the two statements. Because the attribute "is a feminist" is highly representative of women conforming to Linda's description, adding that feature to the unlikely attribute "is a bank teller" increases subjects' probability estimates even though it should decrease them according to extensional principles.

Now consider how a connectionist model might respond to the same two stimuli if it were to be trained on all the people in Linda's college class except Linda. Each person has a gender, a name, a set of attitudes, and a post-college career. Some are feminists and some are not. Some are bank tellers and some are not. A few of the bank tellers are feminists but most are not. The system would come to recognize the various types of people: that is, it would induce patterns corresponding to bank tellers, feminist bank tellers, women, (other) women named Linda, and other superordinate categories that are germane to Linda's cohort.

What would the system do upon meeting Linda? Clearly, Linda's description would activate features associated with feminist bank tellers at least a little since she is described as a female (which most tellers are) and concerned with liberal causes (which most feminists are). The description would activate the features associated with bank tellers less well because there is nothing in the description that would match the pattern for bank tellers except for being female. Moreover, even if the attribute "works with money" were added to Linda's description, the resulting pattern of activation would still more nearly correspond to that representing feminist bank teller than that for teller since the new attribute would be common to both.

But this is just to say that the connectionist system thinks, in some sense, that given the choice between responding to Linda as a bank teller and responding to her as a *feminist* bank teller, the latter course is likely to be more appropriate since it has at least some appreciable support. Obviously, we have taken some liberty in translating the notion of activation level into the linguistic variable of likelihood. But we only mean to say that if the system has come to embody not just patterns, but pattern-action sequences, and if it is somehow restricted to the actions for feminist bank tellers and undifferentiated bank tellers, then it will respond as though the action for feminist bank teller is the more appropriate for application to Linda. To borrow some language from behavioral decision theory, a system that evokes the pattern for feminist bank tellers in preference to the pattern for undifferentiated bank tellers is making a bet about Linda based on its past experience and on the available options.

Now consider two versions of a new problem:

Version 1: In four pages of a novel (about 2,000 words), how many words would you expect to find that have the form _ _ _ _ _ _ n _ (seven letter words with n in the sixth position)?

Version 2: In four pages of a novel (about 2,000 words), how many words would you expect to find that have the form _ _ _ _ _ i n g (seven letter words ending in "ing")? (Adapted from Tversky & Kahneman, 1983a, p. 295).

Subjects given Version 2 of this problem typically produce estimates that are larger than subjects given Version 1 despite the fact that all seven letter words ending in "ing" have n in the sixth position. Tversky and Kahneman attribute this violation of the conjunction rule to reliance on the availability heuristic since the retrieval cue "ending in 'ing'" is, in fact, more helpful in word generation tasks than the retrieval cue "has n in the sixth position."

Now consider how a connectionist word identification model might respond to the two versions of the estimation problem. Given a system trained on all English words of seven letters, the partial pattern _ _ _ _ _ i n g would probably match some words well enough that their activation levels would be amplified by the ogival output function (at least initially; such systems often include mutually inhibitive response nodes that allow the system eventually to settle on a single best-fitting identification). In contrast, the partial pattern _ _ _ _ _ _ n _ might match more words initially but each to such a weak degree that their activation would be attenuated by the ogival function. Thus, the total activation of the system during the beginning of the recognition process would provide a nonextensional measure of the familiarity of partial patterns. In this way, a connectionist system would encode familiarity in availability-like terms even though the system itself could not be said to have a rule corresponding to the availability heuristic. (We note here that Cohen, 1981, made essentially the same point about availability consisting of people's reliance on easily accessible data rather than being a procedure for finding new things out. Since then, Shedler, Jonides, and Manis, 1988, have provided supporting evidence from path analyses showing that availability is often correlated with judgments of frequency and likelihood without being causally related to them.)

The final problem has the following general form:

John P is a meek man, 42 years old, married with two children. His neighbors describe him as mild mannered, but somewhat secretive. He owns an import-export company based in New York City, and he travels frequently to Europe and the Far East. Mr. P was convicted once for smuggling precious stones and metals (including uranium) and received a suspended sentence of 6 months in jail and a large fine. Mr. P is currently under police investigation. Please rank the following statements by the probability that they will be among the conclusions of the investigation. (a) Mr. P is involved in espionage and the sale of secret documents. (b) Mr. P is a drug addict. (c) Mr. P killed one of his employees. (Adapted from Tversky & Kahneman, 1983a, p. 306).

In this problem, subjects given a set of conclusions like those above tend to rank item c as relatively improbable. However, when item c is changed to read "Mr. P killed one of his employees to prevent him from talking to the police," subjects violate the conjunction rule by increasing its ranking.

According to Tversky and Kahneman, the violation comes about because subjects tend to anchor on the simpler-to-assess conditional probability that Mr. P killed an employee given that the employee was going to inform on him. Getting to the requested probability that Mr. P killed an employee *and* the employee was going to inform on him requires adjusting downward in accord with the small probability that Mr. P killed an employee to begin with. Since adjustments are typically insufficient, the probability of the conjunction is overestimated.

Although this is a more complicated situation than we have encountered previously and certainly more complicated than any extant connectionist model is prepared to handle, we can imagine how the problem would appear to a connectionist model that has been trained on hundreds and hundreds of old B movie plots. Over time, the model would discover patterns corresponding to various canonical plot structures, including ones like "Someone kills someone else," "Someone becomes a drug addict," "Criminal kills someone in order to keep from being exposed," and so forth. Given the storyline in the problem, then, it is clear that the plot about the criminal killing an informer will fit better than the general purpose plot about someone killing someone else. Thus, in our connectionist gloss, the system doesn't anchor so much as it fits the story to a previously discovered plot pattern, and adjustment (when it occurs) is just the lessened degree of fit that is achieved if the input story has features that do not match the stored pattern.

Intelligence and Action

The bottom line, then, is that in certain cases, representativeness, availability, and anchoring and adjustment can be seen not as rules that are applied by people consciously seeking answers to problems, but rather as emergent properties of a general pattern recognition process that automatically and continually matches present patterns of stimulation to past patterns induced from the flow of experience. In this sense, we claim that not all intelligent thought embodies the sequential, deductive, deliberate, and conscious character that conceptions of intelligence based on the serial computer require. Although much of the macrostructure of thought does seem to have sequential and deductive structure, we believe that underlying thought processes usually are parallel, inductive, automatic, and, save for their final results, mostly unconscious.

Many of the problems that Kahneman and Tversky use engage us in what Bruner (1986) refers to as the narrative mode of thought. This is the mode that "deals in human or human-like intention and action and the vicissitudes and consequences that mark their course" (p. 13). It is the mode that gives us "good stories, gripping drama, [and] believable (though not necessarily 'true') historical accounts" (p. 13). Unlike the rational (or what Bruner terms "paradigmatic") modes of probability theory and logic that attempt to strip situations down to essentials, the narrative mode steeps us in the richness and verisimilitude of specific detail.

Most of the problem studies discussed in this paper are filled with specific detail that has no formal place in the normative structures that Tversky and Kahneman see as applicable to the problems. Why, then, is it included? Probably for no more nefarious reason than to interest subjects. Nevertheless, once included, the extraneous detail unavoidably engages us in the narrative world of human intention and action.

On first acquaintance, the three heuristic rules that Tversky and Kahneman describe seem to belong to the sequential, deliberate, and conscious world of formal processes (despite their putative deficiencies). At the very least, published processing accounts suggest that subjects *execute* heuristics by *comparing* sample populations or by *attempting to generate* exemplars of the categories to be compared or by *selecting* a starting point and then *adjusting* its magnitude.

Although we accepted these claims ourselves for years, the challenge of thinking about thought in parallel terms has encouraged us to probe our own experiences with these problems to see if the rule gloss rings true. We have discovered, belatedly, that it does not. Even though the processing stories that purport to be explanations for the data make sense in terms of the pragmatic structure of the problems, in most instances, the heuristic answer springs immediately to mind without the kinds of intervening steps that are usually suggested by the rule metaphor.

We claim, then, that the heuristic processing that Tversky and Kahneman see underlying judgments of probability and frequency is intrinsic to connectionist systems and, therefore, implicit in the pattern-based reasoning processes that seem to provide the substrate for human thought. If this is so, the human brain has evolved in a way that is less optimal for handling some kinds of problems than if it had hit on extensional reasoning.

But why *didn't* extensional reasoning arise as a fundamental feature of biological intelligence? It could have, of course, but only if the feedback coming from the environment systematically reinforced extension-action sequences rather than pattern-action sequences.

The fact is, or so it appears to us, that problem solutions are only right if they solve the problem at hand. For most people, the birth order BGBGGB seems more frequent than BGBBBB though it is not. But the cost of this small error made in the lab is sufficiently outweighed by the benefits of generally not giving inordinate weight to rules that only apply to specific unusual situations. For the student in a statistics class, on the other hand, or a sophisticated professional who might place high value on appearing rational in front of his peers, it may pay to practice with this kind of problem so that their characteristic features will be recognized in the future and the specialized probability rule will be the one that is most strongly activated. (This is analogous to the fact that most of us don't worry enough about falling for the old "when a plane crashes on an international boundary, where are the survivors buried?" trick to alter psycholinguistic processes that serve well for comprehending language under naturally adverse conditions. Linguists and lawyers, on the other hand, must refine their skill at bypassing intended meaning and seeing the host of other meanings that might be enabled in altered circumstances.)

The world confronts us with the need for appropriate action, and appropriate action, in turn, requires some degree of specificity (Holland, 1986). Although it is true that Linda is extensionally more likely to be a bank teller than a feminist bank teller, we are more likely to engage her interest in conversation (should that be a goal) if our opening gambit concerns a wildcat strike among discriminated-against tellers in a small town than if it concerns interest rates on passbook savings. If the point is not clear, then consider this: if extensions are relevant in determining appropriate actions toward Linda, then we are best off to treat her as a material object since that, of course, is an even more probable extension.

Obviously, this is not what Tversky and Kahneman have in mind when they contrast "compelling logical rules" with "seductive nonextensional intuitions" (1983a, p. 314). For them, problems are problems and people ought to develop machinery for answering both extensional and nonextensional questions. And so people do when they enroll in probability courses and practice hard at defeating common sense in specialized circumstances. Things would be different if there were clear payoffs for extensional thinking outside the classroom and the lab. But we cannot overlook the fact that formal probability theory developed very late in human intellectual history (Hacking, 1975). Had probabilistic thinking truly been as central to good action as geometry and arithmetic, we believe it would have arisen sooner and would underlie more of natural thought.

We began this chapter with the goal of exploring the differences between intelligence and rationality. We end with the insight that intelligence is a natural product, honed on the stone of experience both individually and evolutionarily. The goal of building artificial intelligence is the goal of comprehending and encompassing not only the hardware and logic that have evolved, but also the compiled accounting of what has been important and what has not.

Rationality, on the other hand, is so much a human artifact that one never even speaks of artificial rationality. Like all things human, this rationality is limited in its scope and usefulness in a complex world. Intelligence is bigger, murkier, and more serious stuff. It has had to keep us alive as a species and as individuals even at the cost of error. Robustness, generality, and practicability are the demands of life itself. They reside in intelligence because they must.

Department of Psychology
University of Wisconsin
Madison, Wisconsin 53706
USA

---

*This chapter is dedicated to the memory of a good friend.

REFERENCES

Allais, M. (1979). The foundations of a positive theory of choice involving risk and a criticism of the postulates and axioms of the American School. In M. Allais and O. Hagen (Eds.), *Expected utility hypotheses and the Allais paradox*. Dordrecht, Holland: Reidel. (Paper published originally in French in 1953).

Bar-Hillel, M. (1973). On the subjective probability of compound events. *Organizational Behavior and Human Performance*, **9**, 396-406.

Baron, J. (1985). *Rationality and intelligence*. London: Cambridge University Press.

Bruner, J. (1986). *Actual minds, possible worlds*. Cambridge MA: Harvard University Press.

Cherniak, C. (1986). *Minimal rationality*. Cambridge MA: MIT Press.

Cohen, L. J. (1981). Can human irrationality be experimentally demonstrated? *The Behavioral and Brain Sciences*, **4**, 317-331.

Berkeley, D., & Humphreys, P. (1982). Structuring decision problems and the 'bias heuristic.' *Acta Psychologica*, **50**, 201-252.

Christensen-Szalanski, J. J. J., & Beach, L. R. (1984). The citation bias: Fad and fashion in the judgment and decision literature. *American Psychologist*, **39**, 75-78.

Edwards, W. (1968). Conservatism in human information processing. In B. Kleinmuntz (Ed.), *Formal representation of human judgment*. New York: Wiley.

Ellsberg, D. (1961). Risk, ambiguity, and the Savage axioms. *Quarterly Journal of Economics*, **75**, 643-669.

Gigerenzer, G., Hell, W., & Blank, H. (1988). Presentation and content: The use of base rates as a continuous variable. *Journal of Experimental Psychology: Human Perception and Performance*, **14**, 513-525.

Hacking, I. (1975). *The emergence of probability*. London: Cambridge University Press.

Holland, J. H. (1986). Escaping brittleness: The possibilities of general-purpose learning algorithms applied to parallel rule-based systems. In R. Michalski, J. Carbonell, & T. Mitchell (Eds.), *Machine learning: An artificial intelligence approach*. Vol. II. Paolo Alto, CA: Morgan Kaufman.

Irwin, F. W., Smith, W. A. S., & Mayfield, J. F. (1956). Tests of two theories of decision in an 'expanded judgment' situation. *Journal of Experimental Psychology*, **51**, 261-268.

Kahneman, D., Slovic, P., & Tversky, A. (1982). *Judgment under uncertainty: Heuristics and biases*. London: Cambridge University Press.

Kahneman, D., & Tversky, A. (1972). Subjective probability: A judgment of representativeness. *Cognitive Psychology*, **3**, 430-454.

Kahneman, D., & Tversky, A. (1979). Prospect theory: An analysis of decision under risk. *Econometrica*, **47**, 263-291.

Kohonen, T. (1984). *Self-organization and associative memory*. Berlin: Springer-Verlag.

Lichtenstein, S., Slovic, P., Fischhoff, B., Layman, M., & Combs, B. (1978). Judgment frequency of lethal events. *Journal of Experimental Psychology*, **4**, 551-578.

Lichtenstein, S., Slovic, P., & Zink, D. (1969). Effect of instruction in expected value on optimality of gambling decisions. *Journal of Experimental Psychology*, **79**, 236-240.

Linsker, R. (1988). Self-organization in a perceptual network. *Computer*, **21**, 105-117.

Lopes, L. L. (1991). *The rhetoric of irrationality*. Theory of Psychology. In press.

Margolis, H. (1987). *Patterns, thinking, and cognition: A theory of judgment*. Chicago, IL: University of Chicago Press.

Massaro, D. W. (1987). *Speech perception by ear and by eye: A paradigm for psychological inquiry*. Hillsdale, NJ: Erlbaum Associates.

McClelland, J. L., & Rumelhart, D. E. (Eds.) (1986). *Parallel distributed processing: Explorations in the microstructure of cognition*. Vol. 2. Cambridge, MA: MIT Press.

Montgomery, H., & Adelbratt, T. (1982). Gambling decisions and information about expected value. *Organizational Behavior and Human Performance*, **29**, 39-57.

Oden, G. C. (1988). FuzzyProp: A symbolic superstrate for connectionist models. *Proceedings of the IEEE International Conference on Neural Networks*, Vol. I, 293-300.

Peterson, C. R., & Beach, L. R. (1967). Man as an intuitive statistician. *Psychological Bulletin*, **68**, 29-46.

Platt, J. R. (1964). Strong inference. *Science*, **146**, 347-353.

Robinson, J. A. (1965). A machine-oriented logic based on the resolution principle. *Journal of the Association for Computing Machinery*, **12**, 23-41.

Rumelhart, D. E., Hinton, G. E., & Williams, R. J. (1986). Learning internal representations by error propagation. In D. E. Rumelhart & J. L. McClelland (Eds.) *Parallel distributed processing: Explorations in the microstructure of cognition*. Vol. 1. Cambridge, MA: MIT Press.

Rumelhart, D. E., Smolensky, P., McClelland, J. L., & Hinton, G. E. Schemata and sequential thought processes in PDP models. In J. L. McClelland & D. E. Rumelhart (Eds.) *Parallel distributed processing: Explorations in the microstructure of cognition*. Vol. 2. Cambridge, MA: MIT Press.

Rumelhart, D. E. & Zipser, D. (1985). Feature discovery by competitive learning. *Cognitive Science*, **9**, 75-112.

Savage, L. J. (1954). *The foundations of statistics*. New York: Wiley.

Schoemaker, P. J. H. (1982). The expected utility model: Its variants, purposes, evidence, and limitations. *Journal of Economic Literature*, **20**, 529-563.

Shanteau, J. C. (1970). An additive model for sequential decision making. *Journal of Experimental Psychology*, **85**, 181-191.

Shanteau, J. C. (1972). Descriptive versus normative models of sequential inference judgment. *Journal of Experimental Psychology*, **93**, 63-68.

Shedler, J., Jonides, J., & Manis, M. (1988). *Availability: Plausible but questionable*. Unpublished manuscript, Department of Psychology, University of California, Berkeley CA 94720.

Slovic, P. & Tversky, A. (1974). Who accepts Savage's axiom? *Behavioral Science*, **19**, 368-373.

Tversky, A., & Kahneman, D. (1973). Availability: A heuristic for judging frequency and probability. *Cognitive Psychology*, **5**, 207-232.

Tversky, A., & Kahneman, D. (1974). Judgment under uncertainty: Heuristics and biases. *Science*, **185**, 1124-1131.

Tversky, A. & Kahneman, D. (1983a). Extension versus intuitive reasoning: The conjunction fallacy in probability judgment. *Psychological Review*, **90**, 293-315.

Tversky, A. & Kahneman, D. (1983b). Judgments of and by representativeness. In D. Kahneman, P. Slovic, & A. Tversky (Eds.), *Judgment under uncertainty: Heuristics and biases*. London: Cambridge University Press.
von Neumann, J. & Morgenstern, O. (1947). *Theory of games and economic behavior*. Princeton NJ: Princeton University. 2nd ed.

*Poznań Studies in the Philosophy
of the Sciences and the Humanities
1991, Vol. 21, pp. 225-249*

Gerd Gigerenzer

ON COGNITIVE ILLUSIONS AND RATIONALITY

The controversy as to whether mathematical probability theory has anything to do with human rationality is as old as the theory itself (Daston 1988). The latest episode of this continuing controversy was sparked by developments in psychology, and has contributed a new, distinctly psychological wrinkle to the time-worn themes: the analogy between that stock-in-trade of perceptual psychology, the visual illusion, and what both sides of the current controversy have termed "cognitive illusions".[1] Amos Tversky and Daniel Kahneman (1974) introduced their research program on judgment under uncertainty with the analogy between biases in probabilistic reasoning and visual illusions, and others chose the generic term "cognitive illusions" for these biases in order "to suggest that these phenomena are quite similar to a variety of perceptual illusions extensively studied by psychologists" (Edwards & von Winterfeldt, 1986, p. 643). L. Jonathan Cohen has challenged the repeated claims that intuitive reasoning is biased and irrational, e.g., that experimental results have "bleak implications for human rationality" and that the human mind lacks "the correct programs for many important judgmental tasks" (Nisbett & Borgida, 1975, p. 935; Slovic, Fischhoff & Lichtenstein, 1976). Nevertheless, Cohen, too, described reasoning fallacies "as cognitive illusions ... to invoke the analogy with visual illusions" (Cohen, 1981, p. 324). Both sides, although diametrically opposed in almost all other respects, share the analogy between cognitive and visual illusions.

Ironically, the analogy originally pointed in the other direction, from visual illusions to statistical reasoning. The visual space is tridimensional whereas the retina, the perceiving organ, is bidimensional. How does the brain get three dimensions out of two?, asked George Berkeley in his *Essay Towards a New Theory of Vision* (1709). From Berkeley to the present, many theories have used reasoning, in particular, statistical reasoning, as an analogy for understanding what perception does. Hermann von Helmholtz spoke of "unconscious inferences", Egon Brunswik suggested parameter estimation by multiple regression as the possible mechanism of perception, and R. L. Gregory compared the inferential mechanism of perception to Fisherian statistical inference and Bayesian probability revision. For these theorists, perception is "betting against reality", and what we perceive is essentially illusory, where illusions are understood as deviations from physical reality. Thus, statistical reasoning is and has been a key analogy for understanding perception and perceptual illusions. Since the 1970s, the direction of the analogy has been reversed. In order to understand intuitive inductive reasoning, in particular so-called errors in probabilistic reasoning, visual illusions have now become the analogy of choice. A paradoxical, a circular, or a fruitful inversion of analogy?

In this paper, it will become clear that the analogy with visual illusions means different things to those who attack and to those who defend rationality in intuitive reasoning. I will first lay out three aspects of visual illusions that are relevant for the analogy. Second, I will argue that two of these aspects hold for neither version of the analogy. Third, I will suggest that the third aspect, which is largely ignored by both sides (though less so by Cohen), has real potential for an understanding of the nature of intuitive probabilistic reasoning. Developing this third point further, I will show how experimental studies can profit from philosophers' conceptual distinctions. This research vindicates Cohen's claim that even untutored intuition is capable of making conceptual distinctions of the sort philosophers make.

Visual illusions

One of the best-known visual illusions is the Müller-Lyer arrow figure (Figure 1). Visual illusions are a subset of perceptual illusions, to be distinguished from purely optical illusions such as the bent-stick-in-water

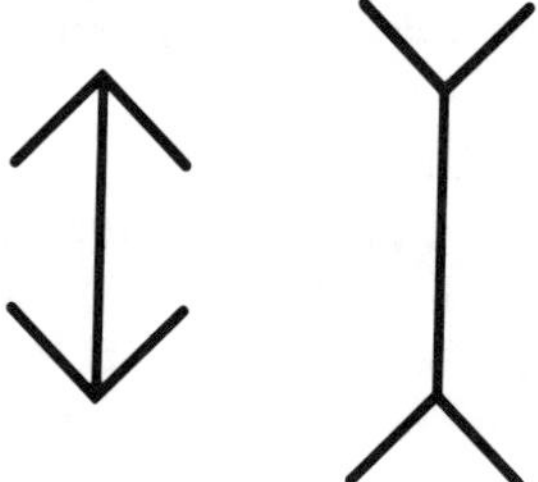

Figure 1: One variant of the Müller-Lyer arrow figure (originally devised by F. C. Müller-Lyer in 1889 in fifteen variants). The two vertical lines are of the same length, although the inward-going and the outward-going arrows make the lines appear smaller and larger, respectively.

effect, and sensory illusions, such as after-images occurring after intense stimulation of the retina. There is a great divide between two major accounts of perception with respect to the nature and the role of perceptual illusions. In the Helmholtzian view, as I mentioned above, only limited and uncertain sensory data is available, and to perceive means to go beyond the information given. Perception interprets incomplete information, and perceptual illusions are both the necessary consequence of this impossible task and the key to understanding the inference mechanism. In the Gibsonian view, in contrast, all the information needed for veridical perception is available in the "ambient array of light", and is simply "picked up" by the observer. Visual illusions are of no or little interest; in fact, there should be no uncertainty and consequently, no illusions. If visual illusions are dealt with, they are attributed to a restricted and unnatural stimulus environment, such as that created by experimenters in traditional laboratory experiments.

For the purpose of this paper, I will list three aspects of the Helmholtzian view of visual illusions which I consider relevant to the analogy with cognitive illusions.

(1) *The "uncontroversial norm" aspect.* In visual illusions, there exists an uncontroversial norm, physical measurement, which defines systematically deviating judgments as illusions. For instance, in the Müller-Lyer arrow figure, this norm is the physical length of the lines.

(2) *The "stability" aspect.* Visual illusions are normally stable. Although visual illusions seem to be a function of early experience and age, the direction and extent of the illusion is relatively stable in humans and probably also in many animals (Gregory, 1974, p. 360). What can be done? Not much, even after having measured the lines, the illusion persists despite better knowledge.[2]

(3) *The "context and prior knowledge" aspect.* Since to perceive means to infer something that is not directly available in the proximal (sensory) information, these inferences have to rely on context information and prior knowledge. This inference mechanism produces both veridical perception (perceptual constancies) and perceptual illusions. According to Gregory (1974), for instance, the Müller-Lyer illusion originates from the way the context (the arrows at the ends of the lines) are taken into account. The arrows automatically generate a three-dimensional representation of the two-dimensional figure, inward-going arrows suggesting the line as the outside (front) corner of a three-dimensional object, and outward-going arrows suggest the line as the inside (back) corner.

Illusions are not due to contingent limitations of the brain, but rather to the nature of the brain's tasks; i.e., to read information from images which at best contain such information only implicitly. Illusions are, therefore, a *necessary consequence of the task*. Gregory (p. 376) conjectures that the brain possesses probably the best perceptual system available and that illusions will be a necessary part of all efficiently designed visual machines.

This is of course a simplified account of visual illusions, relying solely on the Helmholtzian tradition. Since, however, those who originally invoked the analogy have themselves referred to the importance of studying errors and to the betting metaphor of perception (Kahneman & Tversky, 1982b, p. 512), this may be a fair perspective to start with.

Cognitive illusions

Consider the *neglect of base rates* in one of the most well-known demonstrations by Kahneman and Tversky (1973). One group of subjects was given the following problem, known as the Engineer-Lawyer Problem:

A panel of psychologists have interviewed and administered personality tests to 30 engineers and 70 lawyers, all successful in their respective fields. On the basis of this information, thumbnail descriptions of the 30 engineers and 70 lawyers have been written. You will find on your forms five descriptions, chosen at random from the 100 available descriptions. For each description, please indicate your probability that the person described is an engineer, on a scale from 0 to 100.

The same task has been performed by a panel of experts, who were highly accurate in assigning probabilities to the various descriptions. You will be paid a bonus to the extent that your estimates come close to those of the expert panel (p. 241).

Subjects in a second group were given the same instructions but with inverted base rates — that is, they were told that there were 70 engineers and 30 lawyers. Both groups were given the same descriptions, for example:

"Jack is a 45-year-old man. He is married and has four children. He is generally conservative, careful, and ambitious. He shows no interest in political and social issues and spends most of his free time on his many hobbies, which include home carpentry, sailing, and mathematical puzzles" (p. 241).

The authors concluded that base rates were largely ignored by their subjects, since mean probability judgments were about the same in both groups. This base rate neglect was considered a systematic bias in reasoning and was explained by a general heuristic called representativeness: people judge the posterior probability (that a person is an engineer given the above information) largely by the similarity between the description and their stereotype of an engineer. The generality of the heuristic and the neglect of base rates seemed to be confirmed by the fact that the median probability judgments for an uninformative description, "Dick", were also the same (.50), although this description was constructed to contain only worthless information with respect to engineers and lawyers. Dick was described as, *inter alia*, a man of high ability, high motivation, and as well-liked by his colleagues. Only when no description was given, did the probability judgments follow the base rates.

According to the "systematic bias" view, this neglect of base rates satisfies the first two of the above aspects of visual illusions. First, there seems to be an uncontroversial norm that justifies calling a systematic deviation from that norm properly a cognitive illusion, in the sense of a bias in reasoning. In the present case this norm is Bayes' theorem (which Kahneman and Tversky used to calculate the "correct" differences between the probability judgments in both groups). More generally, the norm is referred to as the rules of statistics or "the rules of probability in intuitive judgments" (Tversky & Kahneman, 1977, p. 173). Second, this cognitive illusion seems to be a stable one, at least in this task, it is "a highly robust effect", as we are told by several experimenters (e.g., Tversky & Kahneman, 1977, p. 176; von Winterfeldt & Edwards, 1986, p. 535). Even critics talk about an "overall *tendency* to neglect-base rates" (Adler, 1984, p. 173). Let us now have a closer look at the analogy.

Uncontroversial norms: a single yardstick for rationality?

This first aspect of the perceptual/cognitive analogy maintains: For the reasoning problems posed, there exists a single correct answer just as there does for perceptual problems. A deviation between judgment and normative answer, if systematic, defines an illusion, visual or cognitive. Kahneman and Tversky (1982a, p. 493) explicitly draw this analogy between perceptual and cognitive illusions: "The presence of an error of judgment is demonstrated by comparing people's responses either with an established fact (e.g., that the two lines are equal in length) or with an accepted rule of arithmetic, logic, or statistics". In this analogy, standard probability theory is to inductive reasoning what a ruler is to perceptual judgment. This first aspect of the analogy is at best highly problematic, and, at worst, misleading at three levels.

Level of formal theories: There exist alternatives to standard probability theory (which is based on the Kolmogoroff axioms) as the norm of reasoning under uncertainty. Among others, such non-standard theories have been proposed by Jakob Bernoulli, Johann Heinrich Lambert,

L. Jonathan Cohen, Henry E. Kyburg, and Glenn Shafer. There are good reasons to prefer several alternative theories to a single one, if we want to model what we commonly call "rationality". Among these is the evolutionary reason that it could well be preferable to have a whole toolbox of probability approaches to varied environments, rather than a single specialized tool that works very well only for a few environments. A different kind of reason is that the meaning of "rationality" is notoriously various; it may for instance refer to prediction or to explanation. Standard probability theory does well at defining what it means to minimize errors of prediction in the long run, whereas non-additive probabilities can do well at modeling how the evidence of particular observations bears on several hypotheses.

Level of interpretation: Even within standard probability theory, the question of what belongs to its domain of application is open to interpretation. Is the theory about degrees of belief, relative frequencies, propensities, or about something else? I will argue later that these distinctions are relevant to understanding intuitive reasoning.

Level of representation: Even within standard probability theory and a particular domain of application, the question whether the structure of a particular reasoning problem can be represented or not by a specific statistical model must be carefully examined.

All three levels are important in answering the question of whether there exists exactly one answer to a given problem, or if not, what alternative solutions exist. None of these three complications exist for visual illusions. For measuring the length of two lines, for instance, alternative physical theories, interpretations of domain, and the question of representation play no role. The point Cohen made, and, which I want to follow up here, is that for many reasoning problems, in contrast, more than one resonable answer can be found, even within the narrow confines of a given formal theory and a given interpretation. This holds especially for real-world problems, but also for textbook problems of the real-world kind, and to a lesser degree for problems of the schematic urn-and-balls problems type. In general, the more the urn-and-balls problems are filled with content and context, the more alternative norms can be developed. To make this general statement more precise, I will attempt to state conditions under which we may be justified in claiming the contrary, that is, that a given reasoning problem does indeed have only one correct answer. To make my point stronger, I will consider only the level of representation, assuming consensus about formal theory and interpretation.

I propose that it is justified to claim that a reasoning problem has one and only one correct answer (as is regularly claimed in the literature on biases and heuristics without justification) if and only if two conditions hold, which I call "concept isomorphism" and "structural isomorphism" (Gigerenzer & Murray 1987, chapter 5). I will argue that in order to check these conditions, not only formal structure, but also context and content-related prior knowledge are relevant. I will use Bayes' theorem and textbook problems for illustration.

Concept isomorphism holds if and only if there is a one-to-one mapping between the concepts in a statistical model (e.g., the prior probabilities and the likelihoods in Bayes' theorem) and the *relevant* concepts of a reasoning problem (e.g., base rates mentioned for engineers and lawyers). This means that each statistical concept can be identified with one and only one concept in the problem, and vice versa, and no concept on either side remains. Figure 2 illustrates a situation in which concept isomorphism is violated. For instance, there may be two or more candidates for the prior probability; or some relevant context information in the problem may have no representation in the statistical model.

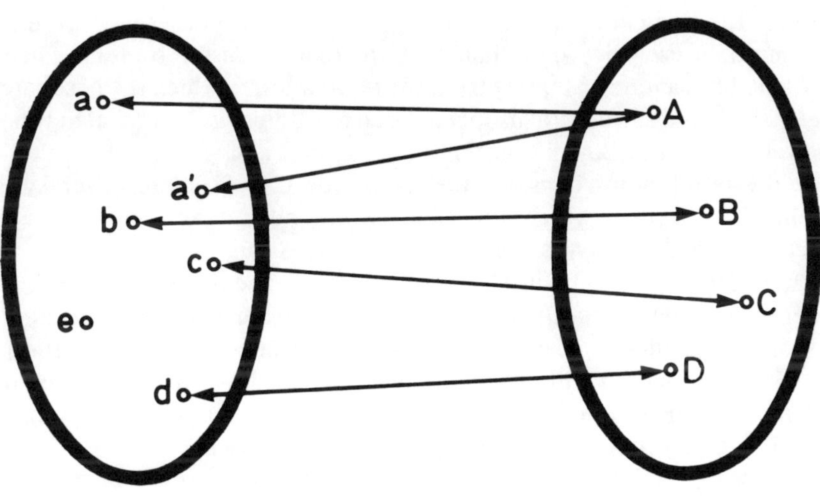

Figure 2: Points in the left circle represent relevant concepts in a reasoning problem, those in the right circle represent concepts of a statistical model. Concept isomorphism is violated in two respects, there exist two different candidates *a* and *a'* for *A*, and the concept *e* is not represented in the statistical model.

To illustrate the latter: In the instructions to the Engineer-Lawyer Problem, subjects were told that a panel of experts "were highly accurate in assigning probabilities to the various descriptions". What can this mean? A subject who considers this information as relevant might infer since being "highly accurate" could only mean to assign probabilities close to zero or to one, this is a hint that the descriptions are highly informative and reliable, and that the solution lies in the descriptions, if only you could decipher them. Thus, this information can be read to imply, in essence, "use probabilities close to zero or one only, if you want to get a large bonus". There is, however, no way to account for this information in the standard application of Bayes' theorem to the problem. The general point is: What qualifies as "relevant" information in the reasoning problem depends on prior knowledge, and is not always exhausted by the concepts in the statistical model.

Structural isomorphism postulates that the structural assumptions underlying a statistical model hold in the reasoning problem (structural homomorphism), and, vice versa, that the structural properties of the reasoning problem are represented by a similar structure in the statistical model. For instance, structural assumptions of the formal model that must hold in the Engineer-Lawyer Problem are that the hypotheses considered are mutually exclusive and exhaustive, and that the descriptions have been obtained by random sampling from the population to which the base rates refer to. Among structural properties of a reasoning problem that may not be represented in Bayes' theorem are temporal and spatial relations. To see the latter point consider the notorious Cab Problem (Tversky & Kahneman, 1980), which describes a hit-and-run accident at night involving a cab and gives the following information:

(i) 85% of the cabs in the city are Green and 15% are Blue.

(ii) A witness identified the cab as a Blue cab. The court tested his ability to identify cabs under the appropriate visibility conditions. When presented with a sample of cabs (half of which were Blue and half of which were Green) the witness made correct identifications in 80% of the cases and erred in 20% of the cases.

Question: What is the probability that the cab involved in the accident was Blue rather than Green? (p. 62).

Tversky and Kahneman look at this problem with the formal structure of Bayes' theorem firmly in mind, and identify the prior probabilities with the numbers given in (i) and the likelihoods p("Blue"/Blue) and p("Blue"/Green) with those given in (ii), .80 and .20, respectively (p("Blue"/Blue) is the probability that the witness says "Blue" if the cab is actually Blue).[3] Inserting these values into Bayes' theorem, they calculate p(Blue/"Blue") = .41, which they consider as the single normative answer. Therefore their subjects' median judgment of .80 appears to be an error of reasoning. Again, the error is identified as neglect of base rates. However, if one looks at Bayes' theorem (as applied) with the structure of the *problem* in mind, one realizes that one relevant structural aspect of the problem is not represented in the Bayesian modeling: the temporal sequence of the witness' two judgments, the first one at time t_1, the night of the accident, and the second one at time t_2, at the court's test. The likelihoods specified in (ii) characterize the witness' performance at t_2, but not at t_1. At issue are, however, the likelihoods at t_1. Several arguments can be made that and how the likelihoods have changed from t_1 to t_2. For instance, assume that both at t_1 and t_2 the witness knew the base rates and judged in a way that minimized judgmental errors, that is, the sum p("Blue"/Green)p(Green) + p("Green"/Blue) p(Blue). Since we are informed in (ii) that the base rates changed from t_1 to t_2, from 15% to 50%, we are then forced to conclude that the likelihoods also changed from t_1 to t_2. Calculating the likelihoods for t_1 and inserting them into Bayes' theorem gives p(Blue/"Blue") = .82, an alternative solution to Tversky and Kahneman's, who ignored the difference between t_1 and t_2 and calculated .41. Incidentally, .82 is close to the median judgment of untutored people (.80), from which latter figure base rate neglect was concluded. Paradoxically, .82 is calculated on the assumption that base rates are taken into account (see Birnbaum, 1983, Gigerenzer & Murray, 1987, pp. 168-173). Of course, there are other reasonable assumptions for a "criterion shift" from t_1 to t_2, which lead to alternative reasonable answers. This one, however, may suffice for making the point.

The example shows that structure that is in the problem but not in the formula matters for the claim that there is a single normative answer. Note that this holds despite the fact that we agreed on a particular formal theory and interpretation. Reasoning problems that are structurally equivalent from the point of view of a statistical model can have different surplus structures.

Conceptual and structural isomorphism are not easily attained for textbook reasoning problems, and this holds *a fortiori* for their real-world versions, such as guessing someone's profession in the cafeteria, or a legal judgment as to whether someone committed a hit-and-run accident. Personal judgment about what are the relevant concepts in a problem situation and about whether isomorphism holds is indispensable, and context and prior knowledge are necessary and legitimate cues in making this judgment. Thus, to speak of a single correct answer in reasoning in analogy to a single measure of length for judging visual accuracy, is to presuppose consensus between experimenters and subjects at all three levels. This shows why the analogy between cognitive and visual illusions is highly misleading with respect to the existence of an uncontroversial norm. We may not legitimately speak of a single norm unless we have first made all our premises, such as random sampling, clear to the untutored layman. The results of the Engineer-Lawyer Problem reported in the next section indicate that such clarification does indeed matter for untutored judgments. The practical consequence of this lack of analogy is: Try to explain people's judgments as they are, not their deviation from some preconceived norm.

Cohen's use of the analogy with visual illusions is different. He challenges standard probability theory, which he calls Pascalian probability, as the only norm for rational reasoning. The alternative norm he proposes is called Baconian probability. Thus, for Cohen at least two alternative theories of probability exist, Pascalian and Baconian, and several interpretations of their domains are possible, but again only one norm for rationality exists. This norm is the intuition of our fellow adults, untutored in probability theory. Given this norm, systematic errors in probabilistic reasoning (i. e., competence rather performance errors) are by definition ruled out — except for the immature, the senile, etc. If, nevertheless, errors do occur, Cohen attributes them either to "cognitive illusions" or to lack of intelligence or mathematical education. Cohen's cognitive illusions are created by the experimenters, not committed by the subjects. How do experimenters do it? Like the Gibsonians, Cohen (1981, p. 324) proposes that such illusions are created by using (i) experimental tasks not representative of the normal conditions of life (e.g., check-ups are excluded), and (ii) unfamiliar or abstract material. For Cohen, the analogy with visual illusions means deception. The experimenter is compared to a conjurer who relies "on the visual inattentiveness of those who are watching, to hold the latter back from obtaining an appropriate additional input to their visual information-processing operation" (p. 325). Note that Cohen's use of the term "cognitive illusion" differs significantly from the notion of systematic biases or fallacies in reasoning. So does his notion of "visual illusion". Cohen is a Gibsonian in so far as illusions are attributed to the experimenter or some special conditions rather than to the very nature of the brain's task. Just as in Gibson's realism the organism simply "picks up" the appropriate information from the ambient array of light, so in Cohen's rationalism the organism "picks up" appropriate rules of reasoning from its untutored competence. Both premises are in principle immune to falsification, it seems to me. But here the analogy ends. A realist is obliged to postulate one and only one physical world, but Cohen does not postulate *one* kind of probability theory as the standard for rationality.

For the Engineer-Lawyer Problem, Cohen (1979) proposes Baconian probability as the norm. The Baconian probability that the person described is an engineer is defined as the inductive reliability of the generalization that all persons described this way are engineers. The inductive reliability is a function of the outcome of tests of the generalization (hypothesis) with an appropriate list of inductively relevant variables. For testing the hypothesis that someone is an engineer, for example, such relevant variables may be the kind of hobbies and political attitudes; in general these are factors with potential causal impact. No measurement function is available for inductive reliability (Cohen, 1979, p. 390), although it may be ranked depending on the tests a hypothesis succeeded in passing. The Baconian probability that the person described is an engineer depends solely on the causal evidence in the description, and not on the base rates of engineers. Therefore, if a description is simply irrelevant to the traits of an engineer or of a lawyer, then the two generalizations "all descriptions of type D are engineers" and "all descritions of type D are lawyers" will both have zero-grade inductive reliability. That is, both of the two Baconian probabilities p_I(engineer/description) and p_I(lawyer/description) may be zero. This reveals a major contrast with conventional or Pascalian probabilities: Baconian probabilities run from nonproof to proof, whereas Pascalian probabilities run from disproof to proof.

To summarize: the "uncontroversial norm" aspect of the analogy with visual illusions is indispensable or the argument that so-called biases in probabilistic reasoning reveal that the untutored mind is running on shoddy software, that is, on programs that work only with a handful heuristics. I have argued, however, that this aspect presupposes consensus on formal theory, interpretation, and isomorphic representation. Therefore, this aspect of the analogy is not valid, unless consensus is established (for a particular situation). Consequently, the frequent joint classification of errors in probabilistic reasoning with errors in memory or in intuitive physics under the heading "cognitive illusions" (e. g., Edwards & von Winterfeldt 1986) is not valid, too, for the same reasons. For Cohen, who does not believe in standard probability theory as the norm of rationality in the first place, this important aspect of visual illusions is either irrelevant, if probability theory is considered as norm, or circular, if intuitive reasoning competence is considered the norm of rationality.

Stability

Stability of cognitive illusions is essential for the view that Cohen attacks. If untutored intuition lacks the correct programs for probabilistic reasoning, and relies on a few general heuristics such as representativeness and availability, then cognitive illusions should be highly stable. Let us consider the Engineer-Lawyer Problem, which seems to me a fair example, since here both sides of the debate seem to accept the stability of the base rate neglect. "Regardless of what kind of information is presented, subjects pay virtually no attention to the base rate in guessing the profession of the target." (Holland et al. 1986, p. 217). Cohen (1979, p. 401), too, refers in his discussion of that problem to an "overall human tendency to prefer reasoning from causal rather than statistical data". How stable in fact is the neglect of base rates? My own research suggests, not very.

First, let us consider the uninformative description "Dick". The puzzling fact reported as evidence for stability is that even if a description is worthless, base rates are ignored. Recall that in addition to "Dick" each subject judged four informative descriptions, which function as context. In a replication of the Engineer-Lawyer study (Gigerenzer, Hell & Blank, 1988) we systematically varied the position of the uninformative description within a set of six descriptions, and separated those cases where untutored intuition encountered the uninformative description first (no context of informative description available) from those where an informative description was given first and "Dick" was in a later position. Base rate neglect largely disappeared in the former case, whereas it was maintained in the latter. Neglecting position and averaging across positions, however, as in the original study, results in an average difference (between base rate groups) which is zero or close to zero, since with six descriptions there are five times as many second-and-later positions than there are first positions.

This tells us that base rates are not automatically neglected when an uninformative description is judged. Quite the contrary. Only if one or more informative descriptions were presented first, did the subject continue to use a representativeness strategy for Dick, too. The fact that a strategy, once adopted, tends to be maintained in subsequent similar problems is well-known as "set" effect or "functional fixedness" from the Würzburg and Gestalt schools of thinking. Furthermore, this result explains the apparently contradictory findings in the litarature on the probability that Dick is an engineer. Studies reporting base rate neglect used *many* descriptions in addition to Dick, those reporting a difference between base rate groups that is about as large as the difference between base rates (Ginossar & Trope, 1980) used only Dick (i.e., always in the first position), and those reporting differences in-between (Wells & Harvey, 1978; Zukier & Pepitone, 1984) used one informative description in addition to Dick, and switched orders randomly. Note that none of these studies analyzed the position effect.

Second, let us see how to eliminate the neglect of base rates for both uninformative and informative descriptions. The interesting question is, "Do untutored people represent the problem as a probability-revision problem?" There is a crucial assumption that must hold in order to use the specified base rates as prior probabilities. The descriptions have to be *randomly* sampled from the 100 descriptions available. If not, the base rates of engineers and lawyers would be irrelevant. Kahneman and Tversky asserted random sampling of descriptions to their subjects, although this was not true. (Descriptions were deliberately constructed, e.g., to fit the American stereotype of an engineer, such as "Jack".). What about the subjects' prior experience? It is not safe to assume that in the subjects' previous experience with guessing a person's profession, these persons were randomly drawn from a population with known base rates or, at least, could be considered to be. For instance, people in several countries watch TV programs in which a panel of experts guesses the profession of a candidate, who answers only yes or no to their questions. Here, the main heuristic strategy is to ask for new information in order to increase what Cohen calls the inductive reliability of a hypothesis, whereas base rates are not important since the candidates were selected and not randomly drawn. In fact, the experts would perform badly if they started with the known base rates of professions and revised them according to Bayes' theorem. I doubt that previous experience with profession guessing suggests to the subjects the same mental representation of the problem as the experimenters had in mind. A mere verbal assertion, the word "randomly" in the instruction, may not be a contextual cue strong enough to revise an already established representation. Note that this issue is not limited to this particular experiment: In all studies on profession guessing I know of which claim a base rate fallacy, either random sampling was merely claimed and never true, or it was not even mentioned (e.g., in Kahneman & Tversky's (1973) "Tom W." problem).

How can we make sure that random sampling gets into the subjects' problem representation? There is an easy way: let the untutored people themselves *do* the random sampling of descriptions. We performed an experiment (Gigerenzer et al. 1988) in which one group of subjects did the random sampling and a second group was tested in a straightforward replication of Kahneman and Tversky's study as a control. With random sampling being explicit by doing it, base rate neglect disappeared for both informative and uninformative descriptions. Average judgments were closer to Bayesian predictions than to what the representativeness heuristic (and Baconian probability) predicts. The control condition, in contrast, largely replicated Kahneman and Tversky's original findings.

This experiment illustrates two points. First, stability is not a general aspect of cognitive illusions. What is considered a general bias of reasoning can be *easily* eliminated, one does not need thousands of repetitions or an upbringing in an entire different environment, as in the case of the Müller-Lyer illusion. Second, the easy elimination indicates the powerful role of contextual information and prior knowledge in cueing a particular mental representation of the problem.

Such results vindicate Cohen's critique of the "systematic biases" view. His concept of cognitive illusions does not imply stability. On the contrary, it means temporary deception by the experimenter. But do our results also vindicate Cohen's explanation of intuitive reasoning? Cohen (1979, p. 397) says "in the experiment about the engineers and lawyers the subjects naturally tended to go by the weight of evidence and use Baconian reasoning". And he asserts that he and Tversky and Kahneman explain intuitive reasoning in the same way: "Representativeness, as Tversky and Kahneman call it [i.e., Baconian probability], is thus not just a heuristic here, as they regard it, but rather the rationally appropriate criterion" (p. 396). I have three comments on this issue. First, although they look alike, representativeness is here not the same as Baconian probability. Consider the uninformative description, where subjects' median probabilities that the person described was an engineer were .50 in the original experiment. That is, the description is no more "representative" of an engineer than of a lawyer, and vice versa. If subjects reasoned in Baconian probabilities, the evidence that the person was either an engineer or a lawyer would be negligible, and they would have assigned probabilities of zero, or close to zero, instead.[4] Second, this shows that even in the original experiment, subjects did not reason consistently in a Baconian way. Probability judgments about informative descriptions were consistent with both Baconian and representativeness reasoning, those about the uninformative description with the latter but not the former, and those without descriptions (base rate information only) were consistent with neither. Our experiment replicates this pattern, and adds more: Judgments about an uniformative description presented first (or alone) are close to Bayesian reasoning (using the base rates specified as priors); and the same holds for all kinds of descriptions if random sampling was performed *by the subject*. Thus, Baconian reasoning is not stable, even with reference to the same task and the same person.[5] Third, Cohen (1979, p. 397) suggests that we "construe at least some such experiments as revealing how subjects do decide between Pascalian and Baconian responses". The one hypothesis, he offers is that non-experts tend to apply Baconian reasoning when causally relevant information is offered in the descriptions. On this hypothesis, real random sampling as opposed to its mere assertion should make no difference, since the amount of inductively relevant evidence remains the same. Our results contradict this particular hypothesis. Nevertheless, Cohen's general suggestion opens a highly valuable research perspective, which I will attempt to develop briefly.

Context and prior knowledge

In the Helmholtzian view, the brain must rely on context and content-dependent prior knowledge in order to make progress in its task, i.e., to infer from proximal cues (such as the bidimensional retinal picture) to the distal world (the tridimensional visual space). Proximal cues are in principle uncertain cues, even for such elementary judgments as the size of an object: The size of the retinal picture of an object is by itself only a poor indicator of its true size. As mentioned in the introduction, the Helmholtzians understand perception via the analogy of reasoning: To perceive means to select a hypothesis about the world, inferred from generalizations based on past experience and context. If content and context are indispensable for improving elementary perception, then this should hold *a fortiori* for probabilistic reasoning. Thus, if we invert the direction of the analogy, an explanation of this aspect should be quite revealing.

Let us look at the Engineer-Lawyer Problem for illustration. How do Kahneman and Tversky and how does Cohen deal respectively with the context and content of the problem? The former look at the problem with the formal structure of Bayes' theorem in mind. No context information and no content-relevant prior knowledge matters, except for that necessary to calculate the likelihoods for the formula (base rates are already numerically specified). Rational reasoning is reduced to analyzing a problem situation solely in terms of a preconceived formal theory; here, in terms of likelihoods and prior probabilities. The Helmholtzian analogy suggests a different kind of rationality. Of primary importance is the presentation of a problem, which mediates, together with prior knowledge, the subjects' representation of the problem. As we have seen above, base rate neglect disappears if subjects do the random sampling themselves, and I conjecture that this change in the presentation of a problem generates a new mental representation of the problem, one that is different from that suggested by prior experience with profession guessing. Kahneman and Tversky, however, did not analyze their subjects' representation of the Engineer-Lawyer Problem and this representation is built up from context and prior knowledge. Their focus is on deviations from formal structures.

In the present case, even their explanation of base rate neglect, representativeness, can be shown to be equivalent to the formal concept of likelihood in Bayes' theorem (Gigerenzer & Murray 1987, pp. 153-155: see also Note 4). Consequently, this "explanation" (intuition relies on likelihoods, not on base rates) is but a redescription of the phenomenon (intuition neglects base rates, but relies on likelihood).

Ironically, here Cohen seems to be the better psychologist. In various places (e.g., Cohen, 1979, 1982, 1983) he has pointed to the role of the subjects' representation of a problem. What he calls the "Norm Extraction Method" is basically a research program to generate interesting psychological explanations (of the after-the-fact kind) about how subjects could have understood a problem, and how this could explain their reasoning. He opposes his "Norm Extraction Method" to the "Preconceived Norm Method" of those he criticizes. The latter presupposes that a given reasoning problem has one single correct answer, based on the (usually implicit) assumption that there is consensus on the levels of formal theory, interpretation and isomorphic representation (see above). The Norm Extraction Method, in contrast, assumes "that, unless their judgment is clouded at the time by wishful thinking, forgetfulness, inattentiveness, low intelligence, immaturity, senility, or some other competence-inhibiting factor, all subjects reason correctly about probability: none are programmed to commit fallacies or indulge in illusions" (Cohen, 1982, p. 251). Cohen is explicit that his motivation for this claim is to protect the foundations of analytical philosophy. If we were programmed to use cognitive heuristics that sometimes generate correct answers and sometimes do not, this would, Cohen fears, "seriously discredit the claims of intuition to provide — other things being equal — dependable foundations for inductive reasoning in analytical philosophy" (Cohen, 1986, p. 150).

Is such a strong defense of human rationality necessary? The analogy with visual illusions may help us to see a third approach, besides the Preconceived Norm and Norm Extraction Methods. This is suggested by the Helmholtzian perspective, in particular, the third aspect of the visual illusion analogy. We have already seen that what Cohen terms the Preconceived Norm Method does not correspond to a Helmholtzian view in this respect. (It corresponds much more to the early Artificial Intelligence notion that perception works by following simple, general rules or algorithms.) But neither does Cohen's own program. Again, it recalls Gibson's program of determining the invariants in the ambient light that enables the realistic perception of the environment. Where the Gibsonians search for invariants, Cohen searches for norms, and both need these to prove convincingly that untutored perception or reasoning has direct access to reality and rationality, respectively. The third view, the Helmholtzian, allows both for optimal cognitive functioning *and* for systematic illusions. Such a view would require altering the Norm Extraction Method in some important ways. I will list a few such alterations.

First, according to the "content and prior knowledge" aspect, systematic reasoning errors are to be expected, and it is not necessary to attribute them to inattentiveness or to "cognitive illusions" created by the experimenter.

Second, such errors are a byproduct of the difficult nature of the task in real-world situations. In order to improve reasoning, the mind must take advantage of contextual information and prior knowledge. For instance, consider the "functional fixedness" or "set effect" mentioned above. On the hypothesis that similar problems demand similar solution strategies, it is optimal to use only the first description for creating a mental representation of the problem and not to repeat this for the second, third, etc. An illusion occurs if this similarity does not in fact hold. In perception we continuously "fill in" by expectation: we watch the first few steps, but we do not have to watch the steps unblinkingly when we walk; we are seldom aware of blinking, and, if I close one eye, there is no blank in the visual world that corresponds to the blind spot in my eye, for the blind spot is filled in using the context and prior experience. Of course, we always can stop and check the steps, as we can check the structure of each new problem. However, it is precisely the fact that we do not have to check and recheck that makes such inference mechanisms so efficient.

Third, the dependence on context and content suggests that we will not get far by posing question of the following, formal kind: Do people intuitively understand the law of large numbers? Is intuitive reasoning Bayesian reasoning? Do people confuse $p(A/B)$ with $p(B/A)$? The Helmholtzian analogy predicts: By choosing a proper content and context, you then will be able to produce reasoning that either does or does not follow a given formal rule, or any point in-between. For instance, when the same subjects who neglected base rates in the above replication of the Engineer-Lawyer study were given further problems of the same Bayesian structure, but with soccer games as the content, their judgments immediately became indistinguishable from Bayesian reasoning. Prior knowledge connects soccer games with statistics, and when judging the probability that a team will win a particular game, subjects combine information about the particular game with base rates of the general performance of the team in the long run (Gigerenzer et al. 1988).

I prefer to interpret such results as indicating that untutored intuition does not consist of formal rules plus a recognition mechanism that abstracts corresponding formal structure from a problem. Even when intuitive reasoning gives the same answer as Bayesian reasoning, as when estimating the probability that a soccer team will win, this does not imply that the untutored mind has the formal rule in its competence. In contrast, I understand Cohen to mean that untutored intuition contains logical and statistical rules. He emphasizes the distinction between a subject's not having a rule and simply not applying it (Cohen 1983, p. 511). For instance, he interprets the subjects' reasoning in the Wason selection task as a failure to apply the law of contraposition, a cognitive illusion in his sense of the term, but rejects the conclusion that the competence of most untutored people does not embrace the law (Cohen 1981, pp. 323-324).

However, as I argued, the Helmholtzian analogy shows that there is at least one alternative to Cohen's premisse of untutored competence in statistical formulae which both supports intuition as basically rational and at the same time allows for systematic illusions.

Can Cohen's premisse of untutored rationality be falsified by observation or experiment? Cohen (1981, p. 330) claims flatly "nothing in the existing literature ... or in any possible future results of human experimental enquiry, could ... establish a faulty competence". But he seems less definite in other places (e. g., Cohen, 1982, p. 252). I think that he can safely use his competence/performance distinction to protect his premisse against empirical falsification. In fact, Cohen takes those contents and contexts where reasoning conforms to probability theory as a proof of lay competence in statistical principles, and dismisses negative instances as due to obstacles like "cognitive illusions". If there exists an empirical argument that could lead him to reconsider his premisse, it is the argument of consistent logical inconsistencies. Today we can survey numerous experimental studies on adult statistical reasoning — from Hofstätter's (1939) pioneering work to the present avalanche. I take the main lesson of those studies to be that untutored reasoning is consistently inconsistent, *if* you look at reasoning from the point of view of logical structure. For instance, if we want to understand the experimental results reported above within a structural-logical conception of the mind, we would have to assume that our typical subject is a Baconian probabilist when she reasons about whether "Jack" is an engineer; uses a representativeness heuristic (but not Baconian probability) when she reasons about the undiagnostic description "Dick"; is a Pascalian probabilist when she encounters only base rates and no description; is an approximate Bayesian if she does the random sampling herself; and is a proper Bayesian if soccer is the issue. This is why I do not believe that some logical structure alone is sufficient to explain how untutored reasoning functions.

The Helmholtzian alternative to both Cohen and the "general heuristics" view of Kahneman and Tversky would be to consider the task of reasoning as coming to terms with a world that presents itself to us only by uncertain cues, and to accept that this uncertainty forces us to rely on context and content (rather than only on formal structure) to improve our bet. But even this aspect of the analogy has its limits. Unlike reasoning, perception goes on to make the same bet again and again, despite better knowledge. As Egon Brunswik once put it, perception is like a stupid animal.

Cueing different interpretations of probability

What would a different research program look like if we took the final aspect of the analogy with visual illusions seriously? Despite Cohen's claim for the in-principle-rationality of untutored reasoning and its competence in applying statistical rules, his work contains the seeds of an interesting answer. In particular, Cohen suggested that the conception of probability, standard or Baconian, frequency or propensity, with which subjects construe their task is cued by the wording and content of the instructions. The question then becomes, what contextual cues and contents cue which notion of probability. This suggests the further question of whether experimenters and subjects have a similar understanding of the task, in particular of key concepts like "probability", "frequency", or "confidence". I consider this suggestion of Cohen's a practical and fruitful one. Curiously, contextually cued "schema" and "frames" are frequently used in cognitive psychology as an alternative view to the application of general systems of mental logic, but rarely to refer to cueing alternative (and equally legitimate) conceptions of "probability".

Let me give an example to show how Cohen's suggestion proves to be particularly fruitful. The "overconfidence" phenomenon is reported to be one of the "notorious biases of human reasoning" (Bar-Hillel & Margalit, 1983, p. 247) that resists numerous "debiasing methods" such as to raise stakes, clarify instruction, warn and inform about the phenomenon, or use different response modes (Fischhoff, 1982). It is classified as "a reliable, reproducible finding" (von Winterfeldt & Edwards, 1986, p. 539). All this makes it look like a good candidate for a stable reasoning error. What is the phenomenon? In a typical experiment, subjects are given two-alternative choice tasks, such as "Which city has more inhabitants? (a) Bonn, (b) Konstanz". After choosing what they believe to be the correct answer, subjects are asked for their confidence that the answer chosen is correct. These confidences were often obtained on a scale from 50% to 100% confident. After many such questions are given to many subjects, the relative frequency of correct answers in each confidence category is determined. A number of studies found that the confidence ratings were consistently larger than the relative frequencies of correct answers, a phenomenon which was called "overconfidence". For instance, when subjects said they were 100% confident, their average relative frequency of correct answers was only about 80%, for 90% confidence it was about 75%, for 80% confidence 65%, and so on. This discrepancy was considered a bias of reasoning, and explained by general cognitive heuristics such as, among others, search for confirming evidence ("confirmation bias"), or general motivational tendencies such as the tendency to overestimate one's knowledge ("illusion of validity").

I want to make two points. First, both the definition of the phenomenon and its interpretation as a bias of reasoning rests on the assumption that relative frequencies are the normative yardstick against which probabilities of individual events (such as that a given answer is correct) are to be evaluated as correct or incorrect. This position has yet to sweep the field of probability and statistics, winning over neither frequentists such as von Mises (1957) nor subjectivists such as de Finetti (1989). Nevertheless, little attention has been paid to this serious conceptual problem in the overconfidence literature, a problem, both for the definition of the phenomenon and its interpretation as a bias (an exception is May 1987). Cohen (1982, p. 256) noticed this widespread neglect of the various interpretations of formal probability in the experimental literature, and claimed that the possibility that subjective probabilities "may differ legitimately in value from the corresponding objective relative frequencies turns out to be vitally important for the interpretation of experimental data on probability judgment". Second, I shall argue that Cohen's conjecture is correct and fruitful for understanding the overconfidence phenomenon. Imagine an experimental task as described above, but where in addition the subjects are asked to estimate the *relative frequency* of correct answers in each of the confidence categories. This allows to compare true relative frequencies with estimated relative frequencies (in addition to comparing true relative frequencies with subjective probabilities for single events).

If the discrepancy called "overconfidence" were a reasoning bias due to limited processing abilities, confirmation biases or general motivational tendencies, then the discrepancy should occur independently of whether the subjects report their degrees of belief that a particular answer is correct, or estimate their relative frequency of correct answers.

If, on the other hand, the philosophers' distinction were relevant and meaningful to the untutored mind, then we might expect the discrepancy to change when we compare estimated with true relative frequencies. We were curious and actually performed such experiments (Gigerenzer, Kleinbölting & Hoffrage, 1990). Subjects were given several hundred two-alternative choice tasks, and were asked after each answer to estimate the probability that the answer was correct, as in previous studies and described above. In addition, after they finished, they were asked to estimate the relative frequency of correct answers in each of the confidence categories. When we compared estimated frequencies with true relative frequencies of correct answers the overestimation disappeared completely. For instance, in those cases where subjects said they were between 70% and 80% confident, only 65% of the answers were correct. But when their task was to estimate the relative frequency of correct answers (among those in which they said they were 70% to 80% confident), the average response was 65%. Different kinds of frequency judgments, such as asking the subject after 50 answers, "How many of the last 50 answers are correct?" gave a similar result. In all cases overestimation completely disappeared, and the discrepancy between estimated and true frequencies was either zero or even turned out to be an underestimation. This result cannot be explained by a general confirmation bias or general motivational tendencies to make ourselves appear better than we are. It suggests, however, that in evaluating one's knowledge, "probabilities for single events" and "relative frequencies" refer to different concepts in the minds of our untutored subjects, as they do in the minds of many philosophers. It also reveals that the previous interpretation of the discrepancy between degrees of belief and frequencies as a *bias in reasoning* meant comparing apples and oranges. Such results open up a quite different theoretical account of this allegedly reliable, reproducible "cognitive error". For one thing, more optimism is in order. Edwards and von Winterfeldt's (1986, p. 656) final word on overconfidence, after they asserted the stability of this "cognitive error", was: "Can anything be done? Not much". My answer is: Keep distinct meanings of probability straight, and much can be done — cognitive illusions disappear.[6]

However, many cognitive psychologists consider Cohen's conjectures of "little relevance to research on judgment and decision making" (Mynatt et al., 1983, p. 507), of no benefit for scientific work (Dawes, 1983) and at best, of merely philosophical interest. Fischhoff (1981, p. 337) even takes Cohen's work as evidence "that psychologists and philosophers would do well to ignore one another", since their pursuits "are somewhat irrelevant to the other". It is ironical that Fischhoff is among those who did the major studies on the overconfidence phenomenon.[7]

Psychologists can learn from philosophers. A sharp eye for conceptual distinctions, supplemented by experiments that can distinguish between them, is a major alternative to the traditional "heuristics and biases" program. It vindicates Cohen's point that even untutored reasoning is capable of making consistent conceptual distinctions of the sort philosophers make.

Institut für Psychologie
Universität Salzburg
Hellbrunnerstr. 34
5020 Salzburg
Austria

FOOTNOTES

[1] The recent debate on reasoning biases and rationality initiated by L. J. Cohen is documented in, among others, *The Behavioral and Brain Sciences* 1981, 1983, 1984, 1987, and *Cognition* 1979, 1980, 1982.

[2] For more on the stability of visual illusions see Robinson (1972).

[3] In fact, the likelihoods p("Blue"/Blue) and p("Blue"/Green) are not unambiguously specified in (ii). What is said is that the two errors, p("Blue"/Green) and p("Green"/Blue) add up to 20%, but this does not determine the size of either of them. But such ambiguities, which could be easily resolved, are not of interest here.

[4] In the present context, the meaning of the term "representativeness" can be reduced to saying that subjects use Bayes' theorem with uniform priors. This can be shown as follows, where UD, E, and L stand for Uninformative Description, Engineer, and Lawyer, respectively. What is to be explained is the subjects median judgment $p(E/UD) = .5$, that is the probability that Dick (the uninformative description, UD) is an eugineer (E). $p(E/UD)/(1-p(E/UD)) = p(UD/E)/p(UD/L)$ is Bayes' theorem with uniform priors. Since $p(UD/E) = p(UD/L)$ (the definition of uninformativeness), we get $p(E/UD) = .5$, which is the actual median answer. Baconian probability, in contrast, should assign $p_I(E/UD) = 0$. It is therefore not equivalent to the notion of "representativeness heuristic" in this context.

[5] Our results also contradict the interesting explanation of intuitive reasoning by reference to conversational maxims. Adler (1983, p. 246) argues that the experimenters' most salient contribution is the biographical sketch, and according to Grice's communicative maxim to "be relevant" it is socially healthy and more cooperative, if the subject judges solely on the basis of the description, which is of greater selective relevance, "rather than (less

cooperatively) integrating it with the base-rate data". Again, randomly drawing the descriptions out of an urn should not matter if the real issue was that conversational maxims such as being cooperative were in conflict with abstraction, as Adler believes. The elimination of base rate neglect by making random sampling explicit, however, indicates that, at least in the present context, cognitive representations are more powerful for probabilistic reasoning than communication maxims.

[6] The same conceptual distinction is crucial for whether people commit the so-called conjunction fallacy (Fiedler, 1988; Tversky & Kahneman, 1983).

[7] Why many psychologists and social scientists neglect conceptual distinctions within probability theory is a different story (see Gigerenzer et al. 1989, chapters 3 and 6, and Gigerenzer, 1987).

REFERENCES

Adler, J. E. (1983). Human rationality: Essential conflicts, multiple ideals. *The Behavioral and Brain Sciences*, **6**, 245-246.

Adler, J. E. (1984). Abstraction is uncooperative. *Journal for the Theory of Social Behaviour*, **14**, 165-181.

Bar-Hillel, M. & Margalit, A. (1983). To err is human. *The Behavioral and Brain Sciences*, **6**, 246-248.

Berkeley, G. (1709). *An essay towards a new theory of vision*. Dublin.

Birnbaum, M. H. (1983). Base rates in Bayesian inference: Signal detection analysis of the cab problem. *American Journal of Psychology*, **96**, 85-94.

Cohen, L. J. (1979). On the psychology of prediction: Whose is the fallacy? *Cognition*, **7**, 385-407.

Cohen, L. J. (1981). Can human irrationality be experimentally demonstrated? *The Behavioral and Brain Sciences*, **4**, 317-331.

Cohen, L. J. (1982). Are people programmed to commit fallacies? Further thoughts about the interpretation of experimental data on probability judgment. *Journal for the Theory of Social Behaviour*, **12**, 251-274.

Cohen, L. J. (1983). The controversy about irrationality. *The Behavioral and Brain Sciences*, **6**, 510-517.

Cohen, L. J. (1986). *The dialogue of reason*. Oxford: Clarendon Press.

Daston, L. (1988). *Classical probability in the enlightenment*. Princeton NJ: Princeton University Press.

Dawes, R. M. (1983). Is irrationality systematic? *The Behavioral and Brain Sciences*, **6**, 491-492.

De Finetti, B. (1989). Probabilism. *Erkenntnis*, **31**, 169-223 (original work published in 1931. New York: Springer).

Edwards, W. & von Winterfeldt, D. (1986). On cognitive illusions and their implications. In H. R. Arkes & K. R. Hammond (Eds.), *Judgment and decision making*. Cambridge: Cambridge University Press.

Fiedler, K. (1988). The dependence of the conjunction fallacy on subtle linguistic factors. *Psychological Research*, **50**, 123-129.

Fischhoff, B. (1981). Can any statements about human behavior be empirically validated? *The Behavioral and Brain Sciences*, **4**, 336-337.

Fischhoff, B. (1982). Debiasing. In D. Kahneman, P. Slovic & A. Tversky (Eds.), *Judgment under uncertainty. Heuristics and biases*. Cambridge: Cambridge University Press.

Gigerenzer, G. (1987). Probabilistic thinking and the fight against subjectivity. In L. Krüger, G. Gigerenzer & M. S. Morgan (Eds.), *The Probabilistic Revolution*. Vol. 2: *Ideas in the sciences*. Cambridge, MA: MIT Press.

Gigerenzer, G., Hell, W. & Blank, H. (1988). Presentation and content: The use of base rates as a continuous variable. *Journal of Experimental Psychology: Human Perception and Performance*, **14**, 513-525.

Gigerenzer, G., Kleinbölting, H. & Hoffrage, U. (1990). Confidence in one's knowledge. A Brunswikean view. Manuscript submitted for publication.

Gigerenzer, G. & Murray, D. J. (1987). *Cognition as intuitive statistics*. Hillsdale, NJ: Erlbaum.

Gigerenzer, G., Swijtink, Z., Porter, T., Daston, L. J., Beatty, J. & Krüger, L. (1989). *The empire of chance: How probability changed science and everyday life*. Cambridge: Cambridge University Press.

Ginossar, Z. & Trope, Y. (1980). The effects of base rates and individuating information on judgments about another person. *Journal of Experimental Social Psychology*, **16**, 228-242.

Gregory, R. L. (1974). *Concepts and mechanisms of perception*. New York: Charles Scribner's Sons.

Hofstätter, P. R. (1939). Ueber die Schätzung von Gruppeneigenschaften. *Zeitschrift für Psychologie*, **145**, 1-44.

Holland, J. H., Holyoak, K. J., Nisbett, R. E. & Thagard, P. R. (1986). *Induction*. Cambridge, MA: MIT Press.

Kahneman, D. & Tversky, A. (1973). On the psychology of prediction. *Psychological Review*, **80**, 237-251.

Kahneman, D. & Tversky, A. (1982a). On the study of statistical intuitions. In D. Kahneman, P. Slovic & A. Tversky (Eds.), *Judgment under uncertainty: Heuristics and biases*. Cambridge: Cambridge University Press.

Kahneman, D. & Tversky, A. (1982b). Variants of uncertainty. In D. Kahneman, P. Slovic & A. Tversky (Eds.), *Judgment under uncertainty: Heuristics and biases*. Cambridge: Cambridge University Press.

May, R. S. (1987). *Realismus von subjektiven Wahrscheinlichkeiten*. Frankfurt/Main: Lang.

Mynatt, C. R., Tweney, R. D. & Doherty, M. E. (1983). Can philosophy resolve empirical issues? *The Behavioral and Brain Sciences*, **6**, 506-507.

Nisbett, R. E. & Borgida, E. (1975). Attribution and the psychology of prediction. *Journal of Personality and Social Psychology*, **32**, 932-943.

Robinson, J. O. (1972). *The psychology of visual illusion*. London: Hutchinson University Library.

Slovic, P., Fischhoff, B. & Lichtenstein, S. (1976). Cognitive processes and societal risk taking. In J. S. Carroll & J. W. Payne (Eds.), *Cognition and social behavior*. Hillsdale, NJ: Erlbaum.

Tversky, A. & Kahneman, D. (1971). Belief in the law of small numbers. *Psychological Bulletin*, **76**, 105-110.

Tversky, A. & Kahneman, D. (1974). Judgment under uncertainty: Heuristics and biases. *Science*, **185**, 1124-1131.

Tversky, A. & Kahneman, D. (1977). Causal thinking in judgment under uncertainty. In R. Butts & J. Hintikka (Eds.), *Basic problems in methodology and linguistics*. Dordrecht-Holland: Reidel.

Tversky, A. & Kahneman, D. (1980). Causal schemas in judgments under uncertainty. In M. Fishbein (Ed.), *Progress in social psychology*. Vol. 1. Hillsdale, NJ: Erlbaum.

Tversky, A. & Kahneman, D. (1983). Extensional versus intuitive reasoning: The conjunction fallacy in probability judgment. *Psychological Review*, **90**, 293-315.

von Mises, R. (1957). *Probability, statistics, and truth*. New York: Dover.

von Winterfeldt, D. & Edwards, W. (1986). *Decision analysis and behavioral research*. Cambridge: Cambridge University Press.

Wells, G. L. & Harvey, J. H. (1978). Naive attributors' attributions and predictions: What is informative and when is an effect an effect? *Journal of Personality and Social Psychology*, **36**, 483-490.

Zukier, H. & Pepitone, A. (1984). Social roles and strategies in prediction: Some determinants of the use of base-rate information. *Journal of Personality and Social Psychology*, **47**, 349-360.

*Poznań Studies in the Philosophy
of the Sciences and the Humanities
1991, Vol. 21, pp. 251-282*

Jonathan E. Adler

AN OPTIMIST'S PESSIMISM:
CONVERSATION AND CONJUNCTION*

1. The lasting significance of L. Jonathan Cohen's critique of recent psychological studies of judgment and reasoning, especially those of Tversky and Kahneman, is threatened by locating it within blunt oppositions like "optimist" and "pessimist" (Cohen, 1977, 1979, 1981, 1986 ch. 3). Optimists are content and pessimists despondent, so that neither is motivated to improve their understanding.

The author (Jungermann, 1986) who suggests these particular labels does so recognizing that it is "somewhat exaggerated". Yet, choice of labels aside, the norm now is to represent the discussion following Cohen's critique in similar terms. "The two camps on rationality" are distinguished this way:

> ... The *pessimists*... claim that judgment and decision making under uncertainty often show systematic and serious errors, due to in-built characteristics of the human cognitive system. Violations of rationality... are interpreted as true deficits of the decision maker. The *optimists* of the other camp claim that judgment and decision are highly efficient and functional even in complex situations. Observed violations of rationality axioms are interpreted as unjustified evaluations based on inappropriate theoretical assumptions of empirical approaches on the part of researcher. (Jungermann, 1986, p. 628).

A better framework for assessing the relevant empirical studies will blunt the opposition between these "camps". I engage this broad theme only in part II. In part I, it is argued in detail that conversational assumptions intrude on subjects reasoning in one experimental paradigm — the conjunction effect studies.

Part II attempts to connect the conversational account with some of the broader issues of rationality raised by Cohen's critique. Conversational accounts are typically optimistic, since they explain (away) alleged errors as due to honest misunderstanding. Nevertheless, our conversational account does not foreclose (cautious) criticism of the subjects reasoning. About the very same judgment task, we can be reasoning appropriately and rationally, and yet still be criticizable. Thus, to return to the theme of part I, the categories of optimist and pessimist must be refined. In part II, an alternative framework that leads to such refinement is proposed. The main claim of this alternative — "Austinian" — framework is that a single dimension of evaluation (correct/incorrect) has been applied to the relevant studies, but that they require a multi-dimensional evaluation.

I

2. One, if not the, central purpose of conversation is to convey information. But in order to convey as much information as we would like under normal constraints of conversation (time, comprehension, interest), audiences must *presume* (and speakers' exploit the fact that audiences so presume) that speakers are trying to be *cooperative* (Grice's Cooperative Principle or CP; Grice, 1975). Presuming a speaker is trying to be cooperative (obeying the CP), we can draw inferences, which he intends us to draw, that go well beyond, but depend upon, what he does say.

The CP is a presumption because it is not supported by evidence, but mandated on procedural grounds. We must avoid having to first investigate the trustworthiness of a speaker before we can accept what he says. So far the parallel with the familiar presumption of innocence in a trial is maintained.[1]

But these presumptions differ in their openness to learning from cases. The extent to which the presumption of innocence has been correct, in the past, is irrelevant to its propriety. Nor is the finding of guilt in any case evidence of the fallibility of the presumption (not, of course, that it is infallible). The presumption of innocence is just a procedural stance, which makes no claim to correctness (as to actual innocence).

The presumption of cooperation is, though, confirmable in a fashion. If it became regularly violated, we would certainly notice, and attempt to apply the presumption more selectively. We are familiar with cases where people are expected to violate the maxims (e. g. used car salesmen). These reveal that the CP has a kind of fallibility. Unlike the presumption of innocence, we learn about new cases from others. We don't (and normatively should not) learn about one person's innocence from anybody else's guilt (leaving aside special cases). But regular disobedience of the CP by members of a representative group do provide some basis for hesitating to apply the CP to the contributions of a new member of that same group[2] (Lewis, 1983). Thus the hold of the CP on us is empirically, not solely prudentially or procedurally, grounded and reinforced.

For the sake of brevity, I will sometimes subsume all of the maxims that Grice holds to follow from presuming the CP, under the single maxim "Be Relevant". For useful information to be conveyed it must be relevant to the interests at hand, truthful, not already known (informativeness) and presented clearly (well mannered). Further, the more we can maximize relevance (as optimizing all of these goals) the better.

The expectation of relevance is reinforced even in contexts where it is likely to be intrusive. Consider, the teaching of a subject like formal logic. Teaching strategies for the construction of proofs requires imposing an artificial limitation: Few, if any, superfluous premises. This restriction, which is extremely strong in light of the vast range of possible premises in proofs, permits students to use strategies such as (a) to work backwards from the conclusion or (b) to apply whatever rules they can, when they find themselves in the dark as to how to reach the conclusion. Neither works well, if at all, when many of the premises are superfluous.[3]

Contributions that satisfy our expectation of relevance confirm that expectation all the more because social learning provides us with a heightened awareness of how much is irrelevant. One of the earliest, most persistent and most frustrating lessons we teach our children is that "It doesn't make a difference". For example, it doesn't make a difference whether the left shoe is tied first or second. Yet children will insist that such things make enormous difference, and they will howl when the pattern they demand is not realized. Further, in an industralized society, we are used to seeing numerous distinct instances of the same brand of product e.g. distinct boxes of Cheerios. Even distinct brands (e.g. of cereal) will often be essentially indistinguishable for the most salient interests (e.g. taste, nutritional value) in that product. (Ullmann-Margalit and Morgenbesser, 1977).

The hold of our expectation of relevance in conversation goes then very deep. Its continued unselfconscious use shows that it is highly confirmed, continuously reinforced, and largely successful. And that expectation receives highlighting from our familiarity with many contrasting instances of irrelevance in other realms of social life. As a result, the basic assumption of the conversational approach to the social psychology experiment is that even persons in the artificial circumstances of the lab, will tend toward treating the experimenter's written and spoken words as conversational contributions. Hence, they will expect the contribution to be cooperative and therefore, relevant. As Kahneman and Tversky note, although it will turn out that they take it less seriously than I do.

> Subjects come to the experiment with lifelong experience of cooperativeness in conversation. They will generally expect to encounter a cooperative experimenter, although this expectation is often wrong. (Kahneman and Tversky, 1982, p. 502).

Of course, as we become more sophisticated language users, we become better adept at recognizing violations, interpreting them so that the violation is removed. Eventually, we are even able to tolerate some violations, where the purpose of the violation is transparent and a strict reading of the contribution is intelligible.

Thus, consider Piaget's classic experiments on class-inclusion, which are structurally identical to the Tversky and Kahneman conjunction effect studies that will be our main concern (Adler, 1984). Children younger than 7 are confronted with an array containing 4 dimes and 3 nickels. They are asked: "Are there more dimes or coins?" The dominant response of these children is "More dimes". We (adults) recognize, however, that the question is a "trick" question, so that we do not treat it as something more like "Are there more dimes or nickels?"

3. Tversky and Kahneman's most dramatic finding is reported at length in a recent article that offers their most considered view. (Tversky and Kahneman, 1983). In this article they present and discuss strtudies of the "conjunction effect" under many conditions. The core study involves a personality sketch "followed by a set of occupations and avocations associated with each of them". (Tversky and Kahneman, 1983, p. 297). The better known one is the following:

> Linda is 31 years old, single, outspoken and very bright. She majored in philosophy. As a student, she was deeply concerned with issues of discrimination and social justice, and also participated in antinuclear demonstrations.

Subjects are then ask to rank the probability of the following statements:

> Linda is a teacher in elementary school.
> Linda works in a bookstore and takes Yoga classes.
> Linda is active in the feminist movement. (F)
> Linda is a psychiatric social worker.
> Linda is a member of the League of Women Voters.
> Linda is a bank teller. (T)
> Linda is an insurance salesperson.
> Linda is a bank teller and is active in the feminist movement. (T&F).

Other conditions include having different groups rank different statements or presenting the options so as to highlight the relation between the conjunction and its conjunct.

The sketch is intended to be representative of a feminist and unrepresentative of a bank teller. The crucial finding is that a high percentage (85-90%) of subjects rate the probability that "Linda is a bank teller and is active in the feminist movement" (T&F) as higher than "Linda is a bank teller" (T). The percentage is little effected by altering conditions so that the logical relation between T&F and T is more obvious. When T&F and T are directly compared Tversky and Kahneman call it a "conjunction fallacy" because a conjunction cannot be less probable than either conjunct. (Tversky and Kahneman, 1983, p. 298).

The finding is especially dramatic since the conjunction law is so simple and basic. Indeed, it is clear in this study, more than in any other related one, that no question of a lack of competence or lack of grasp of a law or rule is involved. When, in fact, the personality sketch is reduced to "Linda is a 31 year old woman" almost everyone rated T over T&F.[4] (Tversky and Kahneman, 1983, p. 305). Everyone will acknowledge the conjunction rule given the right questions. (And, unlike for Meno's slave boy, the questions need not be loaded.) So the main puzzle is why don't subjects apply it to this problem.

4. Gricean presumptions can be hypothesized to intrude on the "Linda" conjunction effect study in two ways, creating distinct, though related, problems. The first, narrow, problem is whether "Linda is a bank teller" (T) is read by subjects as "Linda is a bank teller who is not active in the feminist movement". (T¬-F) The second, broader, problem confronts any study of the influence of (allegedly) "irrelevant" information. Without reference to this specific case, Kahneman and Tversky recognize the general difficulty:

> ...Because the presentation of irrelevant information violates rules of conversation, subjects are likely to seek relevance in any experimental message. (Kahneman and Tversky, 1982, p. 502).

The ascription of greater probability to T&F than T may be a way of rendering relevant what would otherwise be an irrelevant contribution. Since T&F must be (formally) less probable than T, the initial personality sketch would be an uncooperative, because irrelevant, contribution under a purely formal reading. I discuss these two points in turn.

Tversky and Kahneman admit that "Linda is a bank teller" (T) may be taken to implicate "Linda is a bank teller and is not active in the feminist movement". (T¬F) This construal might arise because subjects' consider the task too trivial as stated. There are other inter-related reasons why this (T¬F) reading might arise. First, there is the presence of the option T&F, which as a matter of style and informativeness, encourages us to implicate T¬F from T, thus creating natural contrasts. Second, since the original story points toward Linda's feminism and progressive politics the option of her being a bank teller appears intended by the speaker as dissonant with Linda's being a feminist.

Third, hearers will make assumptions about the speaker's knowledge. As the speaker appears more knowledgeable about Linda, we expect that he would know about her feminism. The reasoning to T¬F can now take two paths. One can reason this way: When the speaker appears to know a good deal about Linda, hearers infer that he has left off "notF" because he takes it as an obvious consequence of the discrepancy between the description and the actual assertion (that she is a bank teller). Alternatively, one can infer from the speaker's saying T, rather than, say, T&F that the speaker must know that T&F is not the case (since otherwise, wanting to be as relevant as possible, the speaker would have said so). Consequently, the implicature is drawn to T¬F. (Hilton, unpublished, p. 8). (This alternative is especially attractive where T&F is already offered as an option. For then, as indicated in the first point above, the speaker implicates that since he knows enough about Linda to offer T&F as an option, he withholds the predicate (F) because he knows it does not obtain).

Both lines of reasoning lead to the same conclusion, so that determining which has greater claim to psychological reality does not much matter to us. However, sorting these out experimentally would complicate even more the question of how to avoid the unwanted implicature. Since the overall purpose is to show that the conversational problem of misunderstanding is a serious one, difficult to control for, complications are welcome. More generally, it is important to emphasize the different routes to the T¬F implicature because of the wide variety of tests that Tversky and Kahneman invoke. On some of these tests, for example, subjects do not view both options T and T&F, and so cannot contrast them. Still, the third reason remains applicable.

Tversky and Kahneman attempt to test for this (T¬F) interpretation by asking "a new group of students (N=119) to assess the probability of T and T&F on a 9-point scale ranging from 1 (extremely unlikely) to 9 (extremely likely)". Tversky and Kahneman write "Because it is sensible to rate probabilities even when one of the events includes the other, there was no reason for respondents to interpret T as T¬F". (Tversky and Kahneman, 1983, p. 299). They found no significant difference from a previous direct comparative evaluation of the probability of T and T&F. However, this variation does not address at all the rationales for the T¬F reading offered above.

There is empirical support for the T¬F implicature. In a study by Morier and Borgida (1984) they tried to induce a strict reading of T by a number of *indirect* means. ("Indirect" because, as just suggested, their variations-such as adding as an option the disjunction "Linda is a bank teller or a feminist"-are not the most straightforward means to cancel or eliminate the alleged implicature.) Although there was little change for the Linda study they found that the "conjunctive error rate" in this debiased form was "significantly lower" for other problems where the personality sketch was less representative of the target occupation property. (Morier and Borgida, 1984, p. 249). More significantly, Hilton (unpublished), relying on a detailed Gricean analysis, found other ways to induce a strict reading of T (i.e. one not [as strongly] implicating T¬F). Under one of his conditions, he found the conjunction effect lowered to "20 and 28 per cent". (Hilton, unpublished, p. 18).

Tversky and Kahneman themselves found that when they substitute for T, T*= "Linda is a bank teller whether or not she is active in the feminist movement". that the violation went down, if I understand them correctly, to 57%. (Tversky and Kahneman, 1983, pp. 298-299). Surprisingly, they find this change neither significant (they abandon the issue at this early stage in their investigations), nor do they actually connect the introduction of T* with the previous discussion of the implication from T to T¬F. Rather, they introduce T* as an independent proposal aiming to make the set-inclusion relation more transparent. It will, however, prove worthwhile to pursue T* as an attempt to cancel the T¬F implicature, since it will help clarify some of the difficulties in that attempt. Tversky and Kahneman hold that these (T and T*) are "extensionally equivalent" by which I assume they mean "logically equivalent", so that they are treating "whether or not she is active in the feminist movement" as conjoining a logically true schema to T (Fx or -Fx).

Notice that if we treat T and T* as logically equivalent then the choice of predicate ("F") is irrelevant. One gets the same logical equivalence by choosing the predicate "is a duck" (Dx or -Dx). Of course, conversationally one of them would be heard as appropriate and the other not. Indeed, given their blatant violation of the maxim of informativeness, tautologies are rarely heard as tautologies[5].

Typically, we use the phrase "whether or not A is a Y" after an assertion of the form "A is an X" as a polite, conventional means to deny a contextually assumed conflict between being an X and being a Y. Statements of the form "A is an X, whether or not A is a Y" contextually seem to mean the same as "A is an X even if A is a Y". Compare "Smith is a crook, whether or not he is class valedictorian" with "Smith is a crook, whether or not he is class valedictorian" with "Smith is a crook even if he is class valedictorian". So "T, whether or not she is active in the feminist movement" would be heard as "T even if she is active in the feminist movement". "Even if p, q" assertions are construed as asserting q together with a speaker's acknowledgement of the hearer's assumption of a conflict between q and p, and the speaker's insistence on q (nonetheless).

If subjects hearing of "T, whether or not she is active in the feminist movement" conflates, as is usual, its semantics and pragmatics, then there is no longer the extensional equivalence with T; and hence, we do not have a clear entailment relationship with T&F. If subjects are meant to treat "whether or not she is active in the feminist movement" as "even if she is active in the feminist movement" the semantics for "even if" also fails to yield the clear entailment relations sought. The dominant view of "even if" is that it is just a conditional plus a pragmatic contrast (Jackson, 1988, Ch. 3). But the probabilistic comparison between $T \rightarrow F$ and T is not at all what is in question[6].

The dilemma, in brief, is this: Either subjects' responses are sensitive at the semantical level or not. If so, then the equivalence is maintained, but the implicature is not cancelled. (As Grice insists, implicatures are attached to the meaning or semantics of statements). If not, then the implicature will be undermined, although probably not fully cancelled, but at this (pragmatic) level of interpretation the equivalence no longer holds. So findings for this condition (using T^*) are inadequate to deny the influence of Gricean implicatures in the original (T) condition.

We standardly cancel implicatures (as well as distinguish between an implication and an implicature) by expanding the original statement, so as to contain a denial of that implicature. (If it were a genuine semantic or logical implication, the result would be a contradiction). "John has 3 children" is shown to implicate, but not imply, that he has exactly 3 children because one could just assert "John has 3 children, and maybe more" (or, "John has at least 3 children"). But if we engage in straightforward cancellation, we cannot study the dependent variable that we want. For the most natural form of cancellation in the Linda study would be either "Linda is a bank teller, and [but?] she may be active in the feminist movement" or "Linda is a bank teller, but for all I (?) know, she is [may be] active in the feminist movement". Neither admits the "transparent" presentation where, in the original studies, T is supposed to be an obvious conjunct of T&F by form alone. But that is symptomatic of the fact that there is no simple and natural way to represent the statements which cancel the proposed implicature in a standard way as either a conjunct of T&F, or a sub-set of T&F. So, whatever interest there might be in studying these in relation to T&F, it would not be a study of the conjunction effect. If this is so, the prospects are poor for finding an unobtrusive means to cancel the unwanted implicature, one that doesn't violate other well inculcated conversational assumptions.

5. The second, broader conversational problem was proposed in an earlier article (Adler, 1984). It is really two difficulties, the second of which is itself complex. First, treating T and T&F in the strict way that experimenters' want, renders the personality sketch completely irrelevant. Whatever the sketch is, T must have at least a high a probability as

T&F. The personality sketch is a useless contribution for the judgment in the same way that the display of coins is useless to the children in Piaget's study, if the children treat the question literally. The similarity is no surprise: Both experiments are, in effect, studies of the class-inclusion relationship. To judge T as less probable than T&F does, however, exploit the personality sketch and thereby render it relevant.

The way the choice of T&F over T renders relevant the biographical sketch, and this is the second difficulty, is by emphasizing Linda's feminism. By so rendering this sketch relevant, subjects' explain or make good sense of or render coherent the experimenter's main contribution of the personality sketch (with the options). These (explanatory) accomplishments are themselves legitimate and penetrate subjects' understanding of probability in everyday reasoning.

Tversky and Kahneman in the main article under discussion consider a related interpretation. They discuss conversational expectations that contributions will have high *expected informational value*; more simply and roughly, that contributions will fit together to make a good story. They observe:

> Consider the task of ranking possible answers to the question, "What do you think Linda is up to these days?" The maxim of value could justify a preference for T&F over T in this task, because the added attribute *feminist* considerably enriches the description of Linda's current activities, at an acceptable cost in probable truth. Thus, the analysis of conversation under uncertainty identifies a pertinent question that is legitimately answered by ranking the conjunction above its constituent. (Tversky and Kahneman, 1983, p. 312).

However, Tversky and Kahneman immediately raise objections to this kind of account:

> We do not believe, however, that the maxim of value provides a fully satisfactory account of the conjunction fallacy. First, it is unlikely that our respondents interpret the request to rank statements by their probability as a request to rank them by their expected (informational) value. Second, conjunction fallacies have been observed in numerical estimates and in choices of bets, to which the conversational analysis simply does not apply. (Tversky and Kahneman, 1983, p. 312).

While I do not claim to have "a fully satisfactory account," Tversky and Kahneman's objections, particularly the first, are questionable. A fundamental reason for their underestimation of the force of the goal of creating a good, coherent story is that they neglect the alternative. The analysis they present is one of merely a ranking of T&F over T. But that ignores the fact that T&F does not merely make a better story than T. Rather, T&F gives us a good, coherent story, albeit rough around the edges (we have to live with Linda as a bank-teller), while T alone offers no coherent story. That is why, the account I defend emphasizes that on the formal analysis the sketch is irrelevant to the judgment.

Let's now turn to their explicit objections, taking the second first. It is not clear to begin with why the conversational model, generally, bracketing the specific notion of "informational value," is inapplicable. For ordinarily we can often substitute "a better bet" for "more probable"[7].

In any event, the reasoning in this second objection even if correct for the "conjunction *effect*" is insufficient for the "conjunction *fallacy*". For the defender of the conversational analysis can deny that the cases (e.g. Linda vs. numerical estimates or choice of bets) are, for normative purposes, relevantly similar.

Consider one of their conjunction effect studies involving betting. (Tversky and Kahneman, 1983, p. 303-304). Given a sequence reporting on the faces of a repeatedly thrown die made up of 2 red (R) and 4 green (G) faces, the following alternatives are offered for comparison:

1. RGRRR
2. GRGRRR
3. GRRRRR

While there is still some temptation to view sequence one as missing a last member, it is not particularly strong. Obviously the first sequence might only include that number of rolls; and, in any case, given that we are convinced that the die is fair, we do not have a basis for taking the speaker as implicating the sixth role one way or another (i.e. R or G). (It is just this lack of implication one way of another that would solve the narrow problem of treating "T" as "T¬F" discussed above).

The more significant point is that these choices are readily construed, unlike for the Linda study, as reporting on, and hence, rendering relevant, the investigator's contribution in throwing of the die. (Formally, of course, the result would be the same without any rolls). Each sequence is initially construed as a report on the outcome of tossings. So, there is no contribution whose purposes need to be inferred, lest it be left a dangling irrelevance.

Why not then hold that in this kind of case there is less ground, conversationally or otherwise, for judging more probable sequence 1 over sequence 2? Betting indicates a special predictive objective with a distinctive criterion of success or failure. As this is a game of chance, there is no differential learning to be found in choosing one sequence over another. The probabilities are fixed once and fall by the nature of the mechanism. Subjects seem to agree to this extent, as Tversky and

Kahneman's data show: An extensional *argument* in the Linda case is much less forceful than an extensional argument (sequence 1 over 2) in the colored die betting problem.[8]

Turning to their more important first objection (that subjects are unlikely to conflate "probability" with cognitive neighbors such as "expected informational value"), Tversky and Kahneman offer no further support for it than this assertion. However, immediately afterward, with regard to a neighbor of probability like "suprise," they note that "Our discussions with subjects provided no indication that they interpreted the instruction to judge probability as an instruction to evaluate surprise". (Tversky and Kahneman, 1983, p. 312). Perhaps the same goes for "expected informational value".

We do not have access to these discussions. Yet it is hard to believe that the "probability" is not conflated with related notions like acceptability, expected informational value, explanatory coherence, support, or increased relevance (to name a few). For as relevance (coherence) is always expected — an expectation typically satisfied in conversation — the ways that purely probabilistic assessments can separate from related epistemic assessments is not salient.[9] "Probability" is ordinarily understood as a term of epistemic assessment, as also are the above mentioned cognitive neighbors, which fall within Grice's broad maxim of quality or truthfulness.

Recognition of the subtle differences among these, while often implicit in our practices, have not been shown to be serviceable in informing our everyday exchanges. Tversky and Kahneman's own illustrations of expected informational value support this claim. If more evidence is needed, try introspection: Don't you find it very easy and natural to substitute for judgments of greater probability (in everyday contexts and usage) judgments of greater worthiness of belief or greater acceptability?

Their own account of the findings is that the biographical sketch is *representative* of being a feminist rather than a bank teller, and representativeness wins out over probability. Now the notion of representativeness has affinities with what I am calling epistemic or cognitive neighbors of probability. The real differences in view are that, first, Tversky and Kahneman want to view representativeness as a second best substitute for probability. My claim is that these various cognitive neighbors do all legitimately apply to the problem as posed, and the evidence does not decisively show one of these to be uniquely applicable. Second, and related, on my view, but not, I believe, Tversky and Kahneman's, these cognitive neighbors penetrate the question of ranking the alternatives according to probability. Subjects can legitimately view the task not as, say, ranking the alternatives according to relative frequency, but, as already suggested, ranking them for acceptability. (Acceptability, conversationally, comes to this: A speaker's assertion that *p* is acceptable if the hearer should not question or doubt it, should treat it as other conversational contributions that it is inappropriate to question, unless he has special reason to do so).

Conversationally, we can draw comparisons with two situations. One is the situation in which two individually implausible contributions become acceptable when conjoined. Thus, if I tell you that Bob is an art major at M. I. T., you may find this implausible given not only the base rates for art majors, but the stereotypes as well. Similarly, if I tell you that Bob is a native of Tahiti. But if I tell you both together, you may find it easy to accept my contribution. For you recall that M. I. T. has a small program to encourage students of art from the Pacific.[10] The second is the situation in which an initially highly implausible contribution, such as the one above claiming that Bob is an art major at M. I. T., is rendered plausible by a further contribution: "... but he is minoring in astrophysics". The former contribution, like Linda's being a bank teller, remains somewhat dissonant, but now acceptable.

These remarks about conversation have analogues in scientific inference, and it is crucial to recognize this if my claim for the real legitimacy of these epistemic neighbors of probability is warranted. However, my comments below are preliminary. Full defense waits upon discussion in part II, especially section 13.

The history of inductive logic for the last sixty years has been marked by the explication of a number of theoretically distinct notions closely related to probability (e. g. Carnap's probability 1 and probability 2, inference to the best explanation, corroboration). Cohen has given extensive evidence that many ordinary uses of probability do not fit the axioms of the probability calculus. Levi, Hintikka, Kyburg, Lehrer, and others have argued that scientists not only assign probabilities to hypotheses, but they *accept* them. That is, a point is reached in inquiry where further experimentation or questioning is not deemed necessary, and the statement is admitted to the corpus of beliefs as true. (For a sample of essays on this subject see Swain, ed. 1970).

Scientific and conversational reasoning are united, specifically, in rejecting at face value one trivial consequence of the conjunction law. The probability of error mounts as the number of conjuncts increases (assuming no extreme values)[11]. Yet, conversations aim at imparting (relevant) information. The more the better. But as we give more information, the probability that something we have said is mistaken increases. Pretty soon the probability of error is greater than the probability of correctness for the contribution as a whole. But it is not standardly a righful ground for mistrusting what someone says that the probability of error has mounted too far. People do make claims, of course, that overreach what they know. But when they do this, it is not on formal grounds that we object. We object because we are aware of their limited evidence for the claim at hand. Indeed, a generalization that many philosophers are inclined to accept is that a doubt or objection is proper only if it shows a *specific* failing or weakness in the assertion or claim being made.

This typical irrelevance of the mounting possibility of error due merely to increasing one's corpus of beliefs is as true for science as it is for conversation. Within the growing body of beliefs, the fact that a conjunction must be less probable than either conjunct is normally put aside. (A technical maneuver of assigning accepted hypotheses probability 1 is often made to avoid difficulties with the conjunction law. But that is a maneuver. The underlying rationale remains that the growth of knowledge in a domain undermines the significance of the mounting probability of error). As a result we can view the increasing comprehensiveness and coherence of our system of beliefs (or contributed information) as further evidence of their truthfulness, rather than as undermining it. Thus we can understand how Newton's Theory can itself gain more credibility than some do its consequences (the laws of Galileo and Kepler) by virtue of uniting them.

6. I conclude that Tversky and Kahneman's objections to this second, broad conversational problem do not succeed. Controlling for these unwanted inferences is even less promising than for the first, narrow problem. The object of the study is to examine the influence of information that is relevant to the assessment we would normally be asked to make, but is rendered irrelevant if a formal analysis is choosen. But, as already mentioned, conversationally we will resist viewing a contribution as irrelevant.

What we expect, in effect, by invoking the presumption of relevance is that, to offer a slogan, "every difference should make a difference". What are from the point of view of logic or semantics, minor variations or other ways of saying the same thing will *initially* not be so understood in the throes of conversation. For each portion of a contribution we seek its point or purpose — we seek its relevance.

Aren't I now exaggerating the problems? After all, conversations are generally successful, even with differences that make no difference. The option T in the Linda study is neat and clear, as is the introductory sketch and task. If conversations loaded with more ambiguity, more unclarity and more non-literal usage work, why not here? By and large

this objection is well founded. The initial expectation that "every difference should make a difference" is readily defeasible. We are familiar
with many cases in which it isn't true. For one thing, we are often
sloppy, adding extraneous material. For another, as a matter of
stylistics, variety in expression is encouraged.

But to overcome this initial expectation usually requires some
explicit cueing. Such cueing might be obtained through pre-experimental training or feedback from the experimenter. Tversky and
Kahneman themselves observe that giving subjects "extensional cues"
— such as, perhaps, to first ask them to estimate the number of feminists, bank tellers and feminist bank tellers in the relevant population-
results in strikingly different results. In one conjunction effect study
"a seemingly inconsequential change in the problem" led to a reduction
in "the incidence of conjunction error from 65% to 31%". (Tversky
and Kahneman, 1983, p. 309).

But the barrier to these kinds of explicit cues is, unsurprisingly, that
they bring attention to a frequency reading of probability and detract
from the everyday reading. If we do not invoke these cues, then within
the confines of normal conversational expectations, the defeasible
assumption that "every difference should make a difference" remains in
force. That assumption represents a special problem for these empirical
studies because the standards that the experimenters require for their
conclusions are much more exacting than the standards for conversational success. Conversations work when we can follow what our
interlocutor says: responding appropriately. If one reflects on actual
conversations, it will be evident that this does not require very precise
agreement on meaning or intentions. But what is sought in these
studies is that subjects in a specialized setting, with no access to the
experimenter's intentions beyond what is presented, obey a sharp distinction between what is said and what is implicated. Yet the whole
reason Grice's work has been so important to philosophers, an
importance increasingly recognized by experimental psychologists, is
that this is a distinction which we do not and, in fact, must not, recognize in on-going conversation. For, presumably, to impose such a distinction would interfere with the drawing of intended inferences, and
so interfere with the smooth interchanges that are the mark of healthy
conversations.

II

7. Two hypotheses positing the intrusion of conversational assumptions
upon the conjunction effect studies have been defended. Both claim
that the intrusion is an attempt to render relevant the fundamental

contribution of the personality sketch, or avoid violations of well founded conversational expectations.

What bearing does the conversational account have to the broader issues of rationality raised by Cohen and others? Conversational explanations for experimental findings are generally invoked as optimistic defenses of subjects reasoning. This is largely, but not wholly, correct for the account defended here. Even if that account is granted, the rationality issue is not settled.

First, the conversationally induced problem of a lack of shared understanding is a subtle one, not due to any blatant verbal trick. It is reasonable to conjecture that the subtlety results in part from subjects limited skill with the rule the experimenter wants to study. To be specific: Our greater pragmatic sophistication alone does not explain the differences in the dominant adult responses to the conjunction efect compared to the Piaget class-inclusion studies. The difference between the number of members of a class and the number of a proper sub-class is so obvious to us that we readily permit the conversationally untoward question — "Are there more dimes or coins?" — at face value. Our greater resistance to the violation of the maxims in the conjunction-effect experiment is partly due, I believe, to a certain lack of either accessibility to or confidence in — though not competence with — the conjunction rule for probabilities. If this is so, then the fact that subjects do not understand the experimenter as he intends his words is itself some evidence of a weakness in subjects understanding of the scope of the conjunction rule in everyday reasoning.

The second limitation of the conversational analysis is that though it is of broad scope it is not deep. Its scope is much broader than here indicated. The conversational analysis bears on the base rate studies which represent the main bulk of the Tversky-Kahneman research paradigm[12]. The natural home of the Gricean view is deductive reasoning, and its relevance to studies of that reasoning has been rightly noted. (For example, by Cohen, 1981, sec. 3 and 1987, p. 152). Indeed, any general account of the social psychology of psychology experiments that involve normal verbal exchanges must give a role to Gricean pragmatics.

But the depth of the conversational approach is limited. Once we get clear on what is meant, we can engage in deductive and inductive reasoning that is not limited to, replaced by, nor necessarily distorted through, conversational expectations. The conversational account should not aspire to the ambitious comprehensiveness of either Tversky and Kahneman's "heuristic" and cognitive econony account or Cohen's Baconian probability and narrow reflective equilibrium view. Conversational reasoning isn't a competitor to these as an account of lay reasoning.

It is a species of inductive reasoning — inference to the best explanation — which is consonant with Cohen's system. Although this theme will not be pursued, some evidence for it is provided above in the discussion of the irrelevance of the mounting probability of error with increasing information.[13]

In the remaining sections, the rationality debate will be addressed with less concentration, except for illustrative purposes, on the conversational analysis or the conjunction effect studies alone. We try to show that even if the pessimists' main premise is granted — that subjects' answers are systematically wrong — that does not settle some of the most important issues about rationality. However, it also turns out that an optimist view of subjects' responses does not foreclose criticism. The issues of rationality are much more complex than the opposition of pessimists and optimists suggest, and it is complex in just the way that renders what I will call an Austinian approach appropriate.

8. J. L. Austin liked to remind philosophers that truth and falsity constitute one among many dimensions of evaluation, and that it is not always the most useful or fruitful. (Austin, 1970). Those of us who tend to run late set our watches ahead to give ourselves helpful, yet false, answers to our question "What time is it?" And true answers are often unhelpful: If I want to know where my daughter is, it doesn't help to tell me that she is on the earth. More pertinently, Austin noted that certain sentences that appear perfectly determinate — e.g. "Italy is bootshaped" — are actually indeterminate until one's purposes are specified, usually tacitly (Austin, 1970, 1963).[14]

Ultimately, I would want to view the experiments on reasoning and judgment, and the subsequent debate, under an Austinian influence. The primary question is for what purposes are subjects answers appropriate or not, costly or beneficial, rather than whether they are right or wrong, correct or incorrect. In this paper, I can only encourage consideration of such an approach. I begin to motivate that alternative perspective by criticizing two views which are paradigmatically pessimist and optimist, respectively.

9. Stephen Stitch criticizing Cohen writes,

...when we judge someone's inference to be normatively inappropriate, we are comparing it to (what we take to be) the applicable principles of inference sanctioned by expert reflective equilibrium. On this account, there is no puzzle or paradox implicit in the practice of psychologists who probe human irrationality. They are evaluating the inferential practice of their subjects by the sophisticated and evolving standards of expert competence. From this perspective, it is not at all that surprising that lay practice has been found to be markedly defective in many areas (Stitch, 1985, p. 133).

Cohen makes a number of telling objections to this account such as that "it is by no means always clear which body of elite opinion is to count as the consensus of relevant experts". (Cohen, 1987, p. 1987). Our criticism from the conversational perspective overlaps Cohen's, but it is directed at the initial stage at which subjects interpret the task and information presented.

In the conjunction effect studies, subjects appear to rate a conjunction as more probable than a conjunct. Stitch wants us to ask the probability theorist whether, presumably, p(T) can be less than p(T&F). The expert, without hesitation, answers in the negative. But the expert's authority for deciding that simple question does not extend to granting him authority on a prior question: Should the task be represented as one of comparing p(T) to p(T&F)? Our conversational account provided a rationale for subjects not viewing the context and task as calling for that purely formal representation and question. The relevant "expertise" to determine whether this interpretation is acceptable is no expertise at all. It just calls for everyday knowledge of conversational propriety.

It is easy to be fooled here into imagining communicative purposes are irrelevant because experts in probability theory (qua experts in probability theory) would quickly favor one judgment.[15] But that is because they have already framed the problem to their interests (expertise). To revert to the Austinian example: If we asked expert cartographers whether Italy is bootshaped, they may similarly say "Of course not". That would not however show that the question is not indeterminate in just the way Austin believes. It would only show that in calling on the cartographer for his expertise, he would, quite properly, take for granted his exacting purposes in mapmaking. Given his purposes, a description of Italy as bootshaped is much too crude.

10. A preferable, though still unacceptable, optimistic view is advanced by Funder in an important recent article (Funder, 1987). He is "... highly critical of the currently dominant paradigm that studies error in social judgment". Funder argues that "... judgments (and judgmental processes) that are incorrect in terms of a limited experimental context may be correct when applied to a wider, more realistic context — a context, in other words, in which judgments are usually made and have real-life consequences". (Funder, 1987, p. 86). His analysis, examples, and references provide compelling evidence in favor of this claim. He takes this claim to show, as with visual illusions, that these studies have no normative implications for improving thinking.

The discomfort I find with Funder's account is that we have higher standards for our thinking than its adequacy for "real life". Consider studies of "lay" or "naive" physics. These are studies of the knowledge of young students about the workings of the everyday physical world, especially motion. A. A. di Sessa (1982) asked students to direct a computer "dynaturtle" to intercept a stationary object at an angle of 45 degrees away from it. They used simple commands of Right, Left, and KICK followed by a number representing the degrees to turn. The KICK command is crucial because "... a dynaturtle never changes position instantly, but can acquire a velocity with a KICK command which gives it an impulse in the direction the dynaturtle is currently facing". (di Sessa, 1982, p. 39). Typically the students in di Sessa's study would give the KICK command when the vertically moving dynaturtle was horizontally parallel to the object. The students were, of course, disconcerted when, in good Newtonian fashion, the dynaturtle resolved the vectors, going over and past the object.

di Sessa successfully explains the reasonableness of students responses — their adopting a kind of Aristotelian physics — in the real world in which they must navigate. So in Funder's sense their errors are no mistakes. Yet surely there is a serious cost. Students take the reasoning underlying their judgments to apply beyond their immediate experience, precisely because they hold it to be the correct account of motion in their immediate experience. The cost is paid in predictive failure in the computer world that they are investigating. Notice that the norm involved (predictive success) is one these students accept: They are attempting to get the dynaturtle to hit the object, and they don't succeed.

In this case we have rationally defensible, yet still criticizable, judgments. It is the compatibility between these that makes it possible to treat the conversational account, as well as other "optimistic" accounts, as rationalizing subjects' judgments without exempting those judgments from all criticism. We argue that claim in sections 13 and 14. In the next two sections we argue to a parallel, but seemingly contrary, position: Even if we grant that subjects' answers are incorrect, that alone is insufficient for significant criticism.

11. The methodological restriction mentioned earlier which excludes pre-experimental training, explicit cues to suspend normal assumptions, or feedback is of the first epistemological importance. For it is evidence that the settings are not primarily testing for the epistemologically central question of how well we modify our judgments with information and experience. As a result what is often tested for, as the conversational approach suggests, is not our reflective, thoughtful, or critical reasoning,

but our initial or immediate thinking.[16] The distinction is one that receives recognition in everyday life. You meet a young child on the street and you say to the parent "He's a lovely boy". The parent responds, "He's a she". It is only the next time you call her a "boy" that your judgment is critically deficient, rather than an honest mistake. Again, subjects may willingly (and uniformly) rate 2 as "more representative" of evenness than 372. Yet, upon a moments reflection, if that, they recognize that the comparison is illicit. (Armstrong, S. L. Gleitman, L. R., and Gleitman, H., 1982).

It is a consequence of our Humean predicament that initial judgments must be strongly biased toward ascribing patterns or regularities. If not — if for any set of events, we assign equal weight to all hypotheses prior to obtaining evidence — then we will never learn from evidence (or experience). Evidence will never raise the probability of a hypothesis because, roughly, of the fundamental inductive truth that any body of evidence for a hypothesis is compatible with the denial of that hypothesis[17]. Thus we must always *begin* any attempt at learning or inductive inference by assuming regularity. A similar point holds for conversation. The assumption of cooperation biases us toward accepting what is said as the result of a sincere intention to communicate (by virtue of what is said). The alternative would be no communication at all if we first required, for each conversational contribution to be accepted, evidence of sincerity.

Robin Hogarth (1986) has noted that many of the judgments that appear mistaken or biased for a discrete (synchronic) test make excellent sense for a continuous enviroment. In considering the so called "adjustment and anchoring heuristic," especially in their role in evaluating joint probabilities, Hogarth observes "In continuous processing, however, a series of adjustment and anchoring responses, all of which may be relatively inaccurate, takes one progressively to the target". (Hogarth, 1986, p. 694). In a continuous enviroment, where there is little *epistemic* cost in conjectural error, initial judgments do not matter much, and will be preferable if stronger[18]. As Popper (1968) would say, the bolder ones will, if true, explain more, and if false, be easier to refute.

Consider a simple, everyday illustration that I will return to. You are driving to a dinner party but you are unsure of the directions. You come to a familiar intersection, vaguely sensing that you should make a right turn. All other things being equal you would turn right. If you were asked at that moment which direction do you believe correct, you would without hesitation say to the right. However, as is typical, things are not all equal and you are not forced at the moment to choose. You recall that if you go left there are lots of stores where you are sure good directions can be had. In this *diachronic* enivironment, you turn left.

Of course, the exigencies of everyday life are often uncaring about our best epistemic strategy. *Practically* the costs of learning from experience are often high. Life is short, emergencies come up (imagine you needed to rush someone to a hospital), so that our errors can harm us, however much we learn from them. That does not undermine the epistemological point. It should only remind us that improving knowledge is not our only concern.

Another reason for the epistemological focus on *diachronic* assessments is that a body of evidence generally permits a number of conflicting inferences. It is (often) only by knowing the *pattern* of modification in judgment with new information or evidence that we can make reasonable assessments. We are in a better position to assess a student, if we observe not only that he now has a C+ average, but the pattern in his grades. He might have tests ranging from D at the start to A currently (assume that all tests count the same). We will have a different evaluation — although we may have to give the same grade — if this same average results from exam grades going in the opposite direction.

12. If subjects view their task diachronically, as part of an inquiry, where feedback and further information is expected, then it must diminish confidence that we can read off what they *believe* on the basis of individual actions (answers, judgments). Yet, it is crucial to the standard treatments of these studies that subjects' judgments represent what they believe in the following way: their judgment is made solely with a concern with being correct; their aim is just to give the right answer.

But already above we observed that where there is limited epistemic cost of error, we are encouraged to make bold inferences. Judgments that on first appearance are claims to believe that a certain conclusion is correct, can actually, though not self-consciously, be offered as good working hypotheses. The strong presumptions of conversation are afterall so successful, to continue our leading analogy, because even when they (mis)lead us to drawing bold inferences that the speaker does not intend, the speaker can just clarify the matter, if he recognizes the misunderstanding.

It is standardly held that to assert p is to represent oneself as believing that p. But even this natural, everyday expression of belief cannot be taken too seriously. The assertions of normal everyday talk (e.g. sincerely uttering "The Yankees will win the pennant game" prior to the game) are evidently much bolder and unqualified than the actual attitudes of the ordinary folks who assert such things. Though assertion implies belief, few, if any, will have full belief that the Yankees will win the pennant game. Conversational norms encourage the expression of belief in an unqualified or unguarded form (rather than being filled with appropriate epistemic qualifiers like "It is highly probable that..."). Our assertions are put forward as beliefs aiming solely to be correct, but this partly serves the social aim of keeping us focused on the main content. What we really believe is then a complex function of what we assert, though assertion is the clearest manifestation of belief.

It is, in fact, a commonplace that it is often quite difficult to tell what a person believes on the basis of any small selection of their acts.[19] Thus consider the drive to the dinner party: Observing only your behavior, most of us would likely conclude that you believe that turning to the left is in the direction of the house. Even when whole groups of persons respond similarly, the attribution of a complex set of beliefs can be problematic. Cohen observes that in the Gambler's Fallacy, we are usually involved in inferring what the gambler believed on the basis of his actions — progressively increasing bets on the same outcome in a game of chance, as that outcome continues not to come up (Cohen, 1981, pp. 327-328 and his replies to comment; Cohen, 1982; and Cohen, 1987, pp. 170-171). We infer that the gambler believes the events are dependent. But Cohen bravely notes that the behavioral evidence is compatible with the gambler's acting on a policy that he fixes ahead of time by reflecting on the law of large numbers[20]. This alternative reading would equally well apply if many gamblers behaved similarly.

Subjects' judgments as behavior are susceptible, I am arguing, to a reading in which their objectives are more akin to those of inquiry or that appropriate to a continuous environment, rather than a multiple choice final exam. In the realistic situations that these tasks suggest, further information, feedback, and the opportunity to modify their judgments is to be expected. Of course, subjects may be told that in these studies such opportunities will not be afforded. But that is unlikely to be sufficient to completely wipe away the expectations we develop from an enormous amount of common experience and learning. This common learning and experience becomes so deeply ingrained that we are unaware of its role in framing the task before us. And well we should be so unaware since one of the purposes of such learning and experience is to develop rapid, automatic adjustments of judgment with new information (e.g. driving a car and the continuous changes in road conditions).

The argument of these last sections is that even if we grant that the subjects' dominant answers are erroneous, it would not follow that their giving these answers are irrational or epistemically defective. This should not be surprising: After all, scientific rationality, which is often held up as the ideal to which we poorly approximate, is realized in science's being a *self-corrective* enterprise requiring a continuous enviroment, and not in the accuracy of its initial hypotheses. The "errors" made whether in everyday or scientific reasoning can be in the service of, and thereby justified by, longer run cognitive of epistemic objectives. If so, further, their judgments cannot be treated as beliefs whose sole objective is truth. To doubt, however, that subjects' judgments are beliefs whose aim is solely the truth is not at all to imply that they are acting insincerely or deceptively. Belief arises from the attempt to satisfy a number of objectives with truth being only the most salient of the cognitives ones. True epistemic criticism in most realistic situations will then be complex (having to take account of the optimization of these objectives) and diachronic, not synchronic.

We turn now to arguing nearly the converse, although the conclusion is similar to one already reached in the discussion of lay physics: The finding of an optimistic rationale for subjects' judgments does not foreclose criticism.

13. If subjects' judgments are partly influenced by the assumption that the enviroment is continuous, then rendering relevant the personality sketch provides a basis for developing expectations about that person, e.g. Linda, should they come to meet her. Such expectations help us develop an initial impression of her, and so to judge how we should behave. It helps us to form positive, rather then null, hypotheses which are useful in gaining more information when we meet her. Bits of action and behavior — e.g. the clothes worn, the mannerisms — can be unified by such hypotheses as that Linda is a feminist. Thus we gain information, and that gain, as we noted above, will not usually be epistemically costly[21].

The judgments that render the personality sketch irrelevant, are likely to surrender such informational benefits. (Likely but not necessarily, assuming that the point of the task is not transparent to subjects, whatever their answer.) If we somehow think it is more worth believing that Linda is a bank teller (than that she is a bank teller and a feminist) then we are less in a position to form expectations and hypotheses about her. Consequently, we have a diminished basis for learning from the experience of meeting Linda because we do not have a good handle on how to interpret her behavior. For the sketch gives us one conception of Linda, while our judgment undermines our confidence in relying on that sketch[22]. (Of course, that will be an advantage if, contrary to expectations, Linda turns out not to fit well with stereotypes for feminists).

However, the cognitive advantages from a bias toward relevance is not without costs for other cognitive aims. Subjects cognitive goals go beyond explanatory and informational gains. Specifically, they also include what I will refer to as *predictive* and *objective* accuracy, although the former term, as we will see, is somewhat misleading.

In contrasting predictive accuracy with explanatory coherence (and related notions), I am refering back to a discussion in the philosophy of science. A well known example of Scriven's (Scriven, 1959, 1962; and the related discussion in Scheffler (1967, Part I, sec. 3-5) is that of paresis which occurs only in those who have had latent, and untreated, syphilis. A person's contracting syphilis promises a good explanation for his manifesting paresis, though the probability of developing paresis, given that he has latent syphilis, is quite low. Positing the etiology of the development of paresis in syphilis has high explanatory value, but low predictive accuracy. As Salmon[23] (1971, pp. 56-57) observes: although the probability of developing paresis given that one has syphilis is low, the positing of latent syphilis does create a narrower reference class within which the probability of paresis is greater than in the population as a whole. It would be bad to bet that a person will develop paresis, given that he has syphilis, though syphilis provides a good (working) explanation for his developing paresis.

Objective accuracy is the goal we are aiming at when we attempt to test our judgments against observations or other (seemingly) neutral evidence. The idea is to let an outside or neutral party — namely nature — decide, and the less we intervene, the more likely the evidence will be untainted by our own interests and biases. Conversationally, it is this goal, although certainly not this goal alone, that leads us to favor treating what others say strictly and literally. The more strict and literal, the less likely we will be imposing our own thoughts on them. (Bach and Harnish, 1979). This goal is somewhat sacrificed in the conjunction effect and similar studies since achieving relevance requires, as we have seen, that what the experimenter says not be taken literally.

It is not unreasonable for subjects to favor the former set of goals rather than the latter. Sacrificing (conversational) objective accuracy allows subjects to continue to view the experimenter as an honest, nondeceptive, participant in a dialogue. The willingness to sacrifice predictive correctness about Linda — just as the positing of the latent syphilis, despite its improbability — improves our ability to exploit the information given and helps us, through hypothesis formation, to gain more pertinent information by guiding and interpreting further investigation or observation. In that way, the explanatory framework promises to vastly improve our predictive control over all[24]. (That is why the term *"predictive* accuracy" is misleading). In themselves, these remain bad bets (e.g. betting on Linda's being a bank teller and feminist over her being a bank teller). But from the point of view of on-going inquiry they are the best bets we have.

Nonetheless, and this is the main point of this section, the fact that favoring one set of goals over others is reasonable does not eliminate the fact that sacrifice of legitimate cognitive objectives has taken place. The objectives sacrificed provide a basis for criticizing subjects. This is especially so if, as with the studies of lay physics, these objectives are the subjects' own. That is the impression gleaned from reports of post-experimental discussions, as well as introspection.

Although these sets of goals can conflict, it is not for that reason that they cannot both be generally maintained. It is not like wanting a healthy marriage and taking a vow of chastity; nor is it even like wanting to be rich and wanting to be a philosopher. For normally, whether in science or everyday life, these objectives go together[25]. Understanding a person's action in terms of explanatory hypotheses (attributing beliefs and desires to them) is an excellent basis for predicting how they will behave[26].

The Tversky and Kahneman studies create, in effect, circumstances that brings these goals into conflict. If we came to appreciate where these conflicts are likely to arise, then these studies should motivate us to figure out ahead of time what the pertinent trade-offs should be. One can then learn to improve one's broadest cognitive goal — to understand the world better — by appreciating the kinds of contexts in which our normal assumptions of regularity or cooperation are likely to prove troublesome.

14. The arguments of the last sections have been aimed at rejecting the idea that the experimental tasks are best evaluated and understood along the dimension of right and wrong; correct or incorrect. Even if subjects answers are wrong, that does not exclude those answers being appropriate on other relevant evaluative dimensions. Correlatively, if there is a legitimate framework that subjects apply, within which their answers are justified, that doesn't exempt them from criticism. The studies of judgment and reasoning are best understood along a number of evaluative dimensions, and it is the basic weakness of the optimist-pessimist opposition to attempt to reduce these to only one.

Any approach that treats the fundamental issue as solely that of right or wrong faces a difficulty that our (admittedly fuzzy) account doesn't. On the former approach, no special illumination is to be found in *ignoring* the rationale that underlies the supposedly wrong answer, from considering and *rejecting* it. That is, no illumination over and above what we would normally learn from rejecting a mistaken view, rather than just ignoring it. But this just cannot be right — subjectively and normatively.

The responses offered by subjects in all the judgment studies do not show subjects *favoring* their answer over the experimenter's preferred alternative. They *ignored* that alternative. But each of the proposed rationales, whether Cohen's Baconian account or the Bayesian one favored by Tversky and Kahneman or our modest conversational hypothesis, provide important insights on these problems. Given those insights we will want, ideally, to be able to look at the problems from all these points of view, offering risky assessments as to which should be favored in specific kinds of context. We will not think it worthwhile, leaving aside practical exigencies, to answer these problems only according to one rationale, ignoring the others. The Austinian perspective does not demand that we reject any of these rationales and, in fact, encourages us to appreciate the illumination each casts on the problems.

If we take seriously that perspective, we should give up as primary the question "Which answer is correct?" It is a misleading question the way "Is Italy **really** bootshaped?" is a misleading question, even though it can always coherently be asked[27]. For it presumes that we can answer the question in abstraction from the different purposes that are served by one judgment rather than another. On the conjunction effect and related studies we should rather ask: What (cognitive or epistemic) purposes are legitimately applied to these problems? If these goals or purposes are in conflict which should dominate? For which of these purposes is one answer more appropriate? What goals are thereby sacrificed?

Pessimists will consider the proposed shift *ad hoc*: The tasks before subjects are self-contained problems, as in an exam, to which there is a right and wrong answer. Subjects are not put under any serious time constraints, and so they give their answers to the best of their abilities. We can thus construe the answers as representing their beliefs. Finally, they will insist, all this purpose — relativity denies the obvious fact that there often is a right and wrong — period.

The attractiveness of this proposal has a source similar to, and similarly as illicit as, that for the question "Is Italy *really* bootshaped?" We imagine that this question is perfectly proper because it is just asking what the objective truth is: "There (the speaker points to the map) is the country Italy, now is it bootshaped or not?" The question isn't imposing any particular purpose or perspective, so that we could answer "Well, for such and such purposes (e.g. drawing) it is, for other purposes (e.g. cartography) it isn't". But it is imposing a perspective or distinct set of purposes namely, ones with high standards of precision. The illicit question succeeds because we acquiesce in identifying greater precision with closeness to objective truth.

If we generalize the lesson, we do not deny that there is often a right or wrong simpliciter. Rather, in many cases where we believe that there is just a flat out right or wrong, our purposes remain, quite properly, tacit. If I ask someone what time it is, and he tells me that it is "around noon" when it is actually 11:10 a. m., then he gave me the wrong answer. That is so even though for some purposes "around noon" would be a correct answer (e. g. where noon time has special significance, and the standards in force are very blunt (morning, afternoon, evening). The tacit purposes in effect in the context I am envisaging does impose standards that require for a correct answer something like "11:10" or "a little after 11", but not "around noon".

The pessimist who insists that these tasks are just self-contained, right or wrong exam type questions is imposing a perspective that he wants to understand as no perspective at all. We can fall under the pessimist's illusion because we have a misleadingly objective perspective on the experiment. We view it from the outside, and then treat the subjects' point of view as our own. We know that the words of the experimenter are to be taken strictly and literally. We know that the experimenter means the task to be fixed as (a) judging probability according to the axioms of probability, (b) representing the alternatives just as they are given, and (c) understanding the problem in terms of no outside, background information. We know that the experimenters want this problem to be viewed as self-contained, as in an exam, and not as a particular moment in an on-going inquiry where our judgments should be strategically governed (e.g. so as to maximize information).

But our objective perspective, like the expert's, is a perspective, and an external one. It is unfair to ascribe it to subjects or to judge them in terms of it, as they must view these tasks from the inside[28]. They have no access to any information as to what experimenter's mean beyond what they are given. And what they are given, so I am claiming, requires them to engage in theory construction. They must construct an account that explains the experimenter's contribution so as to make that contribution appear as rational or reasonable as possible consistent with the evidence offered and their background beliefs and expectations.

Department of Philosophy
Brooklyn College, C. U. N. Y
Brooklyn, N. Y. 11210,
USA

FOOTNOTES

*I am grateful to Catherine Z. Elgin for careful and thorough comments. My thinking on this topic has been greatly improved from discussions and correspondence over a long period of time with L. Jonathan Cohen. A version of this paper was read at the University of Cincinnati colloquium on Human Reasoning, and benefitted from discussion.

[1] Ulluman-Margalit (1983) suggests interpreting Grice's CP as a presumption. I follow her suggestion excepting the disanalogies to be discussed.

[2] I am assuming that though the violations of the CP might be regular they would be contained, and not extensive. We can roughly identify limited non-cooperative groups. Otherwise linguistic communication, as we enjoy it, would probably collapse. On the role of trust in communication see Lewis, (1983).

[3] Discussion of this point in an educational setting is to be found in Adler (1987).

[4] This point is not distinctive of the conjunction effect studies. It is standard in the base rate studies to find that subjects use the base rates as prior probabilities where the sketch is not diagnostic of the pertinent target categories. See the essays in Parts I-III of Kahneman, Slovic, and Tversky, eds. (1982).

[5] For further discussion of the anomaly in such a presentation see Hilton, (unpublished), pp. 19-20; and Adler (1984).

[6] Hilton (unpublished), p. 20 suggests an "even if" treatment but takes it as "T even if is she is *not* active in the feminist movement".

[7] One variant asks subjects to bet on the alternatives. When this is done the conjunction fallacy is lessened. (Tversky and Kahneman, 1983, p. 300).

[8] Compare p. 299 to p. 304 of Tversky and Kahneman (1983). Cohen's recent ideas concerning counterfactualizable vs. non-counterfactualizable probabilities provides principled reasons for treating games of chances separately from normal inductive situations as in the Linda study. See Cohen (1987), sec. 17 and 18.

[9] Cohen's Baconian reading of probability provides further reasons for these multiple associations of the same term. See Cohen (1977), Ch. 1.

[10] Compare this example to the studies on categorization which find that though a guppy might be rated atypical for a fish or for a pet, it may be rated highly typical for a pet fish. For discussion see Osherson and Smith (1982).

[11] Cohen's own conjunction law captures this feature of inductive support. See his (1970), Ch. 1; and (1977), Ch. 5 and 14. Compare the problem of diminished probability with more conjunctions to Hume's probabilistic regress argument in Hume (1975), Book 1 part iii sec. xiii.

[12] For discussion of the base-rate studies see a number of the essays in Parts I-III of Kahneman, Slovic, and Tversky, eds (1982). It is subjects' performance in these studies that has been most strongly defended by Cohen relying upon his own system of inductive probability. See Cohen (1987), sec. 16-18 for his most recent account. A view of the base-rate studies is offered below in section 13.

[13] The basic point of comparison is that inductive inference or support is made to begin with, and thereby facilitated by, very strong assumptions-cooperativeness for Grice and causal regularity for Cohen. For an excellent inferential model of speech communication see Bach, K. and R. Harnish, (1979).

[14] The example is unfairly good for Austin's point (and my use of it) because of the obvious vagueness. But for reasons Austin offers repeatedly, further developed by Lewis (1983b), and for my specific purposes, the example is not atypical. See also sec. 13 below.

278

[15] Yet, as Tversky and Kahneman (1983, p. 300) observe, in the main article we discuss, graduate students in social science at Berkeley and Stanford, with a number of courses in statistics, answer the Linda study in a transparent test with an "incidence of violations" that remains "fairly high". This marks a disanalogy between these studies and the unsurprising lack of knowledge among lay persons in areas such as physics or economics.

[16] For the denial of this claim see Tversky and Kahneman (1983, p. 299); this is also a theme of L. Ross and R. Nisbett (1980) for a broad range of studies. The distinction between levels of thinking that is influencing me, though I remain unclear as to its direct bearing, is that offered by Hare (1983).

[17] This is one way of viewing Carnap's arguments for taking the confirmation function c^* as the proper initial one, rather than $c+$, though he construes the latter as a priori preferable. For the latter does not allow learning from experience. See Carnap (1962), especially sec. 110.

[18] Those inclined toward pessimism will point out that in studies of the modification of belief with new evidence subjects display a high degree of conservatism. See Cohen (1987, pp. 163-164) for doubts about this reading of studies involving artificially constructed sources for sampling (so subjects do not have strong prior beliefs). In other studies where this effect is shown, a number of philosophers have rightly noted that conservatism in a continuous enviroment is desirable. See Sklar (1975); Harman (1986); Goldman (1986). Nisbett and Ross (1980) pp. 191-192 grudgingly admit the point, but do not at all appreciate its force as their final (rhetorical) question makes clear. See also note 28.

[19] For a related and judicious discussion see Edwards and Winterfeldt (1986). Compare, in particular, their p. 649 with Tversky and Kahneman (1983, pp. 308-309).

[20] To engage in such a policy requires that the bettor put himself on "automatic" in increasing bets, but there are other cases – such as Newcomb's Problem – where a refusal to deliberate is sought, if the proper policy can be put in effect. A bettor would lose badly from this policy but that is partly due to the practical constraint about limiting bets. To ignore that constraint or to fail to see its connections with the policy is a mistake, but it is not the colossal Gambler's Fallacy.

In the Gambler's Fallacy we judge the error in terms of a short-run view, not the long-run view inferred from the law of large numbers. Notice a contrast with our response to an example of Lopes' (1986). She imagines a random number generator of 1's and 0's that doesn't seem to be functioning properly. An expert is called in. The expert continually gets 1's, but after each 1 he repeats: Since in a random series 1 and 0 are equally likely I cannot reject the hypothesis that this machine is functioning properly (Lopes, 1986, p. 729). If we find his reasoning bizarre, isn't it because that reaction stems from our viewing this series as a whole (i.e. taking a long run view). While in interpreting the behavior of the gambler (as exhibiting the Gamler's Fallacy) we take a short run view by focusing on an initial sequence of trials.

[21] Tversky and Kahneman's (1983). Examination of the proposal that subjects are seeking "high expected informational value", which we discussed above, does not, notice, consider the proposal in a continuous enviroment.

[22] This same point about the informational advantages of working hypotheses over null

hypotheses in continuous enviroments applies as well to the "fundamental attribution error". For discussion see Nisbett and Ross (1980).

[23] Three observations: First, the question of treatment (or action) complicates matters. The treatment for syphilis is not highly costly or risky compared to the dangers of non-treatment, even if the paresis is not caused by syphilis. So, putting aside direct diagnostic tests, treatment would be advised. Second, there is the question of contrastive classes discussed earlier (inferring T¬F from T). Are we inferring syphilis rather than any other cause or syphilis compared to some other specific illness that is known to be a cause of paresis? Paresis would be explanatory (in Salmon's sense) in the first case, but might not in the second. Normally initial information should be understood as directed at the first question, rather than the second. For the information might not be sufficient for making a discriminatory judgment, while it could be quite adequate for an initial hypothesis. The base-rate studies seem to offer initial information, while asking them to render discriminatory judgments. Third, Salmon's observation adds weight to the question of whether there is a tendency to read the request as for a conditional probability. Tversky and Kahneman op. cit., p. 309 deny this. But the evidence is too limited for a firm judgment.

[24] Where this goal is emphasized there is a significant decrease in the conjunction effect. See Tversky and Kahneman (1983, p. 300). In this section I develop a theme first suggested in a comment Adler (1983). In respect to the base rate studies, see here Cohen's discussion of the different predictive goals of the patient and administrator in Cohen, (1981, p. 329).

[25] The view that appears to emerge from the philosophy of science is that though the symmetry thesis is not true in general, it does hold for many cases.

[26] On the (predictive) success of folk psychological explanations see Fodor (1988)

[27] On how such questions exploit our natural tendencies see Lewis (1983b) and Unger, (1986). It is because of the Lewis-Unger point about our natural tendencies that I reject a tacit point of agreement between Cohen, Tversky and Kahneman, and Nisbett and Ross. They agree that a crucial test of who is right would be subjects' own reflective judgments. They disagree as to what the outcome of that test would be. I believe that Tversky and Kahneman's and Nisbett and Ross' view would be favored, but this would not be a vindication because the preference would arise from a shift in standards of precision and a subtle change in aims. (That is, I disagree also with the shared agreement that a crucial test is worthwhile). So Cohen should not go along with such a test (if he still means subjects' responses to be intuitive). See Kahneman and Tversky (1983a, p. 509). See also Kahneman and Tversky (1982, p. 498); and Tversky and Kahneman (1983, p. 30). (It is puzzling that they hold both that their description is not "evaluative" and yet that subjects "should have known the correct answers"). For Cohen's response see his (1984, p. 736). See also Nisbett and Ross (1980, p. 192).

[28] Goldman (1986, p. 216) observes, in an aside, that in the "debriefing" (conversation) studies subjects told post-experimentally that they had been deceived in the previous experiment, may then come to question whether these same experimenters are now telling the truth. From our external or objective point of view, we know that they are. Do subjects?

REFERENCES

Adler, J. E. (1987). Form, content, and superstition. *Teaching Thinking and Problem Solving,* **9**, 1-2, 10.

Adler, J. E. (1984). Abstraction is uncooperative. *Journal for the Theory of Social Behavior,* **14**, 165-181.

Adler, J. E. (1983). The rationality of the (lay) scientist: Toward reconciliation, *The Behavioral and Brain Sciences,* **6**, 487-88.

Arkers, H. R., and Hammond, K. R. (1986). *Judgment and Decision Making: An Inerdisciplinay Reader.* H. R. Arkes and K. R. Hammond, (eds.) Cambridge: Cambridge University Press.

Armstrong, S. L. Gleitman, L. R., and Gleitman, H. (1982). What some concepts might not be. *Cognition,* **13,** 263-308.

Austin, J. L. (1970). *Philosophical Papers.* 2nd edition J. O. Urmson and G. J. Warnock, (eds.) Oxford: Oxford University Press.

Austin, J. L. (1963). Performative-Constative. In Caton, C. E., (Ed.) pp. 22-54.

Bach, K. and Harnish, R. M. (1979). *Linguistic Communication and Speech Acts.* Cambridge: The M. I. T. Press.

Carnap, R. (1962). *Logical Foundations of Probability.* 2nd. edition, Chicago: University of Chicago Press.

Caton, C. E., ed. (1963). *Philosophy and Ordinary Language,* Urbana: University of Illinois Press.

Cohen, L. J., (1981). Can human irrationality be experimentally demonstrated? *The Behavioral and Brain Sciences* **4**, 317-370.

Cohen, L. J. (1977). *The Probable and The Provable.* Oxford: Oxford University Press.

Cohen, L. J. (1979). On the psychology of prediction: Whose is the fallacy? *Cognition,* **7**, 385-407.

Cohen, L. J. (1982). Are people programmed to commit fallacies? Further thoughts about the interpretation of experimental data on probability judgment. *Journal for the Theory of Social Behavior,* **12**, 251-274.

Cohen, L. J. (1986). *The Dialogue of Reason.* Oxford: Oxford University Press.

Cohen, L. J. (1970). *The Implications of Induction.* London: Methuen.

Cohen, L. J. (1984). Can irrationality be discussed accurately? *The Behavioral and Brain Sciences,* **7**, 736-738.

di Sessa, A. (1982). Students understanding of ordinary physics. *Cognitive Science,* **6**, 37-75.

Feigl, H. and Maxwell, G. eds. (1962). *Minnesota Studies in the Philosophy of Science,* Vol. **III**, Minneapolis: University of Minnesota Press.

Fodor, J. A. (1988). *Psychosemantics,* Cambridge: The M. I. T. Press.

Funder, D. (1987). Errors and mistakes: evaluating the accuracy of social judgment. *Psychological Bulletin,* **101**, 75-90.

Goldman, A. I. (1986). *Epistemology and Cognition.* Cambridge: Harvard University Press.

Grice, H. P. (1975). Logic and conversation. In G. Harman and D. Davidson (Eds.) (1975), pp. 64-75.

Hare, R. M. (1983). *Moral Thinking: Its Levels, Method, and Point.* Oxford: Oxford University Press.

Harman, G. (1986). *Change in View: Principles of Reasoning.* Cambridge: The M. I. T. Press 1986.

Harman, G. and Davidson, D. (Eds.) (1975). *The Logic of Grammar.* California: Dickinson.

Hilton, D. (unpublished). Conversational implicature and the conjunction fallacy: Bank teller' really counts.

Hogarth, R. M. (1986). Beyond discrete biases: Functional and disfunctional aspects of judgmental heuristics. In H. A. Arkes and K. R. Hammond (Eds.), pp. 680-704.

Hume, D. (1975). *A Treatise of Human Nature.* Selby-Bigge (Ed.), Oxford: Oxford University Press.

Jackson, F. (1988). *Conditionals.* Oxford: Basil Blackwell.

Jungermann, H. (1986). The two camps on rationality. In H. R. Arkes and K. R. Hammond (Eds.), pp. 627-641.

Kahneman, D. and Tversky, A. (1982). On the study of statistical intuitions. In Kahneman, D. Slovic, P. and Tversky, A. (Eds.) pp. 493-508.

Kahneman, D. and Tversky, A. (1983). Can irrationality be intelligently discussed? *The Behavioral and Brain Sciences,* **6**, 509-510.

Kahneman, D. Slovic, P. and Tversky, A. (Eds.) (1982). *Judgment Under Uncertainty: Heuristics and Biases.* Cambridge: Cambridge University Press, pp. 493-508.

Lewis, D. (1983). *Philosophical Papers*, Vol. 1. Oxford: Oxford University Press.

Lewis, D. (1983a). Languages and language. In Lewis (1983), pp. 63-188.

Lewis, D. (1983b). Scorekeeping in a language game. In Lewis (1983), pp. 233-249.

Lopes, L. (1986). Doing the impossible: A note on induction and the experience of randomness. In Arkes and Hammond (Eds.), pp. 720-738.

Moirer, D. M. and Borgida, E. (1984). The conjunction fallacy: A task specific phenomenon. *Personality and Social Psychology Bulletin,* **10**, 243-252.

Nisbett, R. and Ross, L. (1980). *Human Inference: Strategies and Shortcoming of Social Judgment.* Englewood Cliffs: Prentice-Hall.

Osherson, D. and Smith, E. (1982). On the adequacy of prototype theory as a theory of concepts. *Cognition*, **9**, 35-58.

Popper, K. (1968). *The Logic of Scientific Discovery.* New York: Harper.

Salmon, W. (1971). *Statistical Explanation and Statistical Relevance.* Pittsburgh: University of Pittsburgh Press.

Scheffler, I. (1967). *Anatomy of Inquiry.* New York: Alfred A. Knopf.

Scriven, M. (1959). Explanation and prediction in evolutionary theory. *Science,* **CXXX.**

Scriven, M. (1962). Explanations, predictions, and laws. In H. Feigl and G. Maxwell (Eds.). *Minnesota Studies in the Philosphy of Science.* Vol. **III**, University of Minnesota.

Sklar, L. (1975). Methodological conservatism. *Philosophical Review,* **84**, 374-399.

Stitch, S. (1985). Could man be an irrational animal? *Synthese,* **64**, 115-135.

Swain, M. (Ed.) (1970). *Induction, Acceptance, and Rational Belief*. Boston: D. Reidel.

Tversky, A. and Kahneman, D. (1983). Extensional versus intuitive reasoning: The conjunction fallacy in probability judgment. *Psychological Review, 90*, 293-315.

Ullmann-Margalit, E. (1983). On presumptions. *The Journal of Philosophy,* **LXXX**, 143-163.

Ullmann-Margalit, E. and Morgenbesser, S. (1977). Picking and choosing. *Social Research,* **44**, 757-785.

Unger, P. (1986). *Philosophical Relativity*. Minneapolis: University of Minnesota Press.

Poznań Studies in the Philosophy
of the Sciences and the Humanities
1991, Vol. 21, pp. 283-301

Tomasz Maruszewski

HUMAN RATIONALITY — FACT OR IDEALIZATIONAL ASSUMPTION[1]

The problem of human rationality is not only empirical but also has important theoretical implications, philosophical ones included. These implications are not limited to anthropology but affect epistemology and the philosophy of science as well. One cannot reject Kelly's thesis (1955, 1970) saying that scientists are human beings and are subjected to the same laws and limitations as they find in the objects of their study.[2] Therefore, different, sometimes even contradictory, theses of the rationality of man should also be applied to scientists. Those who claim that people are irrational should be aware of the consequences of their statement, especially that this statement is made by a representative of the irrational species. To be consistent, they should assume, that they can be irrational themselves or otherwise they would credit themselves with supernatural abilities which make them different from other people.

In the first case we should answer the question 'What are the reasons to make us believe them?' but the second possibility is more complex. The thesis assuming that scientists or philosophers are rational but ordinary people are not, would imply the need to build up different theories explaining the cognitive process in scientists and in other people. In fact, such theories are constructed because the claims of cognitive psychology about human intellectual activity are different from those put forward in the philosophy of science, which aims at the description and explanation of the procedures involved in the process of creation of science. An exception in psychology are those theoretical approaches which are based on the metaphor "man — the scientist in the street".

At the beginning, these conceptions were related with Kelly's theory personal constructs, then, with the increasing popularity of attribution theory, they came into social psychology. The assumption made in one of the most popular conceptions developed within attribution theory was that while reflecting the reasons of other people's behaviour some very simple cognitive procedures are employed, resembling those of inference based on the method of agreement and the method of difference (Kelley,

1973; Laljee, 1981; Maruszewski, 1983). This reasoning was based on the assumption that if ordinary people are similar to scientists and scientists are rational, ordinary people should also be rational. However, the problem is not as simple as it seems to be. Science creates its own prejudices: for example the one saying that a statistical hypothesis should be accepted on the level of .01 or .05 (Skipper, Guenther, Nass, 1970). Another prejudice having practical implications is that children should be born in hospitals although the mortality rate is higher among this group than among babies born at home. We can give many more similar examples.

These and other examples, as well as the difficulty in finding a direct confirmation of Kelley's thesis, make grounds for a revision of the claim that people are rational. Starting from the mid-seventies, a number of papers on cognitive biases have been published; at some time research in this area became epidemic in character. The studies concerned such biases as the fundamental attribution error, false consensus effect, underestimation of the influence of the social role pressures on the human behaviour, and so on (Ross, 1978).

The problem, however, was that it was difficult to decide from what standard these biases are deviating? We should say, of course, that they deviate from a certain theoretical model; however, then we could not clarify from which one. Is there a single universal model of rationality or are there many of them? Are these models independent of culture or relative to it? Are they assumed *a priori* or can they be considered as derived from empirical data? The answer to each of these questions requires fundamental philosophical solutions to be settled and when a certain option is chosen the reference to empirical data should be made.

To conclude this preliminary analysis, I would like to ask the most important question: is human rationality a characteristic which can be discussed with reference to competence or to performance? Acceptance of the second solution would imply that the results of all studies on cognitive biases should be questioned, whereas when the first solution is taken the question arises what kind of data may falsify the thesis about the rationality.

The present article consists of three parts. In the first one I am going to recall some classical approaches to rationality. Roughly speaking, rationality was understood as undertaking such actions in a choice situation so as to ensure the attainment of some (or some class of) goals (Weber, 1949; Simon, 1950). There are some difficulties posed by such an approach. The two following parts of the article will deal with these difficulties. In the second one I will pay attention to the restrictions caused by the language that is used for posing and solving problems concerning rationality. An emerging possibility to deal with rationality on the grounds of some theories of concepts is — in my opinion — quite inspiring but insufficient. There is another difficulty — the most substantial one — in attempts at analysis of rationality, which will be presented in the third part of article. Certain arguments suggesting that it is very difficult to speak about one kind of rationality will be presented. The last, fourth part of the article, will provide a sketch of a solution of the difficulties.

I. On classical approach to rationality

Rationality is a concept that can be derived from a rich philosophical and psychological tradition. Etymology of this term points to the Latin term *ratio* which corresponds to reason or intellect. It suggests that originally rationality might be understood as an ability to make use of one's reason. Thus rationality refers to mental activities rather than to overt behaviour. It can be used in an analysis of various kinds of reasoning, both deductive and inductive. The growing popularity of behaviouristically oriented theories caused some shift in the interpretation of rationality. Instead of speaking of rationality as a feature of mental activity many authors confined this term to rational behaviour only, assuming that the underlying processes are obscure and difficult to study.

As examples of the classical approach to rationality we may cite views of Weber (1949) and Simon (1950). Weber considers rationality from a subjectivistic point of view, claiming that the "... subjective intention of the individual is planfully directed to the means which are regarded as correct for a given end" (1949, p. 34). Thus, if two individuals in the same situation are rational it does not mean that they will act the same. Differences in their behaviour originate from the fact that the two people are not going to attain the same goal. The subjectively rational behaviour is different from "... technically 'correct' behavior" (*op. cit.*, p. 34).

The discrepancy between "rational" and "correct" conduct may originate not only from differences in the goals that the individual tries to attain, and indirectly from the values that order his or her preferences as to the goals. The discrepancy may simply come from differences in recognizing possible actions, that may lead to attainment of a given goal. The goal of a physician and a medicine-man is the same — to cure a patient. But they will act in different ways because they have different knowledge about the mechanisms of functioning of a human body. Thus, on deciding if some action is rational or not, we should take into account the content of knowledge of a given individual.

The last point is strongly stressed by Simon who considers rationality in the context of the decision making theory: "The function of knowledge in the decision making process is to determine which consequences follow upon which of the alternative strategies" (Simon, 1950, p. 75). And further, he gives the following definition or rationality: "Roughly speaking, rationality is concerned with the selection of preferred behavior alternatives in terms of some system of values whereby the consequences of behavior can be evaluated" (*ibidem*, p. 75). The author contends that the term "rationality" should be used in conjunction with appropriate adverbs. As a result we have a great deal of various rational decisions: objectively rational, subjectively rational, unconsciously rational, organizationally rational and personally rational (*op. cit.*, p. 76-77). The variety of these kinds may be confusing: in fact, in each case a decision should be considered from a different point of view. If the above adverbs refer to the independent dimensions of a rational behaviour then the number of combinations of these aspects of rational behaviour may be quite large (for example: objectively, consciously and personally rational; objectively, consciously and organizationally rational, etc. etc.). I do not think that such a richness is the state that would be desirable in science.

One may ask the question, if there is a possibility of avoiding these difficulties and confusion that were an inherited part of these early approaches to rationality. There are two possible options: the first one was proposed by Kmita (1971, 1975), the second stems from ideas developed by Cohen (1982).

I am not going to present them in detail here. I would like to mention that difficulties of "classical" approach may stem from the fact that they use a special kind of language. This language prevents us from carrying out empirical studies on rationality. Thus it would be very interesting to see if a change of the language used in the analysis of the concept of rationality would generate a possibility of a new approach to it.

II. Is rationality a natural concept?

There are three general ways of defining categories. The first consists in enumeration — we try to construct an exhaustive list of all instances of the category. The second involves finding a rule that generates all the instances of the category. The third one consists in finding attributes (perceptual, functional, relational) that are shared by members of the category. In the foregoing part of the article we will pay attention to the last one which is the most typical of sciences except logic and mathematics, where the second one is the most popular (Kihlstrom, Cantor, 1984).

There are many possible interpretations of the way in which clusters of attributes combine to define a category. In these considerations we will confine to two of them, called the "classical" view and the proto-type view.

According to the "classical" view, concepts are defined by specifying a set of conditions which would be both sufficient and necesaary to include a given object to a denotation of a given concept. Such an understanding of concepts originates from Aristotle's ideas and for many centuries philosophers and psychologists have assumed this interpretation to be the only one possible (Smith, Medin, 1981). Attributes of the concept are singly necessary (i.e. every instance possesses every defining feature) and singly sufficient (i.e. every object and event that possesses all the defining features is an instance of the concept) to define a category. It means that the examples of a given category are homogeneous. Another consequence is that there are clear boundaries between different categories. Cantor and Mischel (1979) and Kihlstrom and Cantor (1984) also suggest that concepts are located in a hierarchical system characterized by perfect nesting. It means that the defining features of superordinate categories are also defining features of subordinate categories. For example, all the features used in defining the concept of "animal" may be found among the features defining the concept of "dog".

The "classical" view also assumes that the number of defining features of a given category is finite, although it may be very large. Thus an individual is not likely to be able to list them, although he or she may use them in his or her behaviour. In other words, a knower does not have to know (on the conscious level) all the features that underlie his classification decisions. For example, we can cite the results of Hull (1920) experiment. In the first part of the experiment subjects were shown various Chinese ideographs and asked to guess which of them were "lings" and which were not. After each guess they were told the correct answer. In the second part of the experiment they were shown a number of new ideographs with a request to classify them as "lings" and "not-lings". Although most of subjects responded correctly, they were not able to show the features that were characteristic of all "lings".

However, it appeared that concepts defined by the "classical" view are quite rare. They can be found among scientific concepts or among legal concepts, but they are used very seldom in everyday cognitive practice. This fact provided the basis for formulating the prototype view. According to this view, concepts may be considered as a summary description of the category. But it is impossible to find defining features of the category. "No feature is singly necessary and no set of features is jointly sufficient to define a concept" (Kihlstrom, Cantor, 1984, p. 5).

The features of objects constituting a category are only probabilistically associated with category membership. Therefore researchers no longer speak of defining features, but they assume that all examples constituting denotation of a given concept are characterized by a "family resemblance" (Wittgenstein, 1953). Such concepts are termed natural concepts.

Among the examples of a certain cetegory there are objects that are more or less typical. It was assumed that there are objects more or less round, more or less square or showing "bird-ness" to a greater or lesser degree (e.g. robin seems to be a bird to a greater degree than a penguin or a duck). The objects which are the most typical of a given category are referred to as prototypes (Rosch, 1978). In categorization they are used as reference standards. There has been, however, some trouble in the intepretation of prototypes. Some people treated them as the average of all the known objects belonging to a given category; according to others, the prototype was the mode (i.e., the most frequently met object of the given kind). Another group of authors claimed that the prototype should be a specified object exemplifying a given category and classification of other objects was based on the similarity between them and that specified object. A review of different approaches to the prototype definition is given in the book by Smith and Medin (1981) so we shall not go into details in this paper.

One of the consequences of the lack of defining features is that categories are heterogeneous. Another consequence is the fuzziness of concepts' borders and occurrence of borderline cases. Sometimes it is difficult to say whether a given object falls within the scope of a given category or not. Let us consider, for example, a tomato or a cucumber within the category of "fruit" or an ashtray or a hammock within the category of "furniture". Depending on the context, people can make different classifying decisions. Out of the context no one is likely to classify roller-skates as vehicles but when people are subsequently asked if a car, motor-bike, bicycle, scooter or roller-skates are vehicles, the answer that roller-skates belong to that category is far more likely.

Although philosophers noticed this kind of concepts long time ago, they considered them as deficient in some respects. They supposed that only "classical" concepts are "true" concepts. Beginning from insightful analyses of Wittgenstein (1953) and from empirical studies by Rosch (1978), it appeared that natural concepts are a quite reasonable way of acquiring and storing knowledge.

As was proved by Cantor et al. (Cantor, Mischel, 1979; Cantor, Smith, French, Mezzich, 1980; Kihlstrom, Cantor, 1984) natural concepts are also found among those describing people, their characteristics and behaviour. Thus, we might suppose that the concept of rationality is of the same type. This assumption would solve many problems. First of all we would avoid problems of definition, we would not have to define what rationality is and it would be sufficient to determine the way certain behaviour is classified. Moreover, certain kinds of behaviour could be treated as prototypes of rationality. Such prototypes could, for example, be the behaviour of scientists, of logicians solving a problem within their own field. Other kinds of behaviour could be treated as revealing a lesser degree of prototypicality. As an example we can cite the behaviour of children or people in emotional arousal as well as symptoms of cognitive biases.

The solution proposed above is very tempting as it would make grounds for empirical studies of the way the concept of rationality is understood by different people, how its meaning changes depending on context and so on. According to the results of Cantor et al. (1980) this approach is very helpful in the study of cognitive structure of diagnostic categories used by clinical psychologists.

Probably the approach based on natural concepts might be useful in other domains. However, its greatest disadvantage is that the scientist does not have to investigate the essence of a given concept and its meaning. We all know from everyday experience the category of furniture and there is no need to inquire about its meaning. However, we are left with some discomfort because instead of analysing the meaning of a certain concept which is usually the initial stage of any investigation, we can only learn about classification of different objects by different people. Thus, to continue this example if we treat this problem in a slightly sarcastic way, a researcher, who does not know what furniture is, keeps asking others who know as little as he does. A sum of two or more ignorancies can not make up for knowledge.

Criticism of that kind is of little importance when it is referred to categories well known from everyday experience, and such categories have been studied by those interested in natural concepts. However, the analysis usually concerns such concepts which have had their normative correspondents sharply defined elsewhere. The favourite categories in the study of natural concepts like "fruit", "bird", "animal" have been taken from biology and this is the branch of knowledge where taxonomy is very important. The studies of natural concepts could be then treated as the analysis of projection of scientific knowledge onto everyday knowledge. There is no use in criticising these studies as we know what a bird, fruit or furniture is.

The problem would be much more complex if some natural concept did not have its normative correspondent or if there was a disagreement on the meaning of a given concept. I am afraid that this is the case as far as rationality is concerned. We can not give a set of defining characteristics which would be commonly accepted. According to some data presented below rationality can be understood in many independent ways. Thus, taking up the studies of rationality in the paradigm of natural concepts, we would get some substitutive knowledge: instead of learning of rationality itself we would learn what other people understand by this term. In view of all this, it seems that empirical studies of rationality have to be postponed until a satisfactory agreement in the understanding of this concept is reached on the normative level.

III. Rationality as an ideal type

The above considerations reveal some dangers embedded in a language used in the anlysis of the concept of rationality. The temptation to approach rationality as a natural concept may lead us to undesired direction. It would involve empirical studies instead of a careful theoretical analysis. Therefore, we will try to approach the ideal concept of rationality from the other side. Let us treat rationality not as a feature which usually characterizes people's activity or thinking, but as a state which usually is not reached although people are able to achieve it. In other words, let us treat it as human competence and not as a characteristic of one's performance. In this approach at least two types of solutions are possible. The first type is related to humanistic interpretation in which the assumption of rationality was fundamental (Kmita, 1971, 1975) and the other is related to the Norm Extraction Method (Cohen, 1982). In both cases we can expect some difficulties. These restrictions stem from the fact that there are different forms of rationality. Which particular form of rationality will appear depends on some dimensions of the situation in which a decision is made. This line of reasoning fits better the one proposed by the Norm Extraction Method than the one proposed by humanistic interpretation. However, it may be possible that the latter might be transformed to allow the possibility of the existence of many forms of rationality.

1. Rationality in the context of humanistic interpretation

I will begin the analysis from the first of the above mentioned approaches. Let us consider the explanation provided by humanistic interpretation why a man has chosen a particular behaviour X_1. We assume that a man has to choose one of a few goals which are ordered by the relation of preference. Each of these goals can be reached by different behaviour. According to humanistic interpretation the behaviour X_1 has been chosen because it leads to the most preferred goal. This kind of behaviour may be described as the most rational. This line of reasoning, although very elegant from the theoretical point of view, brings about various problems when one wants to operationalize and use the concept of rationality in empirical studies. First of all, before we start checking which goal is the most preferred, we have to know mutually independent indices referring to the appropriate degrees of preference. We cannot infer about the preferences *post facto*, i.e. on the grounds of the choice of particular activity made by a particular person, since then we are in danger of getting into a vicious circle.

We should bear in mind that an important human need is justification of one's behaviour. After completion of some action one can change one's hierarchy of values (and thus the order of the preferred goals of one's actions) in such a way that one's activity would always be considered rational. In the work of Aronson (1972) we can find a lot of examples of such behaviour. Is it thus possible to avoid a vicious circle in the explanation? Yes, in theory, but it is almost impossible in practice. The theoretical solution would boil down to finding many independent indices ordering preferences of a given man before he turns to action. If the man would then chose the action leading to the most preferred goal, we could consider his behaviour rational. However, the problem gets more complex since an important determinant of choice in the decision-making process is the probability of the goal attainment and not only its value (Kozielecki, 1977; Lopes, Oden, this volume).

Thus, if before taking up an action one finds the probability of reaching the goal low, even a high value of this goal will not compensate for that low probability. In one of the most popular models assuming the tendency to maximize subjective expected utility (SEU), people are assumed to multiply the value of a goal by the probability of attaining it. Thus, the model proposed by humanistic interpretation should be extended to include the degree of difficulty of action which is inversely proportional to the probability of reaching a goal.

In terms of the idealizational theory of science (Nowak, 1980) we would deal with a nomological structure based on two main factors: the goal value and the probability of reaching it. Both factors would be bound by interaction, that is the impact of one of them would depend on the value taken by the other one. To be more exact, this interaction would be multiplicative, because if one of the factors was zero the tendency to act would be also zero, irrespective of the value of the other factor.

It might seem at this point that the model of rational behaviour is homogeneous. From among all alternative actions a man should choose this one for which the SEU is the highest. Yes, it would be all right if a man was using only one method of determination of subjective probability or utility of alternatives. Unfortunately, this is not the case. Therefore, let us consider the second concept of rationality, related to the Norm Extraction Method (Cohen, 1982). As we shall see in the following passage this approach will take into account the aforementioned reservations.

2. Norm Extraction Method — Polymorphous Rationality

The other approach to rationality stems from the Norm Extraction Method proposed by Cohen (1982). His conception is limited to the analysis of subjective probability, although, in my mind, it could be applied in a larger way as well (see also the articles of Nosal and Ścigała, this volume). The analysis of subjective probability, according to Cohen, cannot be started with one specific model of probability (e.g. from the classical calculus of chance) in order to point out the differences between actual behaviour and the model. The Preconceived Norm Method, that assumes one underlying model of subjective probability, may be accusing in character as it focuses on looking for weak points in the subject.

This method is proposed to be replaced by another procedure. Cohen assumes that the subject's behaviour on the level of the ideal type is consistent with a certain model. The main problem, however, is that the model is not given, but should be inferred from the reactions of the subject. That is, the model should be reconstructed or chosen among available ones as the best describing the subject's reactions. This is of course a much more difficult task than looking for deviations from a single assumed model. In the studies of subjective probability two classes of models are possible — one following from the Pascalian conception of probability and the other implied by the Baconian conception.

Before I start specification of the model for various kinds of situations, I would like to point to a certain characteristic feature of decision making situations. There are few situations when one can estimate the probability of success and failure on the grounds of a large series of identical or almost identical events. The use of a Pascalian conception of probability is reasonable only in them. One can estimate one's probability of success from the frequency of certain events one has experienced or on the grounds of *a priori* evaluation. A classical example of situations of that kind are lotteries.

However, a large number of decision-making situations are unique and the Pascalian conception of probability is of no use in them. In such a case one can resort to the Baconian conception of probability. This conception has been presented in many papers and books by Cohen

(1977, 1979, 1980, 1989). One of the most fundamental theses of the Baconian theory reads as follows: "The Baconian probability never depends (except in limiting cases) on the ratio of A that are B to those that are either B or not-B. It depends instead on the extent of causally relevant factors, which are powerless to interfere in any particular case with the A-B connection" (Cohen, 1979, p. 300). This would imply that in situations such as those above one would carry out a mental experiment in which one would reconstruct all factors that may determine the occurrence or absence a given event. The set of factors of possible importance or unimportance is then compared with the set of factors which accompany the occurrence of a given event in a given moment. The smaller the number of factors blocking the occurrence of a given event the higher is its Baconian probability. In terms of the language of methodologists we could say that the value of Baconian probability is estimated from the variance of interfering factors (usually these are the factors included in the variance of error).

It is supposed that in some situation one turns to a Pascalian conception of probability, whereas in others one chooses a Baconian one. The choice between them depends not only on the kind of situation but on the individual himself and on his/her individual preferences as well. These do not have to be the preferences which directly make some people intuitive Pascalians or intuitive Baconians. The present typology of epistemic styles elaborated by Royce (Royce, 1975; Wardell, Royce, 1978) includes the styles determining the appearance of epistemic perspectives favourizing a Baconian or a Pascalian type of reasoning. It is worth noting that the first kind of reasoning is related to looking for a kind of causality and trying to order the surrounding facts and events. Reasoning of this kind exemplifies the rationalistic style. On the other hand the Pascalian type of reasoning is related to recording a great number of facts and looking for a kind of regularity or repeatability of them. This exemplifies the empirical style.

It would be of extreme importance to find which factors favour the use of a particular conception of probability but this question should be answered not only on the grounds of theoretical reflection but empirical studies as well. Postdiction may help in understanding certain facts, in putting forward certain hypotheses and so on but it will never replace prediction. Only by checking the predictions that have been made we can decide whether a conception considered should be accepted or rejected.

According to the Norm Extraction Method there are many possible models of rationality. Detailed characteristic of each of them is beyond the scope of this paper but we shall put forward hypotheses determining the fundamental dimensions of situation which favour the use of different models of rationality. The first of those dimensions was pointed to when analysing the subjective probability which is one of the basic variables influencing man's behaviour in choice demanding situations. This dimension is the kind of situation depending on whether it is a unique situation or a situation met for the first time or it is one of a series of identical or similar situations. The second dimension could be described as affectiveness — descriptiveness of a situation. The dimension was of great interest to social psychologists (Lewicka, 1985). They proved that in situations arousing no emotions people follow logical standards, such as the rule of logical consistency. This would exemplify the traditional conception of rationality restricted to the use of a certain model of reasoning that would satisfy the requirements of logic. In emotion provoking situations human thinking is based on quite different standards. More detailed analysis concerning this problem can be found in the section devoted to the studies of evaluation.

3. What is the difference between rational action or reasoning and rationalization

The analysis I am going to present will be concerned with different kinds of border cases. If we treat rationality as a classical concept, in the Aristotle sense, we can make its meaning more precise by analysing the border cases (Bolton, 1977). The situation is different from the one in the analysis of natural concepts where the meaning of a given concept may be comprehended best on the grounds of the objects taking the central place in a given category. Let us turn now to the analysis of border cases. Both examples that I am going to give concern the activities which are commonly not recognized as rational since they are determined by emotion dependent factors or variables.

a) Defense mechanisms

In a previously published paper (Maruszewski, 1979) I tried to answer the question whether the use of defense mechanisms may be treated as the kind of behaviour failing to satisfy the assumption of rationality. Defense mechanisms are certain psychological manipulations helping to reduce one's level of emotional tension, leaving the situation that has produced this tension unchanged. According to psychoanalysts, defense mechanisms and repression in particular were the factors favourizing development of neuroses or at least formation of an inadequate image of oneself or other people.

It is hard to deny this line of reasoning at least with reference to the cases which were of interest for Freud and his followers. It is obvious that after having resorted to defense mechanisms, people considerably reduced the possibility of obtaining some value or reaching a goal. We should add that the decision whether to use defense mechanisms or not had not been taken following conscious reflection but had had its causes on the unconscious level.

The results of recent investigations of defense mechanisms showed what the changes are in human perception and which personality traits of an individual favour the use of defense mechanisms (Lazarus, 1967; Grzegołowska-Klarkowska, 1986). They also proved that stress-producing situations may be a factor triggering defense mechanisms. The authors pointed to the fact that an essential factor allowing one to answer whether the defense mechanisms are rational, or not, is the structure of a situation of a given person, that is, whether it is an open situation or a deadlock (Lazarus, 1980). When a situation is of the open kind and the individual can do something to avoid psychological harm the use of defense mechanisms should be treated as irrational. In the core of a deadlock the situation is quite different. Let us consider the situation of the man who is incurably ill and knows that he has got only a few months to live. Should then the use of defense mechanisms be treated as irrational? I doubt that. Probably it is the only possible way of coping with such a situation.

In this moment the following question should be raised: what kind of situations should be treated as deadlocks — either those which can be characterized by certain factors or those which are perceived as having no solution? We advocate for the first solution, which, however, brings about some difficulties related to the fact that an individual's behaviour is not determined by objective characteristics of a given situation but by the way it is perceived. Thus, it is possible that a man perceives an open situation as a deadlock. For example, someone has got a toothache but because of his deadly fear of a dentist he cannot force himself to go to him. Such a person removes from his consciousness all possible dangerous consequences of the status quo and turns to using pain killers. Such a behaviour is governed by an irrational defense mechanism which originates from inadequate perception of a given situation.

The problem, however, is not so simple. Defense mechanisms may be treated as rational assuming the change in the hierarchy of values and goals. The goals related to the reduction of fear and maintenance of positive self-evaluation become the most important. However, the question is whether the choice of the goal has been rational. Probably one can use various criteria to evaluate one's goals and these criteria depend on different factors. These criteria, in turn, may change depending on various factors. This way of reasoning may obviously be continued endlessly. Moreover, the fact the goals as well as the criteria of their evaluation and their importance may change, suggest that there is no irrational behaviour. Is thus the thesis about the rationality of man unfalsifiable?

From the point of view of goals accepted by the individual as well as from the point of view of criteria determining the importance of one's goals we can say that human behaviour is rational. If we assumed that a man behaves in such a way that he could not reach the goals he wanted or that the man's behaviour was independent of his goals, we would have much trouble in understanding his behaviour or in predicting it. The attempts to create a theory of behaviour that would not satisfy the assumption of rationality could only concern pathological cases. When we diagnose a limited ability to control one's behaviour — the explanation of psychological mechanisms one uses may sometimes be very sophisticated (*vide* different conceptions used in clinical psychology) but of little practical importance. We cannot predict the behaviour of that individual on the grounds of them and we have to resort to strict control measures of the ill people (limit their physical freedom in a traditional psychiatric clinic, administer medicines to control their behaviour on a chemical level). It is worth mentioning that perception of the lack of rationality in a person significantly changes social interactions with him/her. These interactions are limited to the simplest communications and people around such a person show anxiety.

Rejection of the assumption of rationality inevitably leads to serious theoretical and practical problems. The latter are caused by the fact that we are not able to predict the behaviour of a person in which the lack of rationality is perceived and we begin to put this person under strict control. On the theoretical level the conception of a man deprived of rationality might be only extremely behaviouristic. It would only describe an automat whose reactions would be determined by its history so far of reinforcements, both positive and negative.

b) Valuation of other people's traits and behaviour
A series of investigations have been undertaken to find out whether in evaluating traits and behaviour of others, people follow the Stevens law (Maruszewski, 1988). We also wanted to check if the parameters of the Stevens function are identical independently of the evaluation of features and behaviour of others being positive and negative.

The students who served as subjects were informed about how much someone had contributed to charity, how much money someone had stolen and how many hours per week someone had been working on computer programming. As numerical characteristics of these three types of behaviour were salient and obvious we expected the subjects to respond either according to the Stevens law or in contrary to it. To our surprise a part of students broke off the main procedure. They gave qualitative judgments in the situation that was clearly quantitative. Independently of the sum of money stolen or contributed to charity their evaluation of theft and generosity level was the same. In evaluation of theft it usually was the maximum negative rating whereas in evaluation of generosity the constant rating was in the lower part of the scale. It is worth noticing that the proportion between minimum and maximum amount of money stolen or contributed to charity was as high as 1 to 12.

What is the reason why the qualitative judgment occurred and what are the differences between this kind of judgment and that, which has been expected to be applied in this kind of experiment? In the case of quantitative response the subject referred the evaluated event to a certain mental scale and his task was to find the optimum position of this event on this scale. This procedure requires to compare a particular event or behaviour with different points on the scale. Usually only the extreme points on the scale are taken into account but sometimes for the sake of more accurate reference the middle point is considered. In the case of qualitative response the cognitive procedure is quite different. The subject finds out whether a particular kind of behaviour has exceeded a certain minimum critical point. This point bears certain resemblance to the lower threshold[3]. After it has been exceeded all patterns of behaviour are treated as equivalent. Complete psychophysical interpretatation of this phenomenon has been given elsewhere (Maruszewski, 1988), so we shall not discuss it here. The obvious advantage of the qualitative evaluation is the smaller amount of time required to make a decision than in the case when the subject has to look for the optimum position for a given object on the scale. We should add that in our study the situation like this described above took place in the case of dimensions of a distinct emotional valence, but did not occur as far as the neutral dimension was concerned. At this moment the problem becomes clear. Situations with a distinct hedonic value (positive or negative) require quick decisions — fast decision making is more important than accuracy of differentiation. The situations of that kind require more distinct differentiation of objects when the time of response is essential. It takes much less time to make a decision when the cognitive activity is restricted to categorizational judgment than when the objects within a certain category have to be ordered along a certain continuum. The reasons why decisions in situations with a distinct hedonic value should be quick become clear. We cannot be indifferent to them. They demand on us unhesitating action either towards a given object or away from it. In the case of neutral objects the situation is different. They do not demand any action from us and indifference to them does no harm.

These two kinds of evaluation resemble what Lewicka (1985) referred to as descriptive and evaluative orientation. Both kinds of orientation assume different standards: the former involves the standard of logical consistency and the latter the standard of affective consistency. The application of the standard of affective consistency leads to the maximum polarization of the world image and in the extreme case to "black and white interpretations". Thus, in consequence we would get something resembling qualitative or categorical orientation. When the other standard is used the individual tries to obtain as accurate and objective image of other people and physical objects as possible. If their characteristics are ordered along certain dimensions, the quantitative interpretation occurs. It is a slower process but it provides a better insight into the essence of the matter.

IV. A tentative conclusion

As follows from the above considerations the problem of the rationality of man is far from being solved. The problem does not refer only to the ways of reasoning but, as we have stressed it here, to different forms of emotional behaviour as well.

Analysing man's behaviour in choice demanding situations we pointed to the fact that it is closely related to the process of evaluation. This process is different in the situations of positive and negative valency. These differences have not been analysed in detail in this paper but, for example, the studies of the utility of money revealed that utility of loss is rated in a different way (radical assessment) than the utility of profit (conservative assessment). It is relevant to add here that the studies of positive-negative asymmetry (Lewicka, personal communication) proved that people process in a different way information with positive and negative hedonic value. We have also shown that people can change the criteria of evaluation depending on whether they perceive the situation as open or a deadlock.

The second parameter of the decision situation is also evaluated in different way by decision makers. One can follow either a Pascalian or a Baconian conception of probability. The first conception will be assumed when the individual has frequently met a given situation before, whereas the second will be followed when the situation is treated as unique or when the individual encounters it for the first time.

The last factor differentiating the form of rationality is the kind of situation, i.e. whether the situation is only of descriptive or diagnostic character (when something is to be found out and no solution is demanded) or of affective character (it demands some action to be taken because of the presence of objects of clear valency).

On the grounds of the above remarks we should derive different models of rationality for each of the following situations:

1. Diagnostic situation, unreiterated or unique.
2. Diagnostic situation, reiterated.
3. Situation of positive valency, unreiterated or unique.
4. Situation of positive valency, reiterated.
5. Situation of negative valency, unreiterated or unique.
6. Situation of negative valency, reiterated.

For a complete picture we should distinguish two variants of situations 5 and 6: each of them may be an open situation of a deadlock.

This analysis suggests that the concept of rationality can adopt many more forms than one initially assumes. However, the strategy adopted in this paper may just as well result in the multiplication of beings beyond the necessity. I hope, it does not because all that has been suggested here is on the one hand an example of the application of the Norm Extraction Method (Cohen, 1982) while on the other it may be considered as a concretization of the ideal type of rationality as realized on the level of human competence.

The end of this study marks no more but the point of departure.

Institute of Psychology UAM
Szamarzewskiego 89
60-568 Poznań
Poland

FOOTNOTES

[1] This research was supported by Research Project RPBP.III.29

[2] It is obvious that this argument refers only to the humanities. To a much lesser degree it refers to the studies of inanimate matter (a chemist is restricted by the properties of the substances his organism is built of, but this restriction seems to be less important than those regarding the character of human cognitive processes).

[3] According to Lewicka the decision process involved in checking whether the lower threshold has been exceeded is different in the case of positive and negative events (Lewicka, personal communication).

REFERENCES

Aronson, E. (1972). *Social Animal*. San Francisco: W. H. Freeman & Company.

Bolton, M. (1977). *Concept Formation*. Oxford: Pergamon Press.

Cantor, N., Mischel, W. (1979). Prototypes in person perception. In L. Berkowitz (Ed.), *Advances in Experimental Social Psychology*, vol. 12. New York: Academic Press.

Cantor, N., Smith, E. E., French, R. De. S., Mezzich, J. (1980). Psychiatric diagnosis as prototype categorization. *Journal of Abnormal Psychology*, **89**, 181-193.

Cohen, L. J. (1977). *The Probable and the Provable*. Oxford: Clarendon Press.

Cohen, L. J. (1979). On the psychology of prediction: Whose is the fallacy? *Cognition*, **7**, 385-407.

Cohen, L. J. (1980). Bayesianism versus Baconianism in the evaluation of medical diagnoses. *British Journal of Philosophy of Science*, **31**, 45-62.

Cohen, L. J. (1982). Are people programmed to commit fallacies? *Journal for the Theory of Social Behaviour*, **12**, 251-274.

Cohen, L. J. (1989). An Introduction to the Philosophy of Induction and Probability. Oxford: Clarendon Press.

Grzegołowska-Klarkowska, H. J. (1986). *Mechanizmy obronne osobowości*. [Defense Mechanisms of Personality]. Warszawa: PWN.

Hull, C. L. (1920). Quantitative aspects of the evolution of concepts. *Psychological Monographs* (Whole No. 23).

Kelley, H. H. (1973). The processes of causal attribution. *American Psychologist*, **28**, 107-128.

Kelly, G. A., (1955). *The Psychology of Personal Constructs*. New York: Norton.

Kelly, G. A. (1970). A brief introduction to personal construct theory. In D. Bannister (Ed.), *Perspectives in Personal Construct Theory*. London: Academic Press.

Kihlstrom, J. F., Cantor, N. (1984). Mental representation of self. In L. Berkowitz (Ed.), *Advances in Experimental Social Psychology*, vol. **17**. New York: Academic Press.

Kmita, J. (1971). *Z metodologicznych problemów interpretacji humanistycznej* [On the Methodological Problems of Humanistic Interpretation]. Warszawa: PWN.

Kmita, J. (1975). Humanistic interpretation. *Poznań Studies in the Philosophy of the Sciences and the Humanities*, **1**, 3-8. Amsterdam: Grüner.

Kozielecki, J. (1977). *Psychologiczna teoria decyzji* [Psychological Theory of Decision]. Warszawa: PWN.

Laljee, M. (1981). Attribution theory and the analysis of explanations. In C. Antaki (Ed.), *The Psychology of Ordinary Explanations of Social Behaviour*. London: Academic Press.

Lazarus, R. S. (1967). Cognitive and personality factors underlying threat and coping. In M. H. Appley & R. Trumbull (Eds.), *Psychological Stress. Issues in Research*. New York: Appleton – Century – Crofts.

Lazarus, R. S. (1980). The stress and coping paradigm. In L. A. Bond & J. C. Rosen (Eds.), *Competence and Coping During Adulthood*. London: University Press of New England.

Lewicka, M. (1985). Afektywne i deskryptywne mechanizmy spostrzegania innych ludzi [Affective and descriptive mechanisms of perception of other people]. In M. Lewicka (Ed.), *Psychologia spostrzegania społecznego* [Psychology of Social Perception]. Warszawa: KiW.

Lopes, L. L., Oden, G. C. (1989). The rationality of intelligence, this volume, pp. 199-233.

Maruszewski, T. (1979). On the humanist interpretation in psychology: The case of investigations of the defense mechanisms. *Poznań Studies in the Philosophy of the Sciences and the Humanities*, **5**, 137-154.

Maruszewski, T. (1983). *Analiza procesów poznawczych jednostki w świetle idealizacyjnej teorii nauki* [An Analysis of the Cognitive Processes of an Individual in the Light of the Idealizational Theory of Science]. Poznań: Adam Mickiewicz University Press.

Maruszewski, T. (1988). Psychophysical scaling and the valuation of social objects. Paper presented at International Symposium on Mathematical Psychology. Berlin.

Nowak, L. (1980). *The Structure of Idealization*. Dordrecht: Reidel.

Rosch, E. (1978). Principles of categorization. In E. Rosch & B. B. Lloyd (Eds.), *Cognition and Categorization*. Hillsdale: Lawrence Erlbaum.

Ross, L. (1978). The intuitive psychologist and his shortcomings: distortions in the attribution process. In L. Berkowitz (Ed.), *Cognitive Theories in Social Psychology*. New York: Academic Press.

Royce, J. R. (1975). Epistemic styles, individuality and world-view. In A. Debons & W. J. Cameron (Eds.), *Proceedings of NATO Conference on Information Sciences*. Leyden: Nordhoff International Publishing.

Simon, H. A. (1950). *Administrative Behavior*. New York: Macmillan.

Skipper, J. K., Guenther, A. L., Nass, G. (1970). The sacredness of .05. In R. E. Henkel & D. E. Morrison (Eds.), *The Significance Test Controversy*. London: Butterworths.

Smith, E. E., Medin, D. L. (1981). *Categories and Concepts*. Cambridge: Harvard University Press.

Weber, M. (1949). *On the Methodology of Social Sciences*. Illinois: The Free Press of Glencoe.

Wardell, D. M., Royce, J. R. (1978). Toward a multifactor theory of styles and their relationships to cognition and affect. *Journal of Personality*, **46**, 474-505.

Wittgenstein, L. (1953). *Philosophical Investigations*. New York: Macmillan.

*Poznań Studies in the Philosophy
of the Sciences and the Humanities
1991, Vol. 21, pp. 303-316*

Ireneusz T. Ścigała

ARE PEOPLE PROGRAMMED TO BE NORMAL?
PSYCHOLOGICAL DETERMINANTS OF HUMAN NORMALITY

1. Introduction

It is rather symptomatic that there are times in the life of individuals and communities or even part of societies when interest in the problems concerning pathology versus normality suddenly grows. This is evidenced by an increasing number of specialist papers or monographs as well as by the way of formulation of statements both in professional language and in common parlance. An illustration of the development of interest and of changes in the views on the question of pathology versus normality may be a comparison of dictionary definitions given in old and up to date editions of dictionaries and lexicons. The interesting thing is that the older the edition of the dictionary the fewer lines are devoted in it to the definition of normality when compared to the lines taken by the definitions of different kind of pathologies (individual, social etc.). To be more precise we observed this relation studying dictionaries and lexicons of English published in Great Britain and USA throughout this century (Flower, Thorndike, Levis, Geddie). A similar relation was then found in dictionaries of German (Wahring), French (Gilbert, Laguane, Niobey) and Polish (Doroszewski, Szymczak).

This comparison brings us to a conclusion that the importance of the concept of normality seems to increase systematically when related to the importance of different concepts of kinds of pathologies (social, individual). The above mentioned relation is particularly pronounced when the concept of normality was made in reference to the mental state of humans and when this concept was replaced by the concept of mental health. In such a case it is understood that the best alternative to mental health is mental illness. Disregarding other terminological problems we would like to point out the most important observation that is that the concepts of normality and mental health have appeared rather late as categories in description and explanation of mental state of humans.

Reference to the dictionary definitions is not sufficient to justify our interest in the development of the category of normality in the language of psychology as science. Acceptance of this kind of evidence demands that Humboldt's and Cassirer's thesis of language determinism (Sapir, 1964, Whorf, 1956) be assumed. This kind of argumentation should rather be treated as an inspiration for further research work. Thus we should at last for the time, abandon the important and interesting analysis of the problem what cultural origins of the systematic invasion of the concept of normality into behavioral and social science are (Sowa, 1984).

Trying to answer in this article the question whether we are programmed to be normal, indirectly assumes a thesis to the presented in the introduction about the characteristic for our time interest in the categories not only of pathology and mental disorders but also normality and mental health. In general, I will consider the question whether the presumed cultural shift has its reflection on a subject's state of being programmed to normality. Answering this question I will refer to theoretical analysis of interdisciplinary character. Thus, I shall refer firstly to certain results of both empirical and theoretical works, mainly from the domain of psychology. The problem of normality is usually regarded as belonging mainly to the area of clinical psychology. It doesn't mean, of course, that it will be automatically always possible to speak about normality without any recourses to the knowledge from other domains (as philosophy, culture anthropology). This article is devoted to the subjective or individual determinants of human normality that were suggested by contemporary psychology.

2. Basic meaning of the term normality

Starting detailed analysis of the category of normality I would like to say that I am not going to make use of any more or less complete set of definitions or conceptions concerning this term. The results of many attempts to construct a theoretical model of normality prove that it cannot be done without some preliminary metathoretical observations. This will be the aim of my paper, on the contrary, I am not going to reconstruct any theories of personality or any other psychological theories from the point of view of the concept of normality, which is implied by these theories most often in an implicit way (Ścigała, 1988). Such reconstruction would only extract them out the specific theoretical context. I would like to start the considerations with systematization of important views on the subject of my interest. (Sowa 1984).

The sense of the term normality oscillates among the three commonly recognized overlapping shades of meaning:

1. quantitative — some phenomena are referred to as normal when they are thought as being common, found on a mass scale, or occurring frequently in a given population.

2. conventionalistic — something is considered normal when it is consistent with the accepted conventions, standards, expectations and postulates in the social world; it is not surprising that in the realm of standard — based activities the conventional and the frequent overlap often as both features may be aspects of the same phenomenon.

3. postulative and normative — in this understanding the norm is referred to the postulate — something is normal when it should occur, of course, this normative normality requires further explanation and formulation of a postulate or norm with respect to which a given state of affairs can be considered normal.

As I am, for obvious reasons, disregarding the normality referring to a wider range of phenomena than human psyche, I will refer only to a certain group of norms. Therefore I am not going to consider the set of technical or economic norms with respect to which one decides about the normal character of certain phenomena, processes or states etc. The reconstruction of normality of some economic processes requires at first establishing the right norms for a specific field and analogously for the mental state we should at first lay down the norms for it on the grounds of some theoretical findings. In other words the question of norms should be considered before we are able to say anything about normality.

In trying to establish the mental normality I am directly interested in, one has to remember that in the present time there is a noticeable shift from interest in physiological to interest in social and cultural norms. As we know, the previously mentioned kinds of norms were regarded as criterions of mental pathology. These criterions had given ground for specification of several kinds of mental disorders. At first, of course, several types of physiological norms referring to the functioning of the nervous system, were taken as criterions of pathology versus normality, and only secondarily the social and cultural norms were taken into account. In fact, only recently have the social and cultural sources of mental disorders have been as thoroughly analyzed. To be more accurate today an attempt is being made to synthesize the basic approach toward mental pathology, and to create a new model that would be systematic, integrated and would take into account the different kinds of norms. Not only are physiological norms according to the functioning of nervous system being taken under consideration but also the pattern of functioning of the hormonal system. In this way an attempt is being made to connect all sorts of changes within the normal hormonal and nervous processes with the changes in man's social and cultural environment. (Fann, Karacan, Pokorny, Williams, 1986).

I think however, that this intense attempt to create a new model of mental pathology will not bring the desired results unless it considers even the most fundamental data concering normality. First of all the underestimation of comparative studies on the normal part of the population. It is true, that an attempt has been made to raise the reliability of many of these kinds of theories by equipping them with several empirical procedures. Nevertheless, such a way of raising a theory to a scientific level is typical for the positivistic paradigm. It can lead, therefore, to a delusive reliability of these theories, which in fact refer to quite different theoretical paradigms. It seems to me that this is the case of several psychoanalytical theories, which are now geared with the above mentioned unjustified methodological intervention for the sake of their scientific status. The adopting of statistical approach has similar results. These are treated as a variant for the scientific character of a wide range of problems taken up by different branches. It leads quite often to unconsidered recognition of many theoretical suggestions as scientific. This is also the reason for the strict requirement of carrying out empirical research only on the condition that beforehand necessary references to the theoretical assumptions has been made. However, it might be worth considering, that there is another, essential condition on which empirical research must be carried out, that is a condition much less respected and noticed. The statistical approach is valid only when it refers to a domain where there is a possibility of finding an adequate objective reference. It is true that the proposition presented here, being an essential condition for the carrying out of valid empirical research is not easy to fulfill especially within the domain of cultural and social sciences in which psychology belongs. Nevertheless this shouldn't lead to the giving up of the search of objective references for the particular categories which the above mentioned sciences are interested in. Thus, I consider such a type of methodological requirement as very advantageous, as it would enable one to carry out empirical research where possible and necessary or give it up even when some problems remain unsolved.

I shall try to describe shortly the requirement according to which the existence of objective references is an essential condition for applying the statistical approach by referring to results of an analysis of chosen psychological theories, which I have carried out elsewhere (Ścigała, 1988). Among the theories discussed were also ones that possessed appropriate objective references. For the purpose of this article I will confine myself to only one of the theories subjected to methodological analysis. It occurred as a result of the following in the four factors Eysenck's theory of personality, that only one dimension obtained the quality of empirical and theoretical validity. It has been the intro — extroversion dimension which reflected the theoretical assumptions concerning the functioning of the nervous system. Thus, as it appeared, the basic differences assumed by H. J. Eysenck (1974) in conditioning and learning between extroverts and introverts are being confirmed more and more (Fonberg, 1980, Strelau, 1980, Eliasz, 1986).

In such cases the requirement concerning objective references finds its justification. Nevertheless it is easy to notice that the theoretical validity of only one dimension in Eysenck's theory depends on the possibility of finding the physiological correlate. Thus, the proper construction of this dimension to be conditioned by a thorough knowledge about general norms seems in our case, physiological norms. However even putting aside that on our current level of physiological knowledge, we are not always able to determine the right physiological norms; these norms are also not always useful for establishing mental normality. It should be mentioned that referring directly to physiological norms as objective references for mental normality may lead to extreme reductionism. In a similar way the social and cultural norms which fulfill the conditions of objective references, because of the possibility of objectivization through empirical research, reduce the matter of dispute to the issue of social and cultural normality. This however, in result can blur out the outline between these kinds of normality and mental normality. Thereby, there is a danger of confusing these different kinds of normality with one another, which makes the possibility of solving the problem of mental normality much more difficult. Thus, to avoid the objection of reductionism and at the same time to hold the possibility of empirical verification, we have to examine precisely the problem of psychological norms. It is quite clear that the problem of human normality can not be always reduced to such categories as physiological, social or cultural normality. If so, as far as it concerns mental normality the problem of psychological norms becomes one of outstanding importance. One should however realize that the theoretical status of psychological norms is much less precise than the other kinds of norms (physiological, social, cultural).

3. The theoretical status of psychological norms

As has been said before, the theoretical status of psychological norms seems to be difficult to establish, because of a basic paradox. On the one hand, it is difficult to consider psychological norms as directly dependent on physiological and cultural norms somehow related with them. On the other hand, the above — mentioned norms can deliver the required objective references, which can give the grounds for empirical research. And so, in order to keep out of danger of reductionism and simultaneously of being over — speculative, we should firstly analyze the meaning of empirical research on normality. In case we accept this sense we must realize thet it will be not easy to find objective references for psychological norms. If we succeed by adopting certain theoretical assumptions another important question will emerge: is it possible to assume that the established norms, in the meaning of certain research work, refer to the whole population? Thus it appears that one of the most important problems should be solved before any reliable settlement about normality can be made: are psychological norms to be "preconceived" or rather to be "extracted" while in the research work is being carried out. This question has been formulated for another domain by L. J. Cohen (1982). He analyses thoroughly the most interesting for me kinds of norms understood not as a social standard or according to R. Glaser (1980) as a "criterion referenced" but as a "norm referenced". It should be added that R. Glaser has proposed the above division since he believed that each preliminary psychological investigation gave two kinds of information. One of them concerns the degree to which the problem has been solved and is referred to a more general quantitative standard found in a given population, i.e. is criterion referenced. The other kind of information permits one to order subjects with respect to a relative standard, for example when subject A has found the solution earlier then B, and is norm referenced (Glaser, after Beck 1980).

L. J. Cohen (1982) in his theoretical work has discussed the second kind of information and has proposed some very interesting solutions. Let me remind the reader that this author devoted a lot of work to the detailed consideration of the relationship between the conceptions of probability assumed in science and the conception of subjective probability in which people are programmed. The researcher usually takes it for granted that a problem which the subjects are asked to solve in the experiment may be solved correctly only on the grounds of some conception of probability assumed in science and which he himself has assumed. Therefore he decides whether the subject's behaviour or solution is wrong or right on the grounds of the technique of evaluation appropriate for a given conception. "If some subject's responses seem, when thus evaluated, to be incorrect, their error may be put down by the investigator — in terms of the computational metaphor — either to a standing fault in the programming of ordinary people (in that they are not programmed to apply appropriate valid principles to the task in question or are programmed to apply inappropriate or invalid ones) or temporary malfunction in the running of a valid programme..." (Cohen, 1982, p. 251). This method of interpretation of experimental data concerned with probability estimation the author calls "the Preconceived Norm Method" and sets it against "the Norm Extraction Method". Advocating the latter he argued: ... "So by setting subjects a particular kind of problem — task the investigator may at best expect to discover from their responses what conception of probability they implicitly adopt for solving that kind of problem and how they assess or estimate the size of probability..." (op. cit., p. 251). Thus, according to the norm extraction method some responses that do not include correct answers are not considered wrong as they may be interpreted to be right within the logic of expression assumed by the subject.

Hence, L. J. Cohen assumes that the answers of the subject are basically correct if they are not influenced by temporary or accidental error producing reasons. He adds"... The Norm Extraction Method assumes instead that, unless their judgment is clouded at the time by wishful thinking, forgetfulness, inattentiveness, low intelligence, immaturity, senility, or some other competence — inhibiting factor, all subjects reason correctly about probability: none are programmed to commit fallacies or indulge in illusions..." (op. cit., p. 251).

The interpretation of almost all responses of the subject as basically correct has aroused ambiguous feeling and opinions. On one hand, the fact that people's right to express their own logic has found acceptance in science demands appreciation. But, on the other hand, the liberalism proposed by L. J. Cohen seems to be too great. In this situation we should at first answer the question whether an ordinary man subject to psychological investigation is able to meet the expectations formulated by L. J. Cohen. Although, to tell the truth, the demands made by L. J. Cohen on the people subject to examination are not great as those made on the researchers. Thus, it seems that the relationship between the researcher and the subject has been finally put right (Cohen, 1981). However, there surely are a lot of traditionally thinking people for whom Cohen's proposition would mean the opposite. I would not like to be counted among them, however, I cannot get rid of some doubts about whether the researcher is able to realize the logic of those subject to examination and their individual concept of subjective probability. It is not that I want to question the competence of the researchers but to assume an attitude to the initial assumption of Cohen's proposition that each subject (in practice each normal person) uses a certain logic or conception of probability. At that point the question arises whether the logic of expression within a given conception of probability can be ascribed to those who according to some standards are considered abnormal, e.g. who are suffering from mental illness. Thus, the question about the conception of normality implied by particular theoretical propositions appears again. In Cohen's proposition the range of its applicability is very important, that is whether it may be applied to the whole population or only to the part of this population which according to the assumed normal distribution falls between 2 standard deviation ($x \pm 2\sigma$) so makes about 95, 55% (Wood, 1981). Since L. J. Cohen does not advocate the preconceived norm he will, a fortiori, reject its rigorous version. However, by referring to this extreme of the preconceived norm I wanted to illustrate clearly the problem whose solution is required by each statement of the norm extraction method. This method should assume that the studied population may include people whose logic may be totally understandable for the researcher and who could hardly be said to be using any of the conceptions of probability. I am not concerned at this point with people falling in the range of extreme pathology, as probably they were of interest to L. J. Cohen when he proposed his method.

I would like to consider a large number of reports on the so called healthy or normal majority, which have recently proliferated in the field of social psychology. In particular, I am concerned with the results of the studies which confirmed the earlier reports about permanent tendencies to unintentionally and most often unconsciously distortion of the image of the world (including social world) and the self-image. Taking this into account analyzed people might quite unintentionally express their tendencies to self-deception introducing the researchers into their delusive world. In these circumstances it is even hard to imagine the kind of internal logic of this private world and the conceptions based on it. We could even risk a statement that a deformed vision, conception or logic of perception of the world and the self is met so frequently that one cannot differentiate in a reliable way psychotics and neurotics from other people taking part in everyday life. Of course, if these deformations in social perception and auto-perception were regularly found, they might be treated as regularities and make grounds for different private conceptions which the intersted researcher could take into account while interpreting results produced by a particular person. Unfortunately, the results of contemporary research in the field of social psychology are often contradictory and, due to various theoretical approaches, considerably difficult to synthesize.

Nevertheless for the sake of this study I have undertaken such an attempt concentrating on systematic errors (Kelly, 1973), illusions (Chapman, Chapman, 1981) or recently biases, (Kruglanski, Ajzen, 1983, Markus, Zajonc, 1985). For obvious reasons I shall refer only to some kind of deformations in social perception and auto-perception which are directly related to the basic tendencies of self-deception.

4. Basic tendencies to self-deception

The first attempts to give a more complete characteristic of the problem of self-deception were undertaken by H. Fingarette (1969) and reported in his monograph. The author treats the tendency to self-deception as something quite common and natural as related to the strenuous effort of the individual to cope with one's social and cultural environment which may be highly complex and full of contradictions. He referred to the philosophical ideas concerning this problem, however, he treated them only as a source of inspiration because of his serious objections to the philosophical approaches which were purely descriptive in their character and the lack of precision. Instead, he proposed a more adequate explanation, in his opinion the tendency to self-deception is based on the mechanism of the analysis of something the individual considers as directly threatening.

H. Fingarette assumes that the preliminary analysis of threatening information implies the appearance of a certain deviation in self-conscious (consciousness) processes. It results in producing a fixed way of perception (social perception and auto-perception) which leads in turn to fundamental distortion in the system of knowledge or self-knowledge. According to H. Fingarette the mechanism of analysis of threatening information involves, to a high degree, unconscious processes. However, this term seems to be only conventional since unconsciousness in his approach is understood in a more limited sense than in psychoanalysis and probably refers to preattentive or automatic processes (Greenwald, 1987). Thus a tendency to self-deception depends on the unconscious processes understood in the above specified way which, as H. Fingarette suggested, recognize the information that may threaten the individual and his/her integration and prevent them from reaching the conscious level. It is puzzling why the mechanism described in this way, based on the processes of drawing conclusions and on the representation of at least symbolic and high degree of differentiation which are so characteristic of conscious processes, has been still considered in terms of unconscious processes, has been still considered in terms of unconscious activity. According to A. G. Greenwald (1987) this is the same mistake which has been pointed to in the Freudian approach where unconscious processes are characterized by the attributes of the processes occurring above the threshold of consciousness within the ego.

This has inevitably brought H. Fingarette to recognize the self-deception phenomenon as a paradox as far as intentionally is concerned. Before I tackle this complex problem I am going to present a clear or even simplified solution to the problem of self-deception. It was proposed by C. R. Gur and H. A. Sackeim (1979) on the grounds of the results of a relatively simple experiment. They assumed that the tendency of self-deception is a combined result of conscious and unconscious processes in the meaning pointed out. They claim that the act of conscious recognition of information may be just a verbal declaration, whereas the process of unconscious decoding is revealed on the level of physiological reactions of the organism. An example of such a reaction may be the skin-galvanic effect which they chose to measure.

In the course of this experiment subjects had to recognize among previously recorded voices samples of their own voice. The results proved that the subjects tended to commit rather systematically two kinds of errors. The errors were found from checking the difference between the unconscious reaction (the skin-galvanic effect) and the conscious one (verbal declaration). The situation when a subject did not recognize his/her voice on the conscious level and did not declare it verbally, although he or she recognized it on the unconscious level (galvanometer indication), was treated as a negative error and a manifestation of the self-deception tendency. The same tendency was assumed to be revealed in the case of so called positive error when the subject recognized someone else's voice as his or her own and no skin-galvanic effect was observed. This kind of error is also called narcissistic projection and is related with a tendency to interpret all kinds of stimuli with reference to the characteristics of oneself.

In the terms accepted throughout this paper this reveals also difficulties in decentration and basically proves the so called egocentric orientation in social environment. It is hard to explain negative errors when there is a motivation to block the elements of the structure of the self. Of course, on the grounds of other theories e.g. the theory of psychoanalysis, the origins of the blockings might be found in traumatic events, complexes and so on. However, when we want to remain on the grounds of more up to date theoretical findings related to the psychology of self (Epstein, 1973, Burns, 1979) we should in consistency with Greenwald's (1987) suggestion refrain from treating the system of knowledge as homogeneous. Under such an assumption we find that the structure of the self is particularly pronounced. Although this structure is subject to the same structural restrictions as the whole cognitive system, in the case of situations when the self is directly engaged there is still the possibility of intentional consciousness to intervene. Thus according to A. G. Greenwald (1987) the situation when one does not recognize one's own voice on the conscious level is a manifestation of an intentional defense against threatening information. At this point the question should be asked: which psychological characteristics may be threatening to the individual?

Among the different answers, the most frequent are those referring to the lack of acceptance of oneself, low self-esteem (self-evaluation) occurring simultaneously with a strong need of social acceptance (social approval), increased social fear and so on (Baron, 1974, Jones, 1973). On the basis of this answer we can risk a statement that affective phenomena of that kind may change the already developed congnitive structures that cannot ensure the adequate adaptation of the individual.

In order to compensate for the difficulties in adaptation the interventions of one's consciousness are used. As I have already mentioned this brings about an increased risk concerning the effectiveness of adaptation as the consciousness may increase or reduce the adaptational potential which may result in only apparently adequate action. This kind of activity has much in common with the tendency to self-deception, as the priority of self-esteem and self-acceptation to cognitive functions is at present commonly assumed (Wojciszke, 1989).

Moreover, now we do not refer simply to the levels of self-esteem but to its different forms as well. Of particular interest are the forms of self-esteem distinguished with respect to the need of social approval; they are the genuine self esteem and defensive self-esteem. The criterion of genuine self-esteem is its relation to a small need of social approval while the defensive self-esteem is related to a strong need of social approval. According to the author of the regulative theory of self-esteem C. S. Jones (1973) a low self-esteem immediately produces a considerable increase in the need of social acceptance, this may be compared to the behaviour of water in a closed hydraulic system.

Summing up this consideration we may conclude that both low and great defensive self-esteem can be treated as factors that block effectively the recognition of all information threatening the level of self-esteem and the related social position. In other words, when one wants to maintain one's image approved by others or by oneself, the self-deception tendencies start working. Referring to the kinds of errors related to self-deception from the research of C. R. Gur (1975) and R. A. Wicklund (1979) we might suppose that high self-esteem should facilitate positive errors, whereas low self-esteem facilitate the negative error. This indicates two different sources of self-deception tendency: one related to the need of autoconfrontation, in the case of positive error, and the other related to the need of avoiding the autoconfrontation, in the case of negative error.

In the terms of the objective self-awareness theory proposed by S. Duval and R. A. Wicklund (1975) we can say that self-deception is related either to the avoidance of self-awareness or to a tendency to search for it. In the terms of this theory objective self-awareness is a result of particular focusing of one's attention on oneself but from an external

314

point of view, so from the point of view of an observer. In general it implies that one takes an unpleasant attitude towards looking at oneself from a distance. Therefore, according to the authors of this theory, people in general avoid autoconfrontation although they are apt to talk a lot about themselves. The greater tendency to confront oneself is the wider the divergence between possibilities and ambitions, expectations and aspirations is or in general, between the ideal and the real self-concept. An exception may be the situation directly after a given person has achieved a success.

In general, the motivation to concentrate on oneself depends on affective reaction to the degree one considers oneself as a person of worth or a person showing genuine high self-esteem. If is of importance as far as the estimation of the level of consciousness is concerned. People showing genuine self-esteem should at least potentially, according to S. Duval and R. A. Wicklund (1975), have the strongest motivation to confront themselves and thus should have the highest level of objective self-awareness.

5. Conclusion

To conclude the theoretical discussion it should be emphasized that of all the determinants predisposing to be normal, the category of objective self-consciousness acquires a great importance. As has been shown, people endowed with a profound level of self-consciousness should, in terms of computer language, reveal no serious faults in programming. At the same time we should notice that Cohen's claim that people are not programmed to commit fallacies refers specifically to those people. This claim does not apply to those people who are inclined to commit systematic fallacies such as defensive forms of self-evaluation following from little motivation to confront oneself, as well as excessive tendency to autoconfrontation or permanent self-deception tendencies. Thus in view of the above considerations it is not easy to share Cohen's optimism as to whether the people are not prone to commit fallacies. The same doubt appears when we answer the question whether people are programmed to be normal. The role of the category of objective self-consciousness is very important because it helps to compensate serious errors in the human programming and in consequence helps to compensate difficulties in striving for normality.

Finally, we should also note that the tendency to identify normality with the lack of committing fallacies revealed in the above considerations is equivalent to giving consent to the rationalistic concept of normality. Thus, the problem of human normality becomes simultaneously a problem of human rationality. After all, it is difficult to think of another approach to normality if it is to be considered by any sicience. The approach to the problem of normality presented in this paper, slightly different from to commonly accepted one, would require the integration of knowledge from different fields of science. Moreover this would imply the necessity of changing the opinion that the question of normality belongs to clinical psychology or psychiatry. On the contrary, I would like to point to the fact that this problem becomes more and more important in contemporary social life subject to intensive cultural transformation and should be considered by psychology in a far more general aspect.

In the end, let me make one more final remark. It should be recognized that the way particular concepts or theories treat or understand the problem of human normality may significantly effect the conclusions they bring.

REFERENCES

Baron, P. (1974). Self-esteem integration and evaluation of unknown others. *Journal of Personality and Social Psychology, 30,* 177-197.

Burns, R. B. (1979). *The self-concept.* London: Longman Inc.

Chapman, L. J., Chapman, I. P. (1967). Genesis of popular but erroneous psychodiagnostic observations. *Journal of Abnormal Psychology, 72,* 193-204.

Cohen, L. J. (1981). Can human rationality be experimentally demonstrated? *The Behavioral and Brain Sciences, 4*, 317-331.

Cohen, L. J. (1982). Are people programmed to commit fallacies? Further thoughts about the interpretation of experimental data on probability judgment. *Journal for the Theory of Social Behaviour, 12,* 251-274.

Duval, S., Wicklund, R. A. (1975). *A theory of objective self awareness.* New York: Academic Press.

Eliasz, A. (1986). *Temperament a system regulacji stymulacji* (Temperament and regulatory system of stimulation). Warszawa: Państwowe Wydawnictwo Naukowe.

Eysenck, H, J. (1971). *The structure of human personality.* New York-London: Wiley and Sons.

Epstein, S. (1973). The self-concept revisited, or theory of a theory. *American Psychologist, 27*, 404-416.

Fann, W. E., Karacan, I., Pokorny, A. D., Williams, R. L. (1987). *Phenomenology and treatment of schizophrenia.* New York: Spectrum Publications Inc.

Fingarette, H. (1969). *Self-deception.* London: Routledge and Kegan Paul.

Fonberg, E. (1974). *Nerwice, przesądy a nauka* (Neuroses, prejudices and science). Warszawa: Wiedza Powszechna.

Greenwald, A. (1986). Samowiedza i samooszukiwanie (Self-knowledge and self-deception). *Przegląd Psychologiczny*, **29**, 291-303.

Gur, C. R., Sackeim, R. (1979). Self-deception: A concept in search of a phenomenon. *Journal of Personality and Social Psychology*, **37**, 147-169.

Jones, S. C. (1973). Self and interpersonal evaluations. Esteem theories versus consistency theories. *Psychological Bulletin*, **79**, 185-199.

Kelley, H. H. (1973). The processes of casual attribution. *American Psychologist*, **28**, 107-128.

Kruglanski, A. W., Ajzen, I. (1983). Bias and error in human judgment. *European Journal of Social Psychology*, **13**, 1-44.

Markus, H., Zajonc, R. B. (1985). The cognitive perspective in social psychology. In G. Lindzey, E. Aronson (Eds.), *Handbook of social psychology*. Vol. I. New York: Randon House.

Sapir, E. (1964). Conceptual categories in primitive languages. In D. Hymes (Ed.), *Language in culture and society*. New York: Harper and Row.

Sowa, J. (1984). *Kulturowe założenia pojęcia normalności w psychiatrii* (Culture approach of normality in psychiatry). Warszawa: Państwowe Wydawnictwo Naukowe.

Ścigała, I. T. (1988). Teoretyczne podstawy konceptualizacji kategorii zdrowie psychiczne (Theoretical foundations of conceptualization of category mental health). Unpublished doctoral dissertation.

Ścigała, I. T. (1989). Kreatywistyczny model zdrowia psychicznego (Creativistic model of mental health). (In press).

Whorf, B. L. (1956). Science and linguistics. In J. B. Carroll (Ed.) *Language, thought, and reality: Selected writtings of B. L. Whorf*. Cambridge, Mass.: MIT Press.

Wojciszke, B. (1988). Polimorfizm reprezentacji i wartości obiektów społecznych (Polymorphism of representation and value of social objects). (In press).

Wood, G. (1981). *Fundamentals of psychological research*. Boston – Toronto: Little Brown and Company.

PART V

CONCLUSION

Poznań Studies in the Philosophy
of the Sciences and the Humanities
1991, Vol. 21, pp. 319-342

L. Jonathan Cohen

SOME COMMENTS BY L. J. C.[*]

Contributors to this volume have generated an embarrassingly rich set of ideas which I have much enjoyed studying, and I cannot hope to comment on more than a fraction of these. I shall arrange my comments under three main headings: issues about philosophical method, philosophical issues about inductive or probabilistic reasoning, and issues in psychological theory about human reasoning.

1. Issues about philosophical method

1.1. Maurice Finocchiaro has posed some searching questions, and made a number of important points, about my analysis of philosophical reasoning. I shall take them up in turn.

Finocchiaro is right to distinguish between what he calls "interpretation" and what he calls "evaluation", as two different modes of philosophical analysis. Alongside his own example from the analysis of scientific explanation one could cite for instance, the difference between the interpretative analysis of moral language as in R. M. Hare's *The Language of Morals* (1952) and the evaluative analysis of justice into its component principles as in J. Rawls's *A Theory of Justice* (1972). Of course, many connections and inter-dependencies arise between these two modes of analysis, when one gets down to the details in any area of analytical enquiry. For example, even interpretative analyses often involve normative judgements about the validity or invalidity of relevant inferences. But the distinction between interpretation and evaluation is particularly important from a historical point of view because the post-1945 period of analytical philosophy has been characterised by a growing tendency to practise evaluation alongside interpretation. This development, as I see it, has been one aspect of the more general movement towards including semantic descent as well as semantic ascent within the repertoire of analytical activity — i.e. the movement towards evaluating or describing the various structures that words like "knowledge", "justice", etc. from time to time denote, in addition to analysing the meaning that these terms standardly possess (1986h, pp. 12-34).*

But, while recognising the existence and importance of this movement, I see it as an extension of the recognised problem-area, not as a replacement of that problem-area by a different one. Linguistic analysis, or interpretation as Finocchiaro calls it, remains a vital task (1986h, pp. 51-52). Indeed I take my own account of the metaphilosophical term "intuition" (1986h, pp. 73-82 and 91-97), for example, to be a contribution of that kind (i.e. semantic ascent) even though the way in which philosophical intuitions should be evaluated (1986h, pp. 82-91 and 97-107) is an important issue of a different kind (semantic descent). To say that intuition is a form of belief, not of acceptance, is an interpretative thesis, whereas the claim that singular intuitions are normally more authoritative than general ones is an evaluative thesis.

In this connection Finocchiaro appositely asks how my remarks about inductive and deductive patterns of philosophical reasoning are to be understood. Presumably when I cite Descartes' proof of the existence of an external world as an example of deductive reasoning I am not intending to evaluate it as being actually valid. But then, Finocchiaro asks, what is it that does make an argument deductive rather than inductive or *vice versa*? What are the grounds for preferring the one interpretation to the other? My answer has to be that a piece of philosophical reasoning is to be called "deductive" (or "inductive") in the relevant sense if the author himself — implicitly or explicitly — claims deductive (or inductive) validity for it. That is, the metaphilosopher may *interpret* a philosopher's text in the light of the philosopher's own *evaluations*. Thus at the beginning of part V of his *Discourse on Method* Descartes himself speaks of what he has "deduced" and claims that his deductions are "more clear and more certain than the demonstrations of the geometricians had formerly seemed" (Descartes, 1931, p. 106). Similarly Kant (1907, p. 30), rejected any attempt at an inductive foundation for his moral philosophy when he asserted "Nor could anything be more fatal to morality than that we should wish to derive it from examples". On the other hand, when a philosopher like Williams (1976, p. 117) says that his "procedure in general will be to invite reflection about how to think and feel about some rather less usual situations, in the light of an appeal to how we — many people — tend to think about other more usual situations" he is obviously intending to offer his readers an opportunity for inductive extrapolation. An implicit generalisation that is validated by reference to cases of a familiar type is then applied to cases of a more unusual kind.

It is thus possible to use the terms "inductivism" and "deductivism" in metaphilosophy to connote a philosopher's tendency to prefer making inductive or deductive claims, respectively. And one may then come to notice that, as a matter of fact, inductivist philosophising goes naturally with a localist rather than a globalist orientation, since at its best it avoids jumping to unjustifiably general conclusions. Nor does inductivist philosophising sit easily with sceptical epistemology since any kind of scepticism normally involves the rejection of some commonly respected intuitions about reasons for belief.

Finally, Finocchiaro objects to my characterisation of intuitions as being *a priori* beliefs. But our disagreement on this issue seems to be only a terminological one. As he himself recognizes I do not use the word "intuition" to denote an infallible source of knowledge. Instead it refers to an immediate, unreflective, and untutored inclination to judge that p, where the judgement that p is of a kind that is in principle not checkable by sensory perception (1986h, p. 75). And by characterising intuitions as being *a priori* beliefs I mean no more than this. Philosophical theories may thus be said to be erected on *a priori* foundations. But this does not imply any concession to a rationalist epistemology of the kind that was advocated — in their various ways — by philosophers like Descartes, Spinoza, etc. It implies that philosophical analysis is not an empirical enquiry. But it is not inconsistent with the thesis that natural science **is** such an enquiry. Nor should philosophical intuitions always be confined to the consideration of actual or historical cases rather than imagined or contrived ones. Just as the intuitions of a native speaker can sometimes be usefully exercised on testing the grammaticalness of a string of words that a grammarian has artfully put together, so too our philosophical intuitions can sometimes be usefully exercised on checking the appropriateness of what might be said about imaginary situations like the survival of a brain in a vat.

1.2. Menachem Fisch's stimulating paper provides a useful opportunity for me to emphasise that I have always seen my neo-Baconian analysis of inductive reasoning as a contribution to logic rather than to methodology. The analysis is concerned with a certain kind of timeless relation between propositions, not with a certain temporally ordered set of heuristic procedures. Of course, it is sometimes convenient for stylistic reasons to use the term "hypothesis" to denote the type of proposition that may or may not possess inductive support. But if we do so use this term in the analysis of inductive support we need to make clear that it there lacks the implication — which is appropriate in a methodological context — that it is a person's working assumption (1970e, p. 6). Similarly, if we use the term "confirmation" to denote inductive support, we must make clear that if a is said to confirm b this does not imply, as it would in a methodological

322

context, that *a* is later in time than *b* (1970e, p. 7). Even Mill (1851, bk. III, ch. IX, sec. 6) came to think of his five canons of inductive reasoning as contributions to the logic of inductive proof rather than to the methodology of inductive discovery. And it is in this sense only that what I have called "the method of relevant variables" may be regarded as a unified and graduated systematisation of whatever is sound in Mill's canons (1977b, p. 146).

There seems no *a priori* reason to suppose that a good methodology of scientific discovery can be derived by some simple transformation of the logic of inductive support. Clearly the methodology would *presuppose* some such logic, in that its overriding goal would be the discovery of propositions that were inductively well-supported by accepted truths. But this characterisation of its goal tells us nothing, for instance, about the question whether the supporting propositions should be accepted before or after the supported ones are entertained, about the question whether bold conjectures are preferable to cautious ones, or about any other question of laboratory procedure. Nor is there any *a priori* reason to suppose that the same methodology is appropriate in all branches of scientific enquiry, let alone in normative reasoning, even if inductive support has the same logical structure throughout science, jurisprudence, ethics etc. Indeed the arguments in favour of a particular scientific methodology would presumably need to be empirical, rather than philosophical, ones. Past successes with a particular method would be said to justify future uses of it, while failures with other methods would justify their disuse. So one would like to see how far Fisch's interesting methodology is supported by evidence from the history of science, from experimental tests of its applicability to currently unsolved scientific problems, or from the results of attempts at knowledge engineering in artificial intelligence.

But even without having any details of that evidence one can see at least one major limitation in Fisch's account of induction. He claims that both for logical and for methodological purposes materially similar generalisations (determining an area of inductive comparability) should be identified by the fact that they normally hold true as a result of the operation of the same kind of causal mechanism, rather than — as I have suggested (1977b, pp. 136-137) — by the fact that they exploit a particular vocabulary and share a common terminological structure. Yet this assumption about a common cause would in practice preclude our drawing any kind of inductive comparisons between rival hypotheses about the causation of a particular phenomenon. We could not compare rival theories about what causes a particular kind of cancer, or rival theories about what cures it. Instead of being able to reason inductively about causal issues we should be confined instead to the consideration of purely descriptive generalisations like "All ravens are black" and "All swans are white". And that seems too crippling a limitation to be acceptable.

Fisch also claims that I am wrong to argue that inductive reasoning can not merely approximate, but actually prove, the truth of a universal generalisation correlating two natural kinds or properties with one another. And his reason for saying this is that the method of assessing inductive support for the generalisation is at any given time grounded upon a revisable empirical conjecture concerning material similarity. Now I accept (1977b, pp. 142-143) that even my own criterion for material similarity is associated with a pragmatic approach that acknowledges how the delimitation of particular areas of enquiry requires adjustment in the light of experience. A vocabulary for hypothesis-construction, I say (1977b, p. 143), should pay its way by being fruitful of hypotheses that turn out to be well supported when they are tested against the appropriate series of relevant variables. But in order to know that p a person does not have to know that he knows, and in order to have proved that p he does not have to prove that he has proved it. So a scientist may correctly and for the right reasons hold that p, and believe that he is doing this, even though he may not have conclusive reasons for that belief. Philosophers who argue otherwise are assuming the validity of the so-called "KK principle" — that if a person knows he also knows that he knows. But there are good reasons for not accepting this principle in the context of inductive reasoning. And not the least of such reasons is the frequency with which distinguished modern scientists like Albert Einstein have claimed to "know" this or to have "proved" that.[1] If our analysis of inductive reasoning does not respect the considered judgements of such scientists, it will be very short of premisses on which to rest its conclusions.

2. Philosophical issues about inductive or probabilistic reasoning

2.1. On the subject of idealisation in scientific reasoning, Leszek Nowak's work is outstanding both for the quality of its analyses and for the wide range of applications that these analyses are shown to have. Nevertheless I do not think that he is quite right in what he says about how the method of relevant variables bears on the problem. The fault for that is probably mine, because I see now that I have not always been altogether consistent in what I have said about the matter.[2] So I shall take the opportunity here to restate the central thesis that I take to be correct.

Generalisations in a particular field of research are to be credited with appropriate grades of Baconian legisimilitude — i.e. inductive support —

in accordance with their capacity to resist falsification in, and by, cumulatively richer and richer combinations of the circumstances that constitute relevant variables for that field of research. A relevant variable is a non-exhaustive set of mutually incompatible circumstances, or variants, each of which suffices to falsify, through its causal impact, at least one possible generalisation in the field of research in question. And these special, potentially falsificatory circumstances are to be contrasted with the normal circumstances under which such processes operate unhindered. Thus for a generalisation about the velocity of falling stones one relevant variable might be the various kinds of medium that interrupt the fall of certain objects, as water interrupts falling paper or treacle interrupts falling leaves, while the earth's atmosphere might be taken to be the normal medium through which such objects fall.

Clearly one way to protect a generalisation against being falsified by the presence of a certain relevant circumstance, and thus to raise its legisimilitude, is to introduce an incompatible, non-falsifying circumstance into the antecedent of the generalisation, and this may be either a different variant of the same relevant variable or the normal circumstance that excludes every variant of that relevant variable. Schematically, if V_1 and V_2 constitute a relevant variable, for which N is the normal circumstance, and if

(1) Any R is S

is falsified by V_1, then something of (1) may be saved by hypothesising instead

(2) Any R that is V_2 is S,

Or, where V_2 is also falsificatory, we might hypothesise instead either

(3) Any R that is N is S

or even

(4) In normal circumstances any R is S.

But equally clearly neither (2) nor (3) nor (4) are models or patterns of idealisation. To be termed an "idealisation" a generalisation must be intended to describe a simpler world than the actual one — a world in which some natural process under investigation operates without any interference at all from certain kinds of factors that complicate it in the actual world. Taken instead as a description of the actual world such a generalisation is falsified not only when any of the variants of a relevant variable (like V_1 or V_2) is present but also when circumstances are quite normal in this respect (like N). So the generalisation has to be protected against falsification in a more radical way than (2), (3) or (4) represent. It has to be intended as a description of what the world would be like if it contained none at all of the falsificatory circumstances. The domain of the generalisation must exclude not only any variant of the relevant variable at work but also any other kind of circumstance co-ordinate with those variants — an exclusion which takes the domain right out of the real world. For example, in the case of generalisation about the velocity of falling bodies, it turns out that even the earth's atmosphere complicates the matter — by falsifying a generalisation that has the desired level of simplicity. Hence the domain of the generalisation has to be restricted to objects in a vacuum.

Perhaps someone will object that if an idealised generalisation does not have to be true of the actual world, but only of some much more jejunely furnished non-actual world, there can be no inductive grounds for choosing between one such idealisation and another. No observable fact, he may say, is predictable from or explainable by, any idealisation, and therefore no inductive confirmation or disconfirmation of it is possible. So the method of relevant variables tolerates any idealisation whatever, because no experience in the actual world could ever falsify a generalisation that purports to apply only under conditions that do not occur in the actual world.

The first point that needs to be made in reply to this objection is that generalisations achieve their proper grade of Baconian legisimilitude not by their instantiation or verification in certain circumstances but by their failure to be falsified in them. If an appropriate restriction on a generalisation's domain of application were to succeed in protecting it against falsification in *all* circumstances that are found within the actual world, then the generalisation over that domain would rightly be termed a "law of nature". And the second point is that such a restriction is to be deemed appropriate only if it is required by the evidential facts. The circumstances excluded have to be those that would *otherwise* falsify the generalisation. Thus a desirably neat and simple generalisation about the velocity of falling objects may have to be restricted to objects falling in a vacuum because the presence of some medium through which they fall would create friction that falsifies the unqualified generalisation. But, though the restriction of such a generalisation to objects made of colourless gold would equally protect it from falsification (since colourless gold is as non-existent as a total vacuum), such a restriction would not serve to raise legisimilitude because it is not one that is required by the evidential facts: it is not the composition of the falling objects that falsifies the unqualified generalisation but the presence of a medium through which they fall.

In short, if one thinks in terms of the method of relevant variables — that is, in terms of grading inductive support for a generalisation by its ability to resist falsification under the permutation of relevant variables — there is no difficulty in understanding how appropriate evidence in the actual world may come to support generalisation about an ideal, or correspondingly simplified, world.

2.2. Ryszard Stachowski argues persuasively that some psychological variables exhibit features which exclude conformity to the standard conditions for measurement. Such a variable may conflict, in particular, with the so-called Archimedean postulate. It has features that seem to entail the existence of an actual infinite and to disallow the in principle infinite extendability of the magnitude under measurement. Thus in someone's scale of utilities, for example, life itself might be assigned an infinite value, so that there could be no greater level of value than that assigned to it.

Stachowski makes the further claim that this fact is an anomaly in relation to scientific theory. Just as, say, the perihelion of Mercury conflicts with Newton's theory of motion, so too the assignment of an infinite value to life conflicts with the standard theory of measurement. And, just as the anomaly about Mercury was not taken as a ground for rejecting Newton's otherwise well-supported theory, so too the anomaly about certain psychological and other magnitudes is not to be taken as a reason for rejecting the standard theory of measurement.

But is Stachowski right to assume that the standard theory of measurement is an empirical theory — a theory that is in principle open to falsification by empirical facts, such as those about psychological magnitudes? Obviously it is an empirical matter whether a particular natural magnitude is such that descriptions of it conform to appropriate conditions — specifiable *a priori* — for the interrelationships between these descriptions to obey the laws of arithmetic. Thus in fundamental measurement there must be an operation for combining two systems possessing amounts of the magnitude in question, so that the combined system also possesses some such amount, and the numbers assigned to these amounts must conform to certain principles such as the Archimedean principle (Ellis, 1968, pp. 74-75). It might then be regarded as an empirical matter whether there is an operation of this kind for length, say, or time interval. I.e., it might be regarded as an empirical hypothesis that such-or-such a magnitude is measurable, in the standard sense of "measurable". But it would only be in relation to some abnormally sweeping generalisation, like the hypothesis "Everything is measurable", that the existence of some non-measurable magnitude, like subjective utility, could count as an anomaly. A spelling-out of the conditions that have to be satisfied for a magnitude to be measurable needs to be distinguished from the bold, Pythagorean, unscientific claim that everything is measurable. I am inclined to think, therefore, that our reluctance to reject the standard theory of measurement in the face of psychological facts about contravention of the Archimedean postulate is due rather to the relatively *a priori* state of that theory than to the strength of its empirical standing.

Of course, some people may object here that when we deliberately overlook anomalies in the case of otherwise well-supported scientific theories, like Newtonian mechanics, we in effect treat those theories as *a priori* truths, since we apparently turn a blind eye to the possibility of falsification. Consequently, the objection will run, there is no real difference between the *a priori* status that we assign to a statement of the conditions for measurability and the anomaly-resistant status that some people may wish to attribute to the empirical theory that everything is measurable.

But the objection does not work out. Even if we try to treat a statement of the conditions for measurability as a theory, it is still only the conditional theory that, *if* a magnitude is measurable, *then* the magnitude satisfies the Archimedean postulate and other associated conditions. It is in no sense identifiable with the unconditional theory that everything is measurable.

2.3. The topic of cascaded inference — i.e. transitive inference under uncertainty — is one to which, in a series of papers, David Schum has made a unique contribution. No-one else has investigated its intricate twists and turns so carefully, or has analysed its overall structure so productively. My own comments here are aimed at viewing this topic within the perspective of a reasoned pluralism that I hope Schum will find congenial.

Pascalian conditional probabilities are not transitive. That is to say, we cannot calculate the value for $p(C/A)$ when given values for $p(B/A)$ and $p(C/B)$. So one question is to ask what related values we can calculate from $p(B/A)$ and $p(C/B)$, and another to ask from what related premisses we can calculate $p(C/A)$. Those are just two very simple problems about the mathematics of probability, which become a little bit more complicated when Bayesian methods of calculation from priors and likelihoods are employed.

There is also a more difficult group of problems — philosophical rather than mathematical — about how to assess the strength of an inference from A to C in terms of the inferability of C from B and of B from A. This issue has generally been discussed for the special case in which A, B and C are all singular propositions concerning the same individual person, object or event, as often in the law-courts. But obviously the same kind of problem can arise whatever the logical categories to which A, B and C severally belong.

Now what complicates the latter group of problems is that they essentially involve detachment. Our concern is not with the credibility of "If A then C" as a function of the credibilities of "If A then B" and "If B then C", but with the credibility assignable to C if we assume the truth of A and bear in mind what this implies for the credibility of B. It follows that whatever considerations bear in general on the detachment of monadic statements of credibility will certainly apply in the special case of transitive inference. More specifically, the strength of our title to detach the monadic statement about the credibility of C will depend not only on what A actually says but also on the extent to which A provides all the relevant evidence. So, even if you do treat the credibilities as Pascalian probabilities and are thus able to exploit Bayesian algorithms to the full, you still have to find a measure for the extent to which A provides all the relevant evidence. And there are good reasons (1986a, pp. 263-278, and 1989c, pp. 99-109) why such a measure — called by Keynes a measure of weight — cannot itself be regarded as having the structure of a Pascalian probability. It follows that you will need to answer the question (about the strength of the inference from A to C) in terms of at least two separate and independent evaluations: an assignment of a value, n, to the Pascalian probability $p(C)$ and an assessment of the weight of the argument from A to C or of the weights of the arguments from A to B and from B to C.

Thus in relation to the treatment of inference upon inference, in the analysis of Anglo-American standards of forensic proof, I am inclined to put forward the same line of argument as I have urged about the standard of criminal proof (1986e, pp. 644-645). The Pascalian mode of assessment is inadequate without the Baconian one, but, if we need in any case to bring in a Baconian mode of assessment in order to determine weight, we do not still need to have any requirement spelt out in Pascalian terms. Certainly in criminal cases it will suffice to require at each stage of a cascaded inference the level of Baconian probability that amounts to proof beyond reasonable doubt. And in civil cases the plaintiff will normally require to establish at each stage a superior Baconian probability for his own point over its denial. If, on the facts before the court, B has a superior Baconian probability over not-B, then B is proved — according to the normal standard of proof in civil cases — on those facts (1977b, pp. 270-271). So B may now be detached and used as a sufficiently reliable premiss for the proof of a further conclusion: if, schematically, on the basis of B, C has a superior Baconian probability over not-C, then C too is proved according to the same standard of proof. Of course, *if* the court or the legislature requires a higher standard of proof up to the penultimate stage of cascaded inference, this requirement can be stated in Baconian terms — just as can the requirement for a higher level of proof in criminal cases. But such a requirement would be a special rule of law: it is not imposed by the structure of Baconian reasoning — contrary to what may have been suggested by (1977b, pp. 268-269). Moreover within the overall rules of procedure the defendant is free to bring into court whatever facts he can use to oppose the plaintiff's line of argument, so that the plaintiff's task is not necessarily made easier by his being allowed to win on the balance of Baconian probability.

2.4. Jim Logue has made a refreshingly sophisticated attempt to show how a Bayesian theory of probability might come to terms with a Keynesian theory of weight. Roughly, the weight of the evidence on which a first-order monadic probability judgement has been conditionalised is to be deemed a function of the second-order probability of this judgement, and that second-order probability is itself a function of (i) the resiliency of the first-order judgement within its present conditionalisation and (ii) a measure of how little this resiliency might alter if further conditionalisation took place. With scrupulous intellectual modesty Logue does not claim to be putting forward conclusive arguments in favour of this analysis. But it may nevertheless be useful to have in mind the main difficulty that it does encounter.

Resiliency may vary independently of the amount of evidence on which conditionalisation has taken place. For example, when only a little evidence has been taken into account factors of considerable relevance may not have been considered yet, so that resiliency remains high within the present conditionalisation. On the other hand, if a great deal of evidence has been considered, extensive "wiggles" or fluctuations have had perhaps a better chance of emerging, so that resiliency is low. Accordingly, whether we confine our attention to resiliency within an existing domain of conditionalisation, or extend it to consider future alterations in resiliency, the concept of resiliency does not capture the intuitions that underlie a Keynesian theory of weight. Rather, criteria of weight grade how much of the evidence for calculating resiliency has already been taken into account — though in ranking weight it is reasonable to order the items of evidence in a way that may be expected to create lesser and lesser "wiggles" as more and more evidence is taken into account (1986a, p. 275).

Logue also argues that his type of Bayesianism can meet the objections that I raised against any Pascalian analysis of the standards of proof in Anglo-American law. But, though his arguments are refreshingly novel, I do not think that they succeed.

The first point at issue concerns the suitability of a subjectivist interpretation for the purpose. What is going on, according to a subjectivist analysis, when a probability judgement is uttered? One view is that the speaker is describing his own state of mind — his willingness to accept bets at certain odds, and so on. But this interpretation is hardly appropriate, as Logue recognizes, when two advocates are disputing about a probability in a court of law. So Logue proposes the other interpretation: the speaker is not describing his partial belief but expressing it. And there is a familiar analogy here with a historical shift in the analysis of moral utterances. Instead of treating them as assertions about the speaker's feelings, subjectivists came to think of them as a form of exclamation, commendation or command. But the trouble with all such non-statement-making analyses, as Searle (1969, pp. 136-141) pointed out, is that they run into severe paradoxes when the analysanda occur in the antecedents of conditional sentences, in indirect discourse, etc., and a subjectivist analysis of probability-judgements cannot be rescued from such paradoxes (1977b, p. 29, f. n. 19). No clear meaning can apparently be given to such conditionals, for example, if their antecedents are not truth-value-bearing propositions but mere expressions.

A further difficulty for a subjectivist interpretation arises about exchangeability. Exchangeability is needed if a subjectivist analysis is not going to be open to the objection that it gives its readers inadequate guidance about how probabilities are to be estimated, and Logue's analysis exploits exchangeability accordingly. But the outcomes with which forensic judgements of probability are concerned are rather unlikely to be exchangeable. Suppose, for example, the issue in a suit for breach of contract is whether or not a certain payment was made. It is rather unlikely that a sequence of payments is going to be exchangeable in the requisite sense since their order of occurrence may well affect their several probabilities. Similarly, if the contral issue at the end of a criminal case is whether a certain witness lied, the order in which witnesses testified may be highly relevant to the probability in question.

The second point at issue concerns prior probabilities. I argued that standards of proof should not be expressed in terms of a concept of probability for the evaluation of which — in cases of testimonial corroboration, for example — the trier of fact must know or assume a value for related prior probabilities. Logue's suggestion is that jurors should and do start with any (set of) coherent priors (so long as none is zero) but with a uniform distribution of second-order probabilities — that is, with no commitment to any of these odds. But, if you don't know or assume a value for $p(H)$, how can the process of Bayesian conditionalisation get started?

The third point concerns the interpretation of "proof beyond reasonable doubt". Consider Logue's own example — genetic fingerprinting. However high the Pascalian probability that some particular specimen of blood came from John Smith's bloodstream it will still need to be proved that the victim of the crime was indeed John Smith, that the blood-specimen was found on a particular garment, that the garment was being worn by the accused at the time of the murder, that the accused was not an innocent bystander, and so on. Unless a set of several such facts is established there remains the possibility of reasonable doubt, and a high Pascalian probability about the blood's origin is only one such fact.

The fourth point concerns civil cases where the need to prove a conjunction of independent facts seems to generate a requirement, unknown to the law, that each conjunct be proved at a correspondingly higher level of Pascalian probability than whatever level is appropriate where only a single fact has to be proved. Logue first objects that where there is a plurality of facts to be proved their probabilities are rarely independent of one another. But there are certainly lots of cases in which the Pascalian probability H_1 & H_2 is substantially less than that of H_1 alone, even where $p(H_2) \neq (H_2|H_1)$. Even a partial independence will generate the paradox. Logue's response to this point is that the conjunction-effect, requiring higher than normal levels of probability for each of the conjuncts, is "a perfectly reasonable criterion of decision". Maybe so. But it is a criterion that seems unknown to courts within the Anglo-American system and contrary to the patterns of decision that they actually follow. And my point was not about whether Pascalian probability could in principle provide a foundation for standards of proof in some law courts, but about whether it could coherently be taken to do so in courts of the Common Law tradition (1977b, pp. 50-56 and 116-120).

The fifth point concerns the paradox of the gatecrasher where proof on the balance of probability would seem to allow a rodeo-organiser to recover money from everyone present if the only evidence was that more than half of them were gatecrashers. Logue suggests here that the standard of proof should be interpreted as requiring an appropriate level of second-order probability for the finding that the first-order probability is greater than .5. But in order to secure a reasonable level of resiliency for a first-order probability it would normally be necessary to take quite a lot of facts into account that litigants prefer to ignore. Litigation would become longer, more complex and more expensive. Of course, a higher standard of proof has sometimes been required for certain types of civil cases — a standard somewhere intermediate between the normal civil criterion and the normal criminal one. But this has been done where the effect of losing the case would be particularly serious for the defendant — for example, deportation, illegitimacy, exposure as a swindler, etc. In other cases a single uncontested piece of evidence may be sufficient to establish a plaintiff's case, whatever would be the outcome of a more extensive enquiry. In other words, the plaintiff needs to get the better of the argument against the defendant, in terms of whatever case the defendant chooses to put up. But he does not need to get the better of an imaginary stronger argument, put up by a possible but non-actual defendant. Logue's suggestion might be appropriate for an inquisitorial system, but it does not fit the adversarial patterns of English and U. S. Courts.

3. *Issues in psychological theory about human reasoning*

3.1. Ireneusz Scigała has made some well-justified criticism of what I once called the "Norm Extraction Method" and thought of as a scientific procedure. Czeslaw Nosal's interesting paper also refers to that idea. So I need to reformulate my thesis here too.

In (1982e) I introduced the term "Preconceived Norm Method" as denoting that method of psychological enquiry into probability-judgement whereby the investigator applies his own conception of probability to the subject and hypothesises either faulty programming or adventitious causes of malfunction in order to account for any of the subject's estimates of probability that are erroneous by the standards appropriate to that conception. By the term "Norm Extraction Method" I denoted instead the method of hypothesising about a subject's conception of probability and his mode of assessing it that assumes the subject's estimate to be correct unless affected by adventitious causes of error. And I suggested a number of reasons why the latter method was superior to the former. But in fact there is a good reason for rejecting both methods, so far as scientific psychology is concerned. For in both methods the investigator adopts his own normative views about probability as one of the premises for his theory about his subjects' judgements. In both cases therefore, because probability is — notoriously — a topic for normative controversy, the investigator's method lacks inter-personal objectivity. In practising the Preconceived Norm Method an investigator runs the risk of describing his subject's estimates as erroneous when they are correct by the standards in fact appropriate to the subject's own understanding of the investigator's questions. And in practising the

Norm Extraction Method an investigator runs the risk of describing his subject's estimates as correct when they are erroneous by the standards in fact appropriate to the subject's own understanding of the investigator's questions.

In order to maintain strict scientific objectivity in this area the psychological investigator needs to refrain from making any judgements at all about the correctness or incorrectness of this or that estimate of probability that is made by any of his subjects. So what is required is a Norm Extraction Method Mark II whereby a hypothesis is intended to state merely what rules, procedures or criteria are consciously or unconsciously guiding the answers that a particular type of subject gives to a particular type of question in a particular type of circumstance. Of course, this is quite compatible with the use of a competence-performance distinction, provided it is understood that calling a system of rules, procedures or criteria a "competence" is not a way of asserting that it is a normatively correct or valuable system. Even witchcraft is a competence and subject to the possibility of performance error, and the dialect in which you or I have a grammatical competence may be rather an ugly one. The investigator merely trims, tidies or simplifies his hypothesis about the subject's guiding competence by writing off awkward or anomalous phenomena as defects in performance, the occurrence of which is diagnosed as being due to inattention, laziness, too much or too little self-confidence etc. Further enquiry can then proceed in either of two quite different directions. On the one side there is a scientific, ultimately a neurological, problem about the nature of the mechanism that sustains the competence along with a further problem about how such a mechanism could have evolved. On the other side there is a normative, ulitmately a philosophical, problem about the appropriateness, validity or optimality of applying that particular competence to answering the type of question asked under the circumstances actually prevailing. Sometimes this is called a problem about the "rationality" of the subject. But there are probably several different dimensions of normative appraisal that are relevant here, and it may be that use of the term "rationality" tends to encourage oversimplification by confounding these different dimensions with one another. For example, a subject may be deemed to have judged some probabilities irrationally either because his judgements were inconsistent with one another or because he applied an inappropriate metric in evaluating a Pascalian probability-function or because he was mistakenly thinking in terms of Baconian probability. (See also Adler's remarks on this topic and my comments on them in 3.5. below.).

It follows that the Norm Extraction Method Mark II does not imply or assume that the way a subject normally judges probabilities in a certain type of context is the "rational" way to do this. What is taken to be normal here, and constitutes the competence described by the investigator, need not be identified with what is normatively correct. Instead it is at first just a theoretical construct. The investigator, guided by considerations of systematisation and simplicity, puts down some data as the product of competence-exercising performance and others as the product of competence-breaching performance. An account of the competence is then constructed on the basis of these assignments. Perhaps the existence of the competence can then be confirmed by reference to some idependently warranted fact, such as the existence of the mechanism that sustains it. Or perhaps the investigator's extraction of the subject's "normal" mode of probability-judgement remains more an act of invention than of discovery. Either way the psychologist, *qua* psychologist, confines his enquiries to the description, explanation and prediction of facts, and avoids normative issues (other than issues about how psychological research is to be carried on). On the subject of human "rationality" he is thus neither *a priori* optimistic, as is the practitioner of Norm Extraction Method Mark I, nor *a posteriori* pessimistic, as are most practitioners of the Preconceived Norm Method. He cannot be either, because he avoids the subject altogether (see also Adler's remarks on this issue). Of course, in deciding how to hypothesise about the content of a subject's competence, an investigator may find it suggestive, or even fruitful, to have in mind the various judgemental strategies that he himself might use in such a situation. But, in order not to slide into the Norm Extraction Method Mark I, he must also have in mind the possibility that subjects might not have those strategies at their disposal. And other considerations may also be suggestive or fruitful for the investigator, such as the long-term evolutionary survival-value that one type of strategy may possess even though it is sub-optimal on a particular occasion.

3.2. Tomasz Maruszewski has directed attention to some important issues in the study of human rationality. Perhaps the most useful point that I can add concerns the need to distinguish carefully between certain concepts that may otherwise lose a useful sharpness of definition. But I do not think that the distinction invalidates anything that Maruszewski has said.

Consider the question: is a theory of competence a form of idealisation? An affirmative answer to this question was put forward explicitly by Fodor and Garrett at least as long ago as their (1966, p. 135), and it has tended to persist in the psychological literature as an unchallenged assumption. A grammatical competence in the English language, for example, — i.e. possession of rules that generate all well-formed English sentences — is naturally regarded as constituting part of the ability that an ideal speaker-hearer of the language possesses. In the speaker-hearer's actual performance ungrammatical sentences may sometimes occur and many grammatical sentences never occur. But we abstract away from such deviations or lacunas when we formulate the theory of a person's idiolect, dialect or language. Thus it is quite natural to think of such a formulation as an idealised projection from the person's everyday linguistic activity or intuitions.

Nevertheless there is an important difference between any such description of a psychological competence, on the one side, and, on the other, the theories that are widely accepted as idealisations in the physical sciences. If we take lingustic ability as a paradigm of psychological competence, we have to agree that performance deviations — grammatical howlers, malapropisms, mispronunciations, etc. — are relatively rare. Or at least it must be admitted that innumerable grammatical sentences have actually been spoken. But objects have never fallen to the Earth under frictionless conditions. A competence-theory in linguistics picks out a norm to which ordinary people are in fact capable of conforming quite accurately. But idealisation in physics picks out a law that does not admit of exact exemplification in the actual world because it fits only a world that lacks certain kinds of factor that are present in the actual one.

What lies underneath this difference is the fact that competences are systems of rules that may, or may not, be obeyed on this or that particular occasion to which they are applicable, while idealisations in physics are laws that hold necessarily and exclusively for the simplified domains to which they apply. The two therefore differ substantially in the kind of explanation or prediction that they authorise. A competence that explains or predicts a certain event (such as the use of a singular verb-form in an English sentence that has a singular noun as subject) will also justify that event. But an idealisation that explains or predicts an event (such as the velocity of an apple that has been falling for half a second) does not, in any sense, "justify" that event. In short, the differences between competence-theory and idealisation are considerably more important than any similarities that they may have.

3.3. Gerd Gigerenzer raises some crucial important issues, on most of which I should wish to take the same line as he does. Certainly the analogy betwen visual and cognitive illusions has merely suggestive, rather then explanatory, value. And it looks as though we still have much to learn about the range of factors — in content or context — that can cue certain types of response to questions about probabilities.

Even on the issue about the role of formal rules in untutored intuition my own view may be closer to his than he supposes. What is being implied when a subject is said to have a rule, such as the law of contraposition, and to be capable of either applying it or not applying it as he or she deems appropriate? At least three different implications are possible.

One such implication is that the subject has learned or developed enough formal logic to be able to make conscious use of the law in the spoken, written or tacit reasonings that he conducts in natural language. And obviously that cannot be the implication that fits the present context, where we are concerned with the thought-processes of logically naive subjects. An alternative possible implication is that, though the subject makes no conscious use of the law, it is nevertheless present as an accepted or adopted formula-type in the language of his thought and, correspondingly, tokens of contrapositive inference occur at relevant points in the subject's inner flow of Mentalese that unconsciously underlies his natural language utterances. But I should not wish to be committed to such an implication, because I regard the whole Fodorian story about the language of thought as an ontological extravagance that is not required by the computational theory of mind and has no empirical justification or explanatory value (1986h, pp. 220-231).

There is instead a third kind of sense in which a subject may be said to have, and occasionally to apply, the law of contraposition. In this sense what is implied is not that tokens of contraposition may occur in the running of the subject's Mentalese software, but just that the subject's neuronal hardware behaves as if it had been programmed to generate spoken, written or tacit instances of contrapositive reasoning on occasions that are appropriate in content and context. So on this account an explicit formulation of the law of contraposition is achieved by whoever explains some of the subject's logical intuitions accordingly. That is to say, the explainer cites the rule explicitly when he attributes the relevant intuitions to the subject's unconscious possession of the rule. But the explanation no more requires there to be an explicit formulation of the rule in the inner recesses of the subject's mind than classical mechanics requires there to be a real gravitational force over and above the motion of particles towards one another directly in accordance with their masses and inversely according to the square of their distance from one another. Just as the so-called "realist" analysis is unnecessary in classical mechanics, so too it is unnecessary for the computationalist psychology of intuition. Possession of a naive logical competence does not arise from the mind's housing a discoverable Aristotelian manikin that talks secretly in syllogisms. Rather the description of the competence in terms of rules, content, context, etc. is the way in which the explaining scientist tries to invent a structure for his categorisation of the situation. And it may be that Gigerenzer's preferred account of naive subjects' intuitions follows this "antirealist" pattern, which I too prefer.

Thus in relation to overconfidence, underconfidence and their mean there is no doubt that the heart of the problem lies in correctly identifying the circumstances that affect self-confidence, not in pinning some label on our fellows that imputes a homogeneous mechanism for self-assessment to all of them. Gigerenzer, for example, discovered that on his choice-tasks the subjects estimated their objective frequency of correctness at levels *equalling* their actual frequency of correctness. So he construes many results in which overconfidence is alleged to be manifest as being due to subjects' estimating the strength of their subjective feeling of correctness. On the other hand, J. K. Adams (1987) has demonstrated *underconfidence* in a word-recognition task where subjects were instructed to express each level of confidence in terms of the percentage of responses, made at that particular level of confidence, that they expected to be correct. And S. Oskamp (1965), using Adams's scale, has demonstrated the presence of *overconfidence* within the judgements of clinical psychologists in certain circumstances. Such experiments help to show how different types of content and context affect human self-confidence differently: they do not demonstrate any universally pervasive form of "irrationality".

3.4. Lola Lopes and Gregg Oden are right to insist, in their perceptive paper, that psychologists need to square their treatment of intelligence with their treatment of rationality. Lopes and Oden have made out a strong case for the appropriateness of a connectionist, pattern-processing model in relation to the untutored use of heuristics in everyday problem-solving, as against the appropriateness of a digital, rule-following model for the operation of acadmically sanctioned strategies in order to achieve rationally optimal solutions. The more one insists on the importance of content and context — alongside the nature of the task set — in interpreting the experimental results that are discussed in the "heuristics and biases" literature, the more one has to acknowledge the relevance of a connectionist model's parallel-wise distribution of acquired information across a large collection of continuous weights, as against the abstract and idealised structure of reasoning that can be embodied in the sequential procedures of a digital computer. At the same time it becomes clearer that partisans of connectionist models are just as wrong to insist on the unique virtue of their preferred type of model as partisans of digital models are to make the opposing claim.

Both types of model are needed because the human mind has two strikingly different modes of operation. Without the robustness, generality and practicability of the heuristics that they operate, humans would be unable to resolve their everyday problems satisfactorily. But without the ability to set themselves rules and to monitor their application humans would be unable to organize legislatures and law-courts or to develop science and technology.

An interesting question now arises. Is our own generation the first to become aware of this crucially dual capacity? Has awareness of our two different kinds of approach to problem-solving had to wait until we invented two correspondingly different kinds of artefact for modelling them? Has so central and crucial a feature of our minds remained wholly opaque to them for so many centuries? The answer is both yes and no.

Yes, in that the duality of the human mind had not previously been explicitly stated as a fact about the world. Discovery of this fact will surely be regarded in future histories of science as a major new development that emerged gradually during the 1980's.[3] On the other hand, there is no doubt that the duality of the mind has long been reflected in the system of concepts that we come to use as soon as we learn a language. The terminology of everyday discourse about mental events implicitly presupposes the very same duality of functioning that has come to be explicitly acknowledged in the connexionist-digital debate. In that sense our own generation is by no means the first to become aware of the matter, as a comparison of the everyday concept of belief with the everyday concept of acceptance can make clear.

Such a comparison was attempted initially in (1983d) and expanded further in (1986h, pp. 91-97), and in (1989a). There is more still to be said about the matter, and the outline given in the present comment is necessarily very cursory: see also (forthcoming c).

What is at issue in such a comparison is an analysis of the meanings of the relevant words and phrases, whether in English or in other languages, not a study of the actual events denoted by those words and phrases. But it seems hardly an accident that we have certain frequently used expressions in our language with just those meanings — such as "belief" and "acceptance". And, if our language's ability to express the meanings is to be deemed a piece of "folk psychology", then here at least folk psychology seems to sit well with the emerging consensus of up-to-date academic psychology.

Briefly, "belief that p" denotes a disposition to feel it true that p, whether or not one goes along with that proposition as a premiss. But "acceptance that p" denotes having or adopting a policy of deeming or postulating that p. A person who is said to accept that p is being implied to go along with that proposition (either for the long term or for immediate purposes only) as a premiss in some or all contexts for his own or others' proofs, argumentations, inferences, and deliberations, whether or not he assents and whether or not he feels it to be true that p. Various other differences follow from this. In particular beliefs are involuntary while acceptance is voluntary, so that people can more easily be held responsible for what they accept than for what they believe. And, while people need not necessarily believe the accepted consequences of what they believe, they are committed to accepting all the accepted consequences of what they accept. Of course, most of the time people accept what they believe and believe what they accept. But this does not have to happen. A lawyer may, for forensic purposes, accept what his client states to be the case, even if he does not believe it, and a juror, on the basis of certain facts that he knows, may believe an accused to be guilty even though he restrains himself from accepting this because the facts were not given in court. Again, a scientists who at present accepts a certain theory may do well not to believe it, thus keeping his mind open to the possibility of future counter-evidence. But people who indulge in one form or another of self-deception may be said to believe — deep down — what they refuse to accept.

In these terms we can say that intuitions, which subjects in the heuristic-and-biases experiments reveal, are untutored beliefs, while the rational calculations of the experimenters deploy the premisses and conclusions that they accept. And, while some kind of connexionist model seems needed for the mental features that in everyday life we call "belief", a digital model seems more appropriate for the self-programming that in everyday life we call "acceptance". The duality posited by academic theory is reflected by a corresponding duality in the conceptual system that is embodied in the natural language of modern culture. But the point needs development at much greater length than is possible here.

3.5. Last, but by no means least, I come to my old friend Jonathan Adler's insightful remarks. He deals with two main issues. One concerns the way in which Grice's conventional principles affect the interpretation of experimental evidence about the conjunction effect. The other concerns the diversity of the various cognitive virtues or vices that are attributable on the basis of such experimental results. I shall comment on the issues in that order.

Adler points out that Gricean principles affect the interpretation of the conjunction effect in more than one way. There is a question about the implicature that subjects might assign to the statement "Linda is a bank-teller", in the light of its association with the statement "Linda is a bank-teller and is active in the feminist movement", and there is also a question about the implicature that subjects might assign to the request for a ranking of probabilities in the light of the character-sketch provided.

But it is clear that these two questions are not equally germane to the problem. The object of the experiment is to determine whether exposure to evidence of the kind presented in the character-sketch can make subjects assign higher probabilities to conjunctions than to one or other of their conjuncts. So the character-sketch *has* to be provided and any implicatures that it may have are correspondingly unavoidable. But, though the statement "Linda is a bank-teller" must occur somewhere in the description of the probabilities to be evaluated, it need not occur in such a way that it implicates "Linda is a bank-teller and is not active in the feminist movement". Suppose that reasonably large and well-structured groups of subjects are both presented with the character-sketch, and one group is asked to assign an approximate value to the probability that Linda is a bank-teller while the other is asked to assign an approximate value to the probability that Linda is a bank-teller and is active in the feminist movement. If the conjunction effect still persists, it cannot then be due to an implicature of the conjunct "Linda is a bank-teller". But it might still be the result of the relevance imputed to the character-sketch.

Indeed it is easy to see that, even if it is Pascalian probabilities that are assessed in the light of this evidence, still, so long as what is ranked for comparison is the extent to which probabilities are increased by the evidence, a conjunction may quite legitimately be ranked higher than one of its conjuncts. For example, where E is the character-sketch, we may have $p(T) = .1$, $p(T/E) = .05$, $p(T\&F) = .01$, and $p(T\&F/E) = .015$. That is to say, on the assumption that untutored subjects regard "probability" as a generic term covering a variety of different ways of assessing expectability, it would be easy to relate subjects' answers to a mode of assessment — excess of posterior over prior Pascalian probability — in terms of which their impicit mathematics is impeccable. This mode of assessment is one way of representing the extent of the causal coherence in Linda's circumstances that seems to influence subjects' responses to their task. But there is no reason to suppose that everyone uses the same mode of assessment for such a task.

Adler is undoubtedly right to insist that in commenting on one another's judgements of probability people can and do employ more than just one dimension of evaluation. For example, we can have in mind such questions as:

(i) Is the correct value assigned to the probability (as conceived) on the evidence specified?

(ii) Is the value-assignment coherent with the speaker's other relevant probability-judgements?

(iii) Should other evidence have been taken into account?

(iv) Should a different conception of probability have been employed?

(v) Is the judgement a respectable first-attempt at a task that in every-day life would normally be progressively clarified so as to lead to improved performance?

And so on. Indeed, a philosopher who takes a pluralist line and argues that a variety of different modes of probability-assessment are available must face the fact that this makes the criticism of probability-judgements quite a complex matter. In particular he incurs an obligation to discuss what conceptions of probability are appropriate or optional for what kinds of purpose (1989c, pp. 90-114), just as a quantity of apples may be measured by weight for sale, by size for storage or by number for distribution as dessert. And people who use a conception of probability that is inappropriate or sub-optimal for the context must undoubtedly be recognised by such a philosopher as having committed some kind of error. But certain errors are more serious than others, and what is worrying in some of the psychological literature is the claim that people are innately programmed to commit very serious errors — errors that can, for example, lead to wrong decisions by juries. Correspondingly, if *that* claim is shown to be indemonstrable, we may at least be cheered, even if an overall optimism would be justified only in a world in which untutored subjects always give optimal answers. Perhaps the situation about accusations of irrationality is a little like the one that developed in the Vienna Circle when some of its members accused others of meaninglessness. The shriller cry has to give way to a more tolerant tone, even if the tolerated judgement is not a perfect one.

FOOTNOTES

* Full list of references of L. J. C. which are mentioned here is given on pp. 30-37 (editors' note).

1 Some references are given in (1977b, p. 346, f. n. 1) The general issue of scepticism in the philosophy of science is discussed in (1977b, pp. 345-356).

2 The account given in (1989c, pp. 169-171) and in (forthcoming c) is broadly correct and in line with earlier remarks in (1970e p. 144) and in (1977b, pp. 165-66). But some remarks in (1970e p. 152) and in (1977b pp. 242 and 293) are misleading, because they do not distinguish properly between taking the domain of a generalisation to be objects in normal circumstances and taking it to be objects in idealised circumstances that are never found in the actual world.

3 But see also the earlier literature an hemispheric differences (Cohen, 1983, pp. 224-250).

REFERENCES TO OTHER AUTHORS

Adams, J. K. (1987). A confidence scale defined in terms of expected percentage. *American Jurnal of Psychology, 70*, 432-436.

Cohen, G. (1983). *The Psychology of Cognition*, London: Academic Press.

Descartes, R. (1931). *Discourse on Method*, In E. S. Haldane and G. R. T. Ross (trans.), *The Philosophical Works of Descartes*, Cambridge: Cambridge University Press, vol. 1.

Ellis, B. (1968). *Basic Concepts of Measurement*. Cambridge: Cambridge University Press, pp. 74-75.

Fodor, J. and Garrett, M. (1966). Some reflections on competence and performance. In J. Lyons and R. J. Wales (Eds.), *Psycholinguistic Papers: Proceedings of the Edinburgh Conference, 1966*. Edinburgh: Edinburgh University Press.

Kant, I. (1907). *Fundamental Principles of the Metaphysic of Ethics* (trans. T. K. Abbott). London: Longmans, Green.

Mill, J. S. (1851). *System of Logic*. 3rd edition.

Oskamp, S. (1965). Overconfidence in case-study judgements. *The Journal of Consulting Psychology, 29*, 261-265.

Searle, J. R. (1969). *Speech Acts: An Essay in the Philosophy of Language*. Cambridge: Cambridge University Press.

Williams, B. (1976). Moral Luck, *Proceedings of the Aristotelian Society*, supp. vol. **50**, pp. 115-135.

BIOGRAPHICAL NOTES

Jonathan E. Adler (1948) studied philosophy at Brandeis and Oxford Universities, and is currently Professor of Philosophy at Brooklyn College, C. U. N. Y. Main research fields include: induction, epistemology, ethics and philosophy of psychology. Address: Department of Philosophy, Brooklyn College, C. U. N. Y., Brooklyn, N. Y. 11210, USA.

L. Jonathan Cohen (1923) studied classics and philosophy at Balliol College, Oxford. He now teaches philosophy at the Queen's College, Oxford, and is also a Fellow of the British Academy. For detailed information about his research interests see "From a Historical Point of View" (this volume). Address: The Queen's College, Oxford, OX1 4AW, U. K.

Ellery Eells (1953) studied philosophy and logic at the University of California at Berkeley. He now teaches philosophy at the University of Wisconsin-Madison. His main research fields have been decision theory, inductive logic, probability, causation, and explanation. Address: Department of Philosophy, University of Wisconsin-Madison, Madison, Wisconsin 53706, USA.

Maurice A. Finocchiaro (1942) studied physics at M. I. T. and philosophy at Berkeley, and teaches philosophy at the University of Nevada-Las Vegas. His main research fields are: informal logic and the theory of reasoning, history and philosophy of natural science, Marxism and philosophy of social science. Address: Department of Philosophy, University of Nevada-Las Vegas, Las Vegas Nevada 89154, USA.

Menachem Fisch (1948) studied physics (Bar Ilan University), philosophy and history and philosophy of science (Tel Aviv University). He is now Senior Lecturer in history and philosophy of science at Tel Aviv. His main research interests are: confirmation theory, inductive logics, theories of rationality, nineteenth century physics, mathematics and philosophy of science and the historiography of science and philosophy. Address: The Cohn Institute for History and Philosophy of Science and Ideas, Tel Aviv University, Ramat Aviv, 69978 Israel.

Gerd Gigerenzer (1947) studied psychology and statistics at the University of Munich and is professor of psychology at the University of Salzburg, Austria. His research interests include discovery in the social sciences, in particular how scientists' tools turn into theories of mind; reasoning and rationality, in particular the relevance of content and context in probabilistic judgment; and methodological "rituals" in the social sciences, such as mechanized statistical inference and its link to ideals of impersonal judgment and objectivity. Address: Institut für Psychologie, Universitat Salzburg, Hellbrunnerstr 34, 5020 Salzburg, Austria.

James Logue (1951) studied mathematics and philosophy at Keele University in England, and also holds higher degree from North London Polytechnic and Oxford. He is now college lecturer at Somerville College, Oxford. His principle current research interest lies in developing a quasi realistic account of subjective probability. Address: Somerville College, Oxford, OX2 6HD, U. K.

Lola Lopes (1941) studied psychology at the University of California, San Diego. She is Professor Psychology and Industrial Engineering at the University of Wisconsin and chairman of the Department of Psychology. Her research involves rational decision making under risk and uncertainty. Address: Department of Psychology, University of Wisconsin, Madison, Wisconsin 53706, USA.

Tomasz Maruszewski (1946) studied psychology at the Adam Mickiewicz University, Poznań. He is Associate Professor of Psychology at this University. His research interests include relations between everyday and scientific knowledge, creativity, theories of concept formation and philosophy of psychology. Address: Institute of Psychology UAM, Szamarzewskiego 89, 60-568 Poznań, Poland.

Czesław S. Nosal (1942) studied psychology at the Adam Mickiewicz University, Poznań. He is Professor of Psychology at Wrocław Technical University. His main research interests include cognitive psychology, creativity and typologies of mind. Address: Wrocław Technical University, Institute of Organisation and Management Systems, Smoluchowskiego 25, 50-327 Wrocław, Poland.

Leszek Nowak (1943) studied law at the Adam Mickiewicz University, Poznań and philosophy at the Warsaw University. He is Professor of Philosophy and teaches philosophy at the Adam Mickiewicz University as well as at many European and other universities. His main research include philosophy of science, especially idealizational theory of science, social theory and theory of historic process. Address: Institute of Philosophy UAM, Szamarzewskiego 89, 60-568 Poznań, Poland.

Greg Oden (1947) studied psychology at the University of California, San Diego. He is Professor of Psychology and Computer Sciences at the University of Wisconsin. His research involves the application of fuzzy propositional and connectionist models to natural language processing problems. Address: Departament of Psychology, University of Wisconsin, Madison, Wisconsin 53706, USA.

David A. Schum (1932) studied psychology and mathematics (Ohio State University). He now teaches various courses at George Mason University concerning probabilistic reasoning; he is a Fellow of American Psychological Association. His main research interests concern study of various evidential subtleties brought forth in analyses of inferential networks. Address: Department of Operations Research and Applied Statistics, George Mason University, Fairfax, Virginia, 22030, USA.

Ryszard Stachowski (1937) studied psychology at the Adam Mickiewicz University, Poznań. He is Associate Professor of Psychology and Head of the Department of History of Psychological Thought at this University. His research interests include antique and medieval roots of modern mathematisation of psychology. Address: Institute of Psychology UAM, Szamarzewskiego 89, 60-568 Poznań. Poland.

Ireneusz T. Ścigała (1955) studied psychology and clinical psychology at the Adam Mickiewicz University, Poznań. He teaches psychology at this University. His main research fields involve errors or biases in self-perception and social perception, creativity potential and theories of human normality. Address: Institute of Psychology UAM, Szamarzewskiego 89, 60-568 Poznań, Poland.

JOURNAL OF

Applied

Philosophy

Volume 7 Number 2 1990

This journal provides a focus for philosophical research with a direct bearing on areas of practical concern which are capable of being illuminated by the critical analytical approach characteristic of philosophy, and by direct consideration of questions of value. These areas include law, politics, economics, science-policy, medicine and education, but are not confined to these. The journal, published under the auspices of the Society for Applied Philosophy, aims to foster and promote philosophical work which is intended to make a constructive contribution to problems in these areas and to the identification, justification and discussion of values capable of transcending narrow or local interests.

EDITORS
Brenda Almond & Stephen Clark

EDITORIAL BOARD

Robin Barrow; Andrew Brennan; David E. Cooper; Anthony Duff; Raanan Gillon; James Griffin; R.M. Hare; Martin Hollis; Ian Kennedy; Joel Kupperman; Jeff McMahan; Susan Mendus; Mary Midgley; Anthony O'Hear; John Passmore; Philip Pettit; Amartya Sen; Michael Simon; Peter Singer; T.L.S. Sprigge; Richard Tur; Gerry Wallace.

EDITORIAL ADVISERS
Bernard S. Baumrin (USA)

ASSISTANT EDITOR
Donald Hill

The journal is published twice a year, in March and October. These two annual issues constitute one volume. An annual index and title page is bound in with the October issue. Volume 8—1991. ISSN 0264–3758.

SUBSCRIPTIONS
Subscription rate: one year/volume £72.00, post free. Individuals receiving the journal at a private address for their personal use may subscribe at the reduced rate of £33.00 under the terms of the Personal Subscription Plan. Non-UK prices available on application. Orders should be directed to the publisher: Carfax Publishing Company, P.O. Box 25, Abingdon, Oxfordshire OX14 3UE, United Kingdom or 85 Ash Street, Hopkinton, Massachusetts 01784, USA.

❏ Please enter our subscription to *Journal of Applied Philosophy*

    We enclose £ ....

❏ Please enter my subscription under the terms of the Personal Subscription Plan
(please give private address)

    I enclose £ ....

    Please charge Access/American Express/Eurocard/MasterCard/Visa No ....

❏ Please send an inspection copy

Name ____

Address ____

Signed ________________________________   Date ___________

When completed please send this order form to **Carfax Publishing Company,** P.O. Box 25, Abingdon, Oxfordshire OX14 3UE, UK, *or* 85 Ash Street, Hopkinton, Massachusetts 01748, USA.

PHILOSOPHIA

Philosophical Quarterly of Israel

Editor: Asa Kasher

EPISTEMOLOGY in PHILOSOPHIA

19: 2–3 (1989)
Papers on Alvin I. Goldman's
"Epistemology and Cognition"

Douglas G. Winblad	Skepticism and Naturalized Epistemology
William P. Alston	Goldman on Epistemic Justification
Robert K. Shope	Justification, Reliability and Knowledge
Paul K. Moser	Reliabilism and Relevant Worlds
Robert Almeder and	
Franklin J. Hogg	Reliability and Justification
Paul Thagard	Connectionism and Epistemology
Richard Feldman	Epistemology and Cognitive Science
Michael P. Levine	Epistemology and Congition: Introduction
Jonathan E. Adler	Epistemics and Total Evidence Requirement
Richard Montgomery	Epistemology Reduced to Cognitive Psychology
Kenneth R. Livingston	Concepts, Categories, and Epistemology
Alvin I. Goldman	Replies to the Commentators

*

19:4 (1989) and on
Series on "Error"

In this issue PHILOSOPHIA starts the publication of a series of papers about notions of Error. Papers will include philosophical analyses or presentations of the uses to which some notions of error are put in various fields.
The first two installments are Natika Newton's philopophical paper "Error in Action and Belief" and Gary Jason's survey on "The Role of Error in Computer Science."
PHILOSOPHIA would welcome submitted contributions to its series on "error".

Editorial address:
PHILOSOPHIA, Bar-Ilan University, Ramat-Gan 52100, Israel

or

Prof. Asa Kasher, Dept. of Philosophy, Tel-Aviv University Tel-Aviv
69978, Israel
ASA0425 @TAUNIVM or F22013 @BARILAN

the review of
metaphysics

a philosophical quarterly

ISSN 0034-6632

JUNE 1991 | **VOL. XLIV, No. 4** | **ISSUE No. 176** | **$10.00**

articles

books received

philosophical abstracts

announcements

index

---

philosophical abstracts   *announcements*   *index*

Individual Subscriptions $23.00   Institutional Subscriptions $40.00   Student/Retired Subscriptions $12.00

Jude P. Dougherty, Editor
The Catholic University of America, Washington, D.C. 20064